Topics in Applied Physics Volume 61

W0043933

Topics in Applied Physics Founded by Helmut K. V. Lotsch

Volumes 1–56 are listed on the back inside cover

Photorefractive Materials and Their Applications I

Fundamental Phenomena

Edited by P. Günter and J.-P. Huignard

With Contributions by
A. M. Glass   P. Günter   J.-P. Huignard   M. B. Klein
E. Krätzig   N. V. Kukhtarev   J. F. Lam   R. A. Mullen
M. P. Petrov   O. F. Schirmer   S. I. Stepanov   J. Strait
G. C. Valley

With 134 Figures

Springer-Verlag Berlin Heidelberg GmbH

Professor Dr. *Peter Günter*

Institut für Quantenelektronik, ETH Hönggerberg
CH-8093 Zürich, Switzerland

Dr. *Jean-Pierre Huignard*

THOMSON-CSF, Laboratoire Central de Recherches, Domaine de Corbeville, B.P. 10
F-91401 Orsay Cedex, France

ISBN 978-3-662-30927-8   ISBN 978-3-540-47881-2 (eBook)
DOI 10.1007/978-3-540-47881-2

Library of Congress Cataloging-in-Publication Data. Photorefractive materials and their applications /
edited by P. Günter, J.-P. Huignard; with contributions by A. M. Glass ... [et al.]. p. cm. – (Topics in
applied physics; v. 61–) Bibliography: p. 1. Electrooptics – Materials. 2. Photorefractive materials.
3. Holography. 4. Image processing. I. Günter, P. (Peter), 1944–. II. Huignard, J.-P. (Jean-Pierre),
1944–. III. Glass, A. M. (Alastair M.) IV. Series: Topics in applied physics; v. 61, etc. TA1750.P47 1988
621.367 – dc19 87-32242

© Springer-Verlag Berlin Heidelberg 1988

Originally published by Springer-Verlag Berlin Heidelberg New York in 1988
Softcover reprint of the hardcover 1st edition 1988

Typesetting, printing and binding: Brühlsche Universitätsdruckerei, 6300 Giessen
2153/3150-543210

Foreword

In recent years the photorefractive effect has become the nonlinear optical mechanism of choice for optical image processing. This is primarily because the effect is easily applied to real-time parallel processing using low-power lasers and existing bulk materials. No reticulation of pixels nor other complex material processing is required. The information-processing capacity is determined only by the diffraction limit of the optics and three-dimensional nature of the optical beam interactions within the crystal volume.

The photorefractive effect has a particularly interesting history. A little over 20 years ago, soon after the mystery of "optical damage" was discovered in $LiNbO_3$, this effect was used to advantage as a holographic optical memory with huge storage capability. At that time, no basic understanding of the origin of the effect had been discovered. The interest in memory was not long-lived, however, mainly because the stored information degraded both during read-out and (over longer time scales) in the dark. For a while only the challenge to understand the basic mechanisms giving rise to this effect, with a view to eliminating the problem of wave-front distortion in useful electro-optical materials, drove the research. More and more materials were found to be photorefractive. New electro-optic interactions were discovered. The fields of nonlinear optics, optical spectroscopy, paramagnetic resonance, ferroelectrics, electronic transport and Fourier optics were brought together to develop a reasonably complete understanding of the complex microscopic mechanisms involved.

Just as this field was reaching scientific maturity in the mid-1970s, it was recognized that the very characteristic which destroyed the optical memory was exactly what is required for information processing and phase conjugation: namely, a memory which could be rapidly refreshed for real-time interferometry. With the knowledge base available, this field has progressed rapidly and more applications are being demonstrated. More precise theory has been developed which reveals new and interesting aspects of photorefractive interactions not previously recognized. The need for improved materials has become urgent and it is apparent that different applications require different materials properties. Some require fast materials response times, some require high diffraction efficiency. Since the magnitude of the nonlinear index change is proportional to absorbed energy, the usual trade off of speed versus sensitivity applies in photorefractive processing. With the present momentum, it is certain that improved materials and improved device performance will be realized. It will be clear to the reader, however, that for each material the ultimate performance is predictable from a knowledge of the electro-optic properties of the host crystal. The optimized performance within this limit is achieved by introducing appropriate defect states into the crystal. The ability to tailor the important parameters over very wide ranges is indeed an advantage of photorefractive materials and will provide a range of alternatives for future device design.

Murray Hill, NJ, November 1987   *A. M. Glass*

Preface

The photorefractive effect was originally discovered as undesirable optical damage in nonlinear and electro-optical crystals. Light-induced changes of the refractive index limited the usefulness of crystals such as $LiNbO_3$ with large electro-optic and nonlinear optical coefficients, because the index changes give rise to decollimation and scattering of laser beams in devices such as modulators and frequency doublers. Subsequently, materials exhibiting such an optical damage effect – later called the photorefractive effect – were applied as holographic recording materials.

The optically induced refractive index changes are generally explained in the following way: When light of a suitable wavelength is incident on a crystal, photoelectrons are generated, migrate in the lattice and are subsequently trapped at new sites. The resulting space charges give rise to an electric field strength distribution in the material which changes the refractive index via the electro-optic effect.

This volume gives a comprehensive treatment of photorefractive effects in crystals. It reviews our present understanding of the fundamental origins of the effect in a variety of materials from ferroelectrics to compound semiconductors. The various materials parameters which enter into the design of optimized materials are described and the beam interactions which form the basis of applications are discussed. The discussion prepares the reader for the detailed review of applications contained in a forthcoming volume.[1]

The two volumes have been prepared for researchers in the field as well as for students of solid-state physics and engineering. The chapters contain sufficient introductory material for those unfamiliar with the topic to obtain a thorough understanding of the photorefractive effect, as well as providing a useful reference source for researchers already involved in this field.

In this volume the basic properties of photorefractive materials are treated. The companion volume deals with the applications of photorefractive materials in dynamic holography, light amplification, optical phase conjugation, optical signal processing, and with devices such as spatial light modulators and integrated optics components. The content of this volume can be summarized as follows.

In the second chapter, P. Günter and J.-P. Huignard give an introduction to the different photorefractive effects observed in electro-optic crystals. The discussion starts with a phenomenological description of photo-induced space-charge fields.and the different changes of optical properties related to them. The relevant physical properties of photorefractive materials are reviewed and

---

[1] Topics in Applied Physics, Vol. 62

some of the applications of photoinduced changes of refractive indices are discussed briefly.

After the introductory chapter on photorefractive effects, G. Valley and J. Lam present a detailed description of the photorefractive effects in dielectric crystals. They describe photorefractive effects for the general case where photogenerated electrons and holes contribute to the photoconductivity of the electro-optic material solutions of the nonlinear wave equation for steady-state and picosecond pulse excitation. Experiments on degenerate four-wave mixing and self-pumped phase conjugation via nearly degenerate backward stimulated two-wave mixing are outlined.

Chapter 4 by N. Kukhtarev summarizes the present status of theoretical work on the photorefractive effect by the rather successful group in Kiev. Starting from the band transport model, the author of this chapter also treats special effects such as optical self-diffraction from photoinduced gratings and light propagation in optically active photorefractive crystals such as $Bi_{12}(Si, Ge, Ti)O_{20}$. In addition, the appearance of optical bi- and multi-stability is predicted theoretically.

The centers that are relevant for the photorefractive effect are treated in Chap. 5 by E. Krätzig and O. F. Schirmer. Of particular interest is the information on the centers that supply charge carriers, on trapping centers, and on optical excitation and transport processes. In order to facilitate an understanding of the experimental results, a short survey of the experimental methods used for investigating photorefractive defects is given.

Chapter 6 by R. A. Mullen summarizes the measurement of physical parameters, such as the densities and cross sections of the impurity centers, diffusion lengths, mobilities and recombination coefficients, using photo-refractive techniques. The results of a charge hopping model are compared with the results of the usual band transport model.

In Chap. 7, M. B. Klein outlines the present findings on the photo-refractive effect in $BaTiO_3$, the material with the largest known photoinduced refractive index changes. The large nonlinearities observed in this material give rise to self-pumped phase conjugation, phase conjugation by four-wave mixing with gain for the phase-conjugate beam, and coherent light amplification with a large amplification factor. Applications of these effects will be discussed in the forthcoming volume.

In Chap. 8, A. M. Glass and J. Strait deal with photorefractive effects in semiconductors. The photorefractive centers in GaAs and InP are discussed and it is shown that, because of the large carrier mobilities observed in these materials, the recording times for photoinduced gratings are substantially shorter than, for example, in oxygen octahedra ferroelectrics.

The final chapter by S. I. Stepanov and M. P. Petrov deals with transient and moving photoinduced gratings and their applications in coherent beam amplification and optical phase conjugation.

Zürich, Orsay   *P. Günter*
November 1987   *J.-P. Huignard*

Contents

Contributors

Glass, Alastair M.
AT & T Bell Laboratories, 600 Mountain Avenue
Murray Hill, NJ 07974, USA

Günter, Peter
Institut für Quantenelektronik, ETH-Hönggerberg
CH-8093 Zürich, Switzerland

Huignard, Jean-Pierre
THOMSON-CSF, Laboratoire Central de Recherches
Domaine de Corbeville, B.P. 10, F-91401 Orsay Cedex, France

Klein, Marvin B.
Hughes Research Laboratories, 3011 Malibu Canyon Road
Malibu, CA 90265, USA

Krätzig, Eckhard
Fachbereich Physik, Universität Osnabrück, Barbarastraße 7
D-4500 Osnabrück, Fed. Rep. of Germany

Kukhtarev, Nicolai V.
Institute of Physics, Academy of Sciences of the Ukrainian SSR
Prospect Nauki, 144, SU-252650 Kiev 28, USSR

Lam, Juan F.
Hughes Research Laboratories, 3011 Malibu Canyon Road
Malibu, CA 90265, USA

Mullen, Ruth A.
Hughes Research Laboratories, 3011 Malibu Canyon Road
Malibu, CA 90265, USA

Petrov, Mikhail P.
A. F. Ioffe Physical Technical Institute of the USSR Academy of Sciences
SU-194021 Leningrad, USSR

Schirmer, Ortwin F.

Fachbereich Physik, Universität Osnabrück, Barbarastraße 7
D-4500 Osnabrück, Fed. Rep. of Germany

Stepanov, Sergei I.

A. F. Ioffe Physical Technical Institute of the USSR Academy of Sciences
SU-194021 Leningrad, USSR

Strait, Jefferson

Williams College, Department of Physics
Williamstown, MA 01267, USA

Valley, George C.

Hughes Research Laboratories, 3011 Malibu Canyon Road
Malibu, CA 90265, USA

1. Introduction

Peter Günter and Jean-Pierre Huignard

With 1 Figure

Photorefractive materials have been investigated extensively in the past. The possibility of obtaining high optical nonlinearities at milliwatt power levels makes them particularly attractive for applications in optical signal processing, dynamic holography, and phase conjugation.

The photorefractive effect has been observed in a variety of electro-optic materials including $LiNbO_3$ [1.1, 2], $LiTaO_3$, $BaTiO_3$ [1.3], $KNbO_3$ [1.4], $K(TaNb)O_3$ [1.4, 5], $Ba_2NaNb_5O_{15}$ [1.6], $Ba_{1-x}Sr_xNb_2O_6$ [1.7], $Bi_4Ti_3O_{12}$ [1.8], $Bi_{12}(Si, Ge)O_{20}$ [1.9], KH_2PO_4 [1.10], CdS [1.11], Rb_2ZnBr_4 [1.12] and $(Pb, La)(Zr, Ti)O_3$ ceramics [1.13], $GaAs$, InP [1.14] and other compound semiconductors, so that it may be considered a general property of electro-optic materials. Depending on the band gap of the material and the energy levels of the donor or acceptor levels of the impurity ions involved, the photorefractive effect may be induced by ultraviolet, visible or infrared radiation.

The energy required to obtain an appreciable photorefractive effect in $Bi_{12}SiO_{20}$ and $KNbO_3$ is close to the photographic sensitivity of silver halide layers. Holographic gratings are mostly recorded with visible cw lasers (Ar, Kr, He-Cd, He-Ne). The storage times of the recorded holograms or grating decay times range from milliseconds for $KNbO_3$ [1.15], hours for KTN to months and years for $LiNbO_3$ [1.2] if a fixing process is applied. The holographic gratings can be erased by uniform illumination.

Using pulsed lasers with high energy densities, short writing and erasing times can be obtained. Experiments with frequency-doubled Nd : YAG lasers at 530 nm show that optical writing, reading and erasing of refractive index gratings are possible on a nanosecond time scale or even faster [1.16–1.18].

In the years since the discovery of the photorefractive effect, a great deal of effort has been devoted to identifying the microscopic details of the mechanism in order to optimize materials for either memory or nonlinear optical applications [1.19]. A wide range of physical studies have been undertaken, including optical spectroscopy of color centers, traps and transition metal ions, controlled crystal preparation, investigations of transport properties, electro-optics, holography, integrated optics, nonlinear absorption, and transient spectroscopy. These studies have also led to the discovery of additional new effects in electro-optic materials, namely the bulk photovoltaic effect and excited state polarization. At present, we have a reasonably good (though still incomplete) understanding of the photorefractive effect, enabling the preparation of materials having a high photorefractive sensitivity with holographic recording sensitivities comparable with high-resolution silver halide emulsions, as well as materials in which the photorefractive sensitivity is very low or even unmeasurable.

The possibility of using photorefractive materials as storage media in holographic memory systems was first proposed by *Chen* et al. [1.20]. Erasable read-write optical memories allowing, in principle, diffraction-limited storage densities (capacity limit typically 10^{12} bits/cm^3 for three-dimensional storage) should be possible with photorefractive crystals. Their use for read-write applications (because they are optically "erasable"), and for read-only applications (because they are fixable [1.21, 22]) make these materials the most versatile of all storage media. For comparison, photographic emulsions are suitable only for read-only applications; magneto-optic materials require high optical power density in short times; and thermoplastics, which require processing, cannot be used in real time. Photochromic processes, closely related to photorefractive effects through the fluctuation-dissipation theorem, have lower photosensitivities since they require changes in the electronic configuration or crystallographic structure. Photorefractive effects, on the other hand, result from a polarization change of the electro-optic crystal as a result of a photoinduced charge transfer. Volume phase holograms (hologram thickness much larger than the light wavelength) recorded by the photorefractive effect also offer the possibility of attenuation-free read-out with high angular selectivity.

The volume nature of thick holograms permits the interference of an incident light beam with its own diffracted beam inside the recording medium [1.23]. This effect causes the continuous recording of a new grating that is nonuniform throughout the thickness of the material; this may be phase shifted with respect to the initial grating and may interfere constructively or destructively with it. The phase shift between the fringe pattern and the recorded grating leads to a dynamic energy redistribution between the two recording beams [1.24]. This self-diffraction exists because the writing and reading of the grating occurs simultaneously and in a self-consistent manner. The refractive index change induced by the radiation gives rise to a phase and intensity redistribution of the interference field and this, in turn, is reflected in the spatial distribution of the refractive index changes. All this determines the complicated dynamic nature of self-diffraction, including transient effects and the establishment of a stationary state, yielding considerable differences between dynamic gratings and the static gratings obtained, for example, with recording in photographic emulsions. The beam coupling effect via the written hologram can be useful for coherent light amplification of time-varying light beams. In the stationary case maximum energy transfer (gain) is obtained if the holographic grating is shifted by a quarter of the fringe spacing with respect to the interference pattern. Since the grating shift depends on the recording mechanism, measurements of the energy transfer can also give additional information about the nature of the photorefractive effects. Conversely, with a detailed knowledge of the photorefractive recording mechanisms it is possible to optimize the grating phase shift by a suitable choice of physical parameters (applied electric field, fringe spacing, etc.) and to control the gain in energy transfer experiments with these parameters. Transient dynamic self-diffraction in electro-optic crystals is analogous to nonlinear optical effects such as self-focusing, self-defocusing, stimulated scattering and other multiple light beam interactions in media with nonlinear polarizabilities. The difference is that the nonlinear refractive index change is of electronic origin and this occurs instantaneously, whereas the refractive index changes in photorefractive crystals are due to electro-optic effects driven by space-charge fields. The time required for these space charges to build up depends on the efficiency of the photoinduced charge transport. This inertia in the "nonlinear" response of photorefractive materials prevents the phase shift between the dynamic grating and the light field instantly adjusting itself to the configuration in which energy transfer does not occur.

Nonlinear optical phase conjugation by four-wave mixing in photorefractive crystals and other dynamic recording media has undergone tremendous development in the last few years [1.25–30]. Much of the interest in this application of laser-induced gratings arises from a fascination with the idea of time reversal of an electromagnetic wave and the demonstration of restoration of severely distorted optical beams to their original, nonaberrated state [1.25–30].

The generation of the phase-conjugate wave occurs through nonlinear mixing of optical waves in a dynamically recording medium. One possible scheme, which is directly related to the recording scheme used in holography (Fig. 1.1), is four-wave mixing. It has been shown in [1.25, 26] that a conjugate wave is generated by mixing a signal wave A_0 to be conjugated with two antiparallel pump waves A_R and A_{R1} (Fig. 1.1) (degenerate four-wave mixing). Therefore any dynamic recording medium with photoinduced changes of refractive index or absorption constant can be used as a phase conjugator in this configuration.

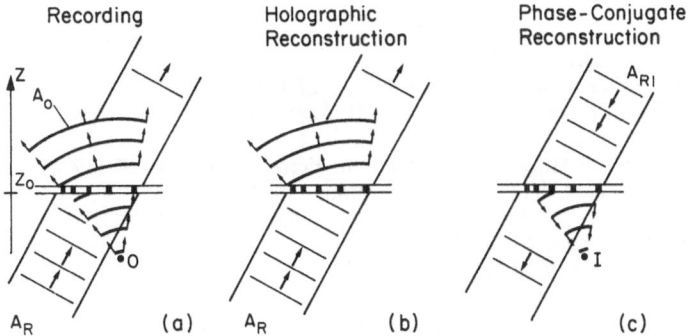

Fig. 1.1a–c. Hologram recording and read-out (schematic diagram of principles) demonstrated for recording the spherical wave front of an object point source. (a) Recording; (b) "normal" read-out by the reference wave A_R produces a virtual image of the object; (c) "phase-conjugate" read-out by illuminating the hologram with a second reference wave A_{R1}, antiparallel to the original reference wave A_R produces a phase-conjugated object wave and a real image of the object point

Phase-conjugate waves have been generated by this technique in photo-refractive $Bi_{12}SiO_{20}$, $Bi_{12}GeO_{20}$, $LiNbO_3$, $LiTaO_3$, $KNbO_3$, $BaTiO_3$, and $Sr_{1-x}Ba_xNb_2O_6$ [1.31–36]. In materials with large electro-optic coefficients cw phase-conjugate wave generation with simultaneous amplification at the signal wave can be achieved. Using the holographic approach to four-wave mixing [1.26] the appearance of the fourth wave can be interpreted as a consequence of the diffraction of one of the pump beams at the refractive index grating recorded by the other two beams.

In photorefractive materials, phase conjugation by four-wave mixing can occur without additional pump beams. White et al. [1.37] have demonstrated that photorefractive phase conjugators can be self-pumped if a pair of mirrors are aligned to form a resonator cavity containing a photorefractive $BaTiO_3$ crystal. Both pumping beams are generated from the input wave by two-wave mixing of the input wave with scattered light propagating towards the mirrors. It was shown later that one of the mirrors can be removed and the phase conjugator still operates if it is started with a weak beam [1.38].

In photorefractive crystals two different self-pumped phase conjugator configurations have been found that need no external mirrors and which are both self-starting [1.39, 1.40]. The pumping beams are self-generated from the incident wave and are internally reflected at the crystal faces. Since the pumping beams do not leave the crystal, the device is compact, relatively insensitive to vibration, and completely self-aligning. The quality of phase conjugation by such a device was demonstrated by focusing complicated images into the crystals and observing the faithfulness of reproduction even through the image-bearing input and output beams passed through a severe phase aberrator.

References

1.1 A. Ashkin, G.D. Boyd, J.M. Dziedzic, R.G. Smith, A,A. Ballmann, H.J. Levinstein, K. Nassau: Appl. Phys. Lett. **9**, 72 (1966)
1.2 F.S. Chen, J.T. LaMacchia, D.B. Frazer: Appl. Phys. Lett. **13**, 223 (1968)
1.3 R.L. Townsend, J.T. LaMacchia: J. Appl. Phys. **41**, 5188 (1970). See also M.B. Klein, Chap. 7 of this volume
1.4 P. Günter, U. Flückiger, J.P. Huignard, F. Micheron: Ferroelectrics **13**, 297 (1976)
1.5 R. Orlowski, L.-A. Boatner, E. Krätzig: Opt. Commun. **35**, 45 (1980)
1.6 J.J. Amodei, D.L. Staebler, A.W. Stephens: Appl. Phys. Lett. **18**, 507 (1971)
1.7 J.B. Thaxter: Appl. Phys. Lett. **15**, 210 (1969)
1.8 L.H. Lin: Proc. IEEE **57**, 252 (1969)
1.9 M. Peltier, F. Micheron: J. Appl. Phys. **48**, 3683 (1977)
1.10 V.M. Fridkin, B.N. Popov, K.A. Verkhovskaya: Appl. Phys. **16**, 313 (1978)
1.11 A. Ashkin, B. Tell, J.M. Dziedzic: IEEE J. QE-3, 400 (1967)
1.12 T. Nakamura, V. Fridkin, R. Magomadov, M. Takashige, K. Verkhovskaya: J. Phys. Soc. Jpn. **48**, 1588 (1980)
1.13 F. Micheron, A. Hermosin, G.B. Smith, J. Nicolas: C.R. Acad. Sci. 8 (Dec. 1971)
1.14 Chap. 8 of this volume
1.15 P. Günter: Ferroelectrics **22**, 671 (1978)
1.16 N.J. Berg, B.J. Udelson, J.N. Lee: Appl. Phys. Lett. **31**, 555 (1977)

1.17 C.T.Chen, D.M.Kim, D. von der Linde: IEEE J. QE-**16**, 126 (1980)
1.18 G.C.Valley, A.L.Smirl, M.B.Klein, K.Bohnert, T.F.Bogess: Opt. Lett. **11**, 647 (1986)
1.19 A.M.Glass: Opt. Eng. **17**, 470 (1978)
1.20 F.S.Chen, J.T.LaMacchia, D.B.Frazer: Appl. Phys. Lett. **13**, 223 (1968)
1.21 D.L.Staebler, J.J.Amodei: Ferroelectrics **3**, 107 (1972)
1.22 F.Micheron, C.Mayeux, J.C.Trotier: Appl. Opt. **13**, 784 (1974)
1.23 H.Kogelnik: Bell. Syst. Tech. J. **48**, 2909 (1969)
1.24 V.Markov, S.Odulov, M.Soskin: Opt. Laser Technol. 95 (April 1979)
1.25 H.J.Eichler, P.Günter, D.W.Pohl: *Laser-Induced Dynamic Gratings*, Springer Ser. Opt. Sci. Vol. 50 (Springer, Berlin, Heidelberg 1986)
1.26 A.Yariv: IEEE J. QE-**14**, 650 (1978)
1.27 R.W.Hellwarth: J. Opt. Soc. Am. **67**, 1 (1977)
1.28 C.R.Giuliano: Phys. Today 27–34 (April 1981)
1.29 J.P.Woerdman: Opt. Commun. **2**, 212 (1971)
1.30 B.Ya.Zel'dovich, N.F.Pilipetsky, V.V.Shkunov: *Principles of Phase Conjugation*, Springer Ser. Opt. Sci., Vol. 42 (Springer, Berlin, Heidelberg 1985); B.Ya.Zel'dovich, N.F.Pilipetskii, V.V.Shkunov: Sov. Phys.–Usp. **25**, 713 (1982)
1.31 P. Günter, F. Micheron: Ferroelectrics **18**, 27 (1978)
1.32 P. Günter: Phys. Rep. **93**, 199 (1982)
1.33 N.V.Kukhtarev, V.B.Markov, S.G.Odulov, M.S.Soskin, V.L.Vinetskii: Ferroelectrics **22**, 949 (1979)
1.34 J.Feinberg, D.Heiman, A.R.Tanguay Jr., R.W.Hellwarth: J. Appl. Phys. **51**, 1297 (1980)
1.35 J.P.Huignard, J.P.Herriau, G.Rivet, P.Günter: Opt. Lett. **5**, 102 (1980)
1.36 A.Marrakchi, J.P.Huignard, P.Günter: Appl. Phys. **24**, 131 (1981)
1.37 J.O.White, M.Cronin-Golomb, B.Fischer, A.Yariv: Appl. Phys. Lett. **40**, 450 (1982)
1.38 M.Cronin-Golomb, B.Fischer, J.O.White, A.Yariv: Appl. Phys. Lett. **41**, 689 (1982)
1.39 J.Feinberg: Opt. Lett. **7**, 486 (1982)
1.40 P.Günter, E.Voit, M.Z.Zha, H.Albers: Opt. Commun. **55**, 210 (1985)

2. Photorefractive Effects and Materials

Peter Günter and Jean-Pierre Huignard

With 31 Figures

The light-induced changes of refractive indices in electro-optic crystals are based on the spatial modulation of photocurrents by nonuniform illumination. The generation of photocurrents at low light intensity depends on the presence of suitable donors, because most of the crystals of interest are intrinsically transparent in the visible. The electrons or holes, which are excited from the impurity centers by light of a suitable wavelength, are upon migration, retrapped at other locations, leaving behind positive or negative charges of ionized trap centers. The photoexcited charges will be reexcited and retrapped until they finally drift out of the illuminated region and are trapped. The resulting space-charge field between the ionized donor centers and the trapped charges modulates the refractive indices via the electro-optic effect. Uniform illumination erases the space-charge fields and brings the crystal back to its original state (which process is known as optical erasure).

2.1 Generation of Charge Carriers

For a large photorefractive efficiency, suitable donors or traps and efficient charge migration are necessary [2.1]. In undoped crystals the traps are provided by small traces of impurities. In most of the oxygen-octahedra ferroelectrics that have a pronounced photorefractive effect, Fe impurities are the most important donor and trapping centers. This dopant acts as a donor-acceptor trap via intervalence exchanges such as $Fe^{2+} \rightleftharpoons Fe^{3+}$ [2.2–4]. The effect of iron doping and oxidation and reduction treatment on the optical absorption of $KNbO_3$ crystals is illustrated in Fig. 2.1a. The intrinsic absorption edge located at about 3.2 eV is due to interband transitions from the valence band (based on oxygen $p\pi$ orbitals) to the niobium $d\varepsilon$ conduction band. After Fe doping and reduction treatment a new and characteristic band at about 2.55 eV is seen, which gives rise to photoconductivity upon optical excitation. This band can be assigned to intervalence charge transfer: an electron from a $d\varepsilon$ orbital of an Fe^{2+} ion is transferred to the niobium orbital (see also Fig. 2.3). Intervalence charge transfer can be regarded as photoionization of the Fe^{2+} ion: the final state is a Fe^{3+} ion and a mobile electron in the conduction band.

Upon oxidizing the crystals the intervalence transfer band decreases and the crystal becomes transparent at $\lambda \simeq 490$ nm when all impurities are trivalent. The susceptibility to light-induced index changes is directly correlated with the

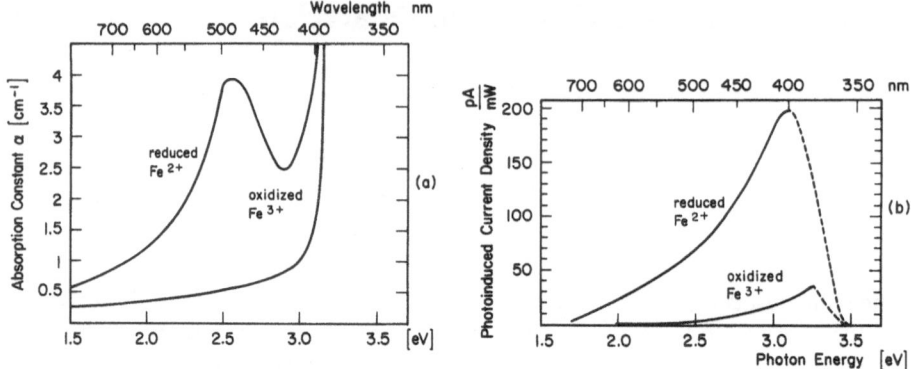

Fig. 2.1. Absorption constant α (**a**) and photoinduced current density j^{ph} (**b**) of reduced (Fe^{2+}) and oxidized (Fe^{3+}) $KNbO_3$ vs photon energy [2.5]

strength of the intervalence band and thus with the Fe^{2+} concentration. The trivalent ions are important, in that they act as electron acceptors, and conduction electrons can be trapped by Fe^{3+} ions, which so convert to Fe^{2+} ions. In the case of oxidized $LiNbO_3$: Fe it has also been shown [2.6] that for low Fe^{2+} concentration the conductivity is p-type and thus holes are photoexcited from Fe^{3+} impurities. In this case Fe^{2+} ions act as traps. We see that in this model the photorefractive effect in Fe-doped oxygen-octahedra ferroelectrics is based on $Fe^{2+} \leftrightarrow Fe^{3+}$ interconversion, and the space-charge set up by the action of light is attributed to a charge redistribution between divalent and trivalent impurities. In recent years it has been shown that the simple model of $Fe^{2+} \leftrightarrow Fe^{3+}$ interconversion cannot explain all phenomena connected with photorefraction and that also other defects such as oxygen vacancies, self-trapped electrons and color centers play an important role [2.7–13]. More details on the connection of the defect structure and the photorefractive effect are given in Chap. 5. Other multivalent transition metal ions have also been used as activators, but Fe ions are the most efficient centers in most of these materials. It has been shown by *Megumi* et al. [2.14] that Ce-doped strontium barium niobate has an efficiency which is nearly four orders of magnitude larger than in the undoped material. Undoped samples of $Bi_{12}SiO_{20}$ and $Bi_{12}GeO_{20}$ [2.14, 15] are very efficient photoconductors with low dark conductivities which allow the buildup of large photoinduced space charges. Photoelectrons are excited in these materials from an extrinsic center (a Si or Ge vacancy associated with an oxygen vacancy) and trapping occurs at an empty photoluminescence center.

Periodic grating-like excitation is particularly well suited to experimental observation of the photorefractive effect and also for its mathematical description. The spatial modulation of the light intensity gives rise to a corresponding modulation of the electron and ionized donor densities (e. g., Fe^{3+} in $LiNbO_3$). Initially the negative and positive electrical charges of the electrons and ionized donors compensate so that there is no net space charge. However, the electrons move by diffusion, under the action of an external field or due to the photovoltaic effect. The electrons are subsequently trapped by empty donor centers. Because of the movement of the electrons, there is a spatial difference in the excitation rate of the ionized electric charge. An electric field builds up which modulates the refractive index via the electro-optic effect.

2.2 Transport of Charge Carriers

The movement of the photoexcited free carriers can be effected by three different mechanisms: diffusion, drift (when an electric field is externally applied) and the photovoltaic effect, as will be discussed in the following.

2.2.1 Diffusion

The development of the space-charge field E_{sc} is sketched in Fig. 2.2. The light intensity I excites ionized donors and electrons. In addition, a spatially homogeneous background electron density may be present. The electrons diffuse so that the spatial amplitude of the electron density is reduced when compared to the spatial amplitude of the ionized donor density. This amplitude difference gives rise to a space-charge distribution modulated in phase with the light intensity. The resulting electric field distribution E_{sc} is shifted by a quarter grating period $\Lambda/4$ against the light intensity.

For the further development of the electron density distribution, the influence of the space-charge field has to be considered in addition to diffusion. If diffusion dominates, the stationary electron density modulation is smoothed out completely. If, on the other hand, the space-charge field dominates the electron movement, the field prevents further smoothing of the electron distribution. In photorefractive crystals, the latter case is assumed and used for mathematical modeling.

2.2.2 Drift

Displacement of the electron distribution can also be achieved by a static electric field. If the ionized donor excitation rate is proportional to $\cos Kx$, the drifted electron excitation rate is proportional to $\cos(Kx + \phi)$ (Figs. 2.2b,c), where K is the grating wave vector and ϕ the phase shift. For sufficiently small displacement the difference of these distributions, i.e., the space-charge density, is proportional to $\sin Kx$. The resulting electric field is proportional to $-\cos Kx$, which corresponds to the intensity distribution except for the negative sign. An alternative way to understand the buildup of a spatially modulated field starts from the current produced by the applied dc voltage. In the stationary state, the current density j is constant as indicated in Fig. 2.2b. The conductivity of the

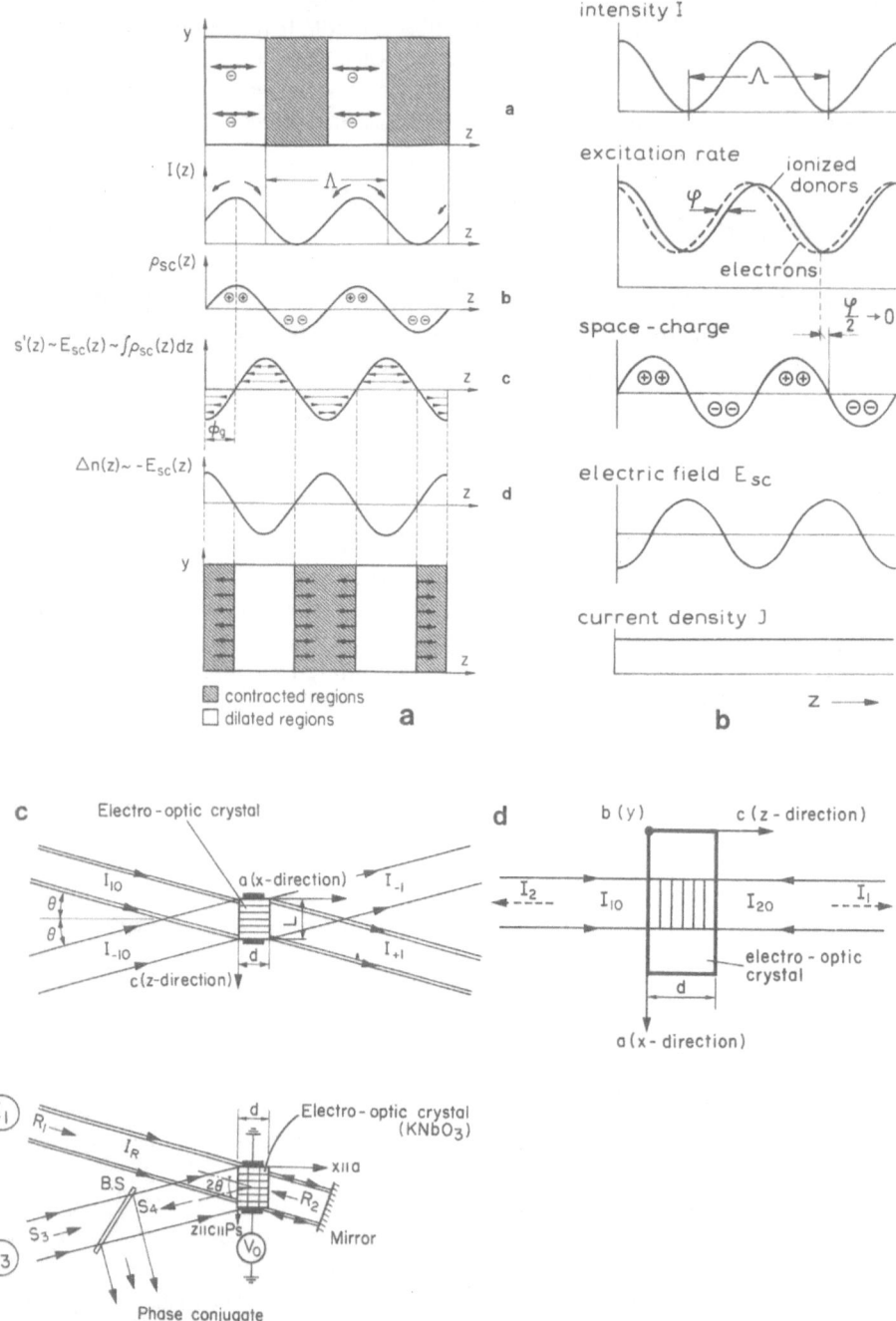

Fig. 2.2a–e. Caption see opposite page

material is spatially modulated due to the modulation of the free carrier density. The electric field E_{sc} therefore also becomes spatially modulated.

Drift and diffusion are sufficient to explain the photorefractive effect in paraelectric crystals such as $K(Ta_{1-x}Nb_x)O_3$, $Bi_{12}SiO_{20}$, $Bi_{12}GeO_{20}$, GaAs, and in highly photoconducting ferroelectrics like $KNbO_3$, where the photovoltaic currents to be discussed below can be neglected.

2.2.3 Photovoltaic Effect

In all electro-optic crystals a photocurrent can also be produced without an applied voltage. Photoelectrons are excited into the charge transfer band with a preferential direction of the velocity along the direction of the polar axis. Additional current contributions due to anisotropic electron trapping and ion displacement are also possible. In total, the photovoltaic current density j^{ph} is given by

$$j^{ph} = \beta_{ijk} E_j E_k^* , \tag{2.1}$$

where E_j, E_k are electric field strength components of the light wave (E_k^* is the complex conjugate of E_k) and β_{ijk} is the third-rank photovoltaic tensor which obeys $\beta_{ijk} = \beta_{ijk}^*$ and which has nonzero components only in media lacking a center of symmetry [2.16–19]. Gratings in electro-optic crystals which are formed by photovoltaic charge transport mostly make use of the tensor component β_{333}. For this configuration

$$j^{ph} = -\beta_{333} E_3 E_3^* = -\beta_{333} I = -\kappa\alpha I , \tag{2.2}$$

where I is the light intensity, α the absorption coefficient for light polarized along the x_3 direction and κ a constant characteristic of the crystal and doping.

The relation $j^{ph} = -\kappa\alpha I$ is valid only for short electron transport lengths [2.18] where the electron transport length is defined as the mean distance the electrons travel before randomization of the velocity. The generalization to an arbitrary electron transport length will not be discussed here.

Glass et al. [2.19] have explained these zero-field currents by a photovoltaic effect due to an asymmetric charge transfer process and Frank-Condon shifts of the excited ions along the polar axis of pyroelectric crystals. Directional intervalence transfer from a substitutional cation in a pyroelectric host will occur if the overlap of defect orbitals with the host cations in the $\pm$ polar directions

Fig. 2.2. (a) Buildup of a space-charge field E_{sc} by diffusion. (b) Buildup of a space-charge field E_{sc} by an external dc voltage (Fig. 2.2c) or by the photovoltaic effect. (ϕ: shift of the excitation rate; j: steady state current density due to the applied voltage). (c) Photorefractive grating recording by means of two-wave mixing. (d) Configuration for recording reflection gratings in photo-refractive crystals. (e) Photorefractive grating recording and read-out by four-wave mixing differ due to the local asymmetry [2.20]. Since the asymmetry has the same sense at all equivalent defects there is a net electronic current upon excitation:

$$j_1 = e\dot{n}_1(p_+l_+ - p_-l_-) = \kappa_1 \alpha I ,\qquad (2.3)$$

where $\dot{n}_1$ is the number of absorbed photons per unit volume and time, p_+, p_- and l_+, l_- are probabilities and mean free paths, respectively, for electrons (with charge e) emitted in opposite directions.

Following excitation, the ionized impurity will in general be displaced along the polar axis due to Frank-Condon relaxation (coupling with asymmetric modes), yielding an additional contribution to the photocurrent:

$$j_2 = e\dot{n}_1 \Delta l_i = \kappa_2 \alpha I ,\qquad (2.4)$$

where Δl_i is the ionic displacement.

Recombination is also asymmetric, but in general the recombination current will not cancel the excitation current since different states are involved ($p_\pm$ and $l_\pm$ are different). The total steady-state current due to electronic and ionic motions is thus given by

$$j^{\mathrm{ph}} = j_1 + j_2 = en_1 L_{\mathrm{ph}} = \kappa \alpha I ,\qquad (2.5)$$

where $L_{\mathrm{ph}} = p_+L_+ - p_-l_- + \Delta l_i$ is a mean effective drift length along the polar axis. This photovoltaic transport length L_{ph} can be regarded as the "mean distance traveled before randomization of the velocity".

Therefore, the photovoltaic current depends only on this mean effective drift length while the photoconductivity depends on the entire history of the excited carrier. The spectral variation of the photovoltaic current need not be the same as that of the absorption constant α, since the former depends not only on the transition probability, but also on the details of the local potential and mean free path. It has also been shown that anomalous bulk photovoltages can be expected even in pure ferroelectrics or electro-optic materials if electron-phonon coupling is considered (e.g., collective Frank-Condon relaxation model [2.21, 22]).

The photovoltaic effect produces a shift in the spatial distributions of the electrons and the ionized donors in a similar way to an applied steady field (Fig. 2.2b). The electric field grating therefore is in phase with the excitation intensity distribution, except for the unimportant factor -1.

It has been shown in [2.23] that gratings can also be recorded by using nondiagonal components of the photovoltaic tensor. In these cases spatially oscillating photovoltaic currents can be produced by perpendicular polarization (ordinary and extraordinary) of the excitation beams. For such a light wave with

$$E = e_o E_o e^{ik_o r} + e_e E_e e^{ik_o r}\qquad (2.6)$$

(e_o and e_e are the polarization unit vectors, E_o and E_e the complex amplitudes of the ordinary and extraordinary waves) an oscillating current appears in the crystal (e.g., along the x-direction if $\beta_{113} \neq 0$):

$$j_1^{ph} = -\beta_{113}^s E_o E_e \cos(q \cdot r) - \beta_{113}^a E_o E_e \sin(q \cdot r)$$

$$= (\beta_{113}^s)^2 + (\beta_{113}^a)^2 \, E_o E_e \cos\left(qr - \arctan \frac{\beta_{113}^s}{\beta_{113}^a} \right) \, . \tag{2.7}$$

This current consists of two parts, which are attributed to the symmetric and antisymmetric components of the photovoltaic tensor.

2.2.4 Charge Transport Equation

In different materials, depending on the electronic properties and experimental conditions, any one of the transport processes mentioned above can dominate. Since the transport of charge carriers upon illumination is of basic importance for the photorefractive effect, we first consider the photocurrents in both the absence and presence of external electric fields in electro-optic crystals, including bulk photovoltaic effects but neglecting transient terms arising from polarization changes when the light is turned on (pyroelectric currents and excited-state dipole effects [2.24, 25].

The transport equation for monochromatic excitation along the polar z-axis of the configuration shown in Fig. 2.2c is then given by

$$j(z, t) = e\mu n(z, t)\left[E_{sc}(z, t) - \frac{V}{L} \right] - \mu k_B T \frac{dn(z, t)}{dz} + e L_{ph} \dot{n}_1(z, t) \, , \tag{2.8a}$$

or

$$j(z, t) = [\sigma_0 + \mu b I(z, t)]\left[E_{sc}(z, t) - \frac{V}{L} \right] - Db \frac{dI(z, t)}{dz} + \kappa \alpha I(z, t) \, , \tag{2.8b}$$

where V is the externally applied voltage between the electrodes on the c-faces of the crystal, L the electrode spacing, μ the mobility of the charge carriers, k_B the Boltzmann constant, T the absolute temperature, D the diffusion constant, $E_{sc}(z, t)$ the electric space-charge field component along the z-axis, and $n(z, t) = n_d + n_1(z, t)$ the total concentration of charge carriers, where n_d is the thermally excited free carrier concentration in the dark and $n_1(z, t)$ the excess free carrier concentration due to illumination. The latter is given by $n_1(z, t) = \Phi \dot{n}_1 = \Phi \alpha I(z, t)/h\nu$, where α is the absorption constant, Φ the quantum efficiency for exciting a charge carrier, $h\nu$ the photon energy, and $I(z, t)$ the light intensity distribution, $\sigma_d = e\mu n_d$ is the dark conductivity and $\mu b I(z, t)$ the photoconductivity with $b = e\Phi\alpha/h\nu$.

The first two contributions in (2.8) are due to drift and diffusion of photoexcited charge carriers, and the last one to the photovoltaic effect. Note that according to (2.8a) the photogenerated carrier density n_1 and the photoinduced space-charge field and, according to (2.8b), the light intensity and the photoinduced space-charge field are nonlinearly coupled and can affect each other, especially, for example, in the steady-state stage of hologram formation.

It will be shown below that in different materials different terms of the transport equation can dominate. For instance, in $LiNbO_3$ and $LiTaO_3$ for homogeneous illumination the photovoltaic term dominates, whereas in photoconducting materials, such as $Bi_{12}SiO_{20}$, $Bi_{12}GeO_{20}$, $K(NbTa)O_3$, and reduced $KNbO_3$, the photoconductivity term is dominant even for relatively small fields. The diffusion term dpends on the mobility as the only materials parameter and is important only for intensity patterns with high spatial frequencies.

The effect of each of these terms in the charge transport and photorefractive effect of $KNbO_3$ has been considered in [2.5]. In this material the different terms can have similar orders of magnitude and, depending on the experimental situation, the relative importance of each can be changed. The characteristic parameters of the three material-dependent terms – photovoltaic effect (ΦL_{ph}), photoconductivity and dark conductivity ($\Phi \mu \tau$ and $e n_d \mu$, respectively) – can be deduced from current-voltage measurements under illumination. The experiments have to be made with the crystal in thermal equilibrium with the radiation to avoid transient contributions. A typical result for an oxidized $KNbO_3$ crystal with 650 ppm Fe is plotted in Fig. 2.3a. The drift length ΦL_{ph} can be calculated from the current densities for zero applied field. The steady-state photovoltaic fields as a function of the light intensity are plotted in Fig. 2.3b. They show a behavior which is expected from (2.8) for $j = 0$. The photovoltage increases linearly for small light intensities (dark conductivity prevails) and saturates in the regime where photoconductivity dominates. The photorefractive effect shows a similar intensity dependence to the photovoltage (see, for example, Fig. 2.7).

The space-charge fields responsible for hologram formation by the photorefractive effect can be derived from the photocurrent data by taking into account the spatial modulation of photocurrents by the nonuniform illumination. In the following we will assume a material with a single donor (concentration N_D) and trap (concentration N_A) center (total concentration $N = N_D + N_A = $ const) and electrons (concentration n) being excited from the donors to the conduction band. It has been pointed out that photoexcited holes may also be contributing to the photoconductivity [2.26–28]. For these special cases the transport equation (2.8) has to be modified in order to include contributions from photoexcited free holes. The situation for the excitation and recombination of the electrons is illustrated in Fig. 2.4.

The rate per unit volume for electron excitation into the conduction band is $[SI(z,t) + \beta]N_D$, and for trapping, $\gamma_R n N_A$, where S is the cross section of photoionization, γ_R is the recombination constant and β is the rate of thermal electron excitation. Ionized donors act as acceptors and other compensative acceptor levels are assumed not to be involved in the phototransition.

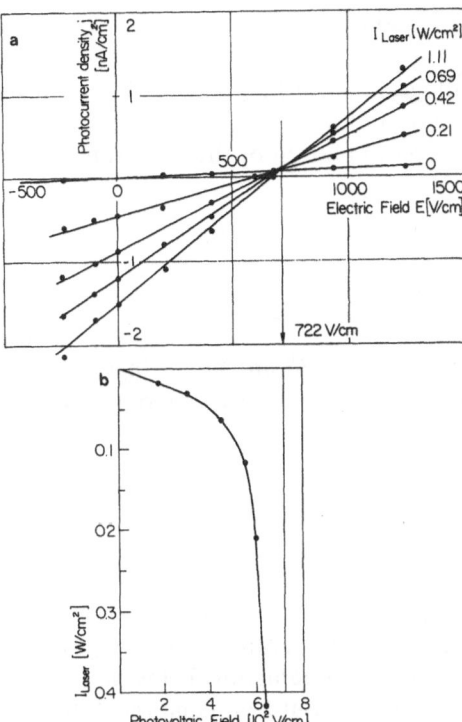

Fig. 2.3. Photocurrent density vs applied electric field (**a**) and intensity dependence of photovoltages (**b**) in $KNbO_3 : Fe$ 650 ppm (oxidized) for $\lambda = 488$ nm

The total change of n resulting from photoionization of the donors, charge transport, and trapping is determined by the continuity equation

$$\frac{dn(z,t)}{dt} = \frac{dN_A(z,t)}{dt} - \frac{1}{e}\frac{\partial j(z,t)}{\partial z} \; , \tag{2.9}$$

where

$$\frac{dN_A(z,t)}{dt} = [SI(z,t)+\beta][N(z,t)-N_A(z,t)] - \gamma_R n(z,t) N_A(z,t) \; . \tag{2.10}$$

In the following we also assume that only a small fraction of the donors will be ionized ($N_A \ll N$). Introducing further the free carrier lifetime $\tau = 1/(\gamma_R N_A)$, assumed to be independent of n (linear recombination), and $\beta N = n_d/\tau$ (in thermal equilibrium: thermal excitation rate equal to recombination rate), (2.10) becomes

$$\frac{dN_A(z,t)}{dt} = g(z) - \frac{n(z,t)-n_d}{\tau} \; , \tag{2.11}$$

where $g(z)$ is the generation rate, proportional to the light intensity.

2.3 Photoinduced Space-Charge Fields

The mechanism of photorefractive hologram recording will be discussed by considering the formation of an elementary holographic grating as realized by the setup shown in Fig. 2.2c. The interference pattern of the hologram can be described by

$$I(x,z) = I_0 \exp\left(-\frac{\alpha x}{\cos\theta}\right)(1 + m\cos Kz) \ , \tag{2.12}$$

where $K = 2\pi/\Lambda$ is the spatial frequency and $\Lambda = \lambda/(2\sin\theta)$ the fringe spacing; $I_0 = I_{+1} + I_{-1}$ is the sum of the incident intensities of the two interfering beams with modulation index

$$m = 2\frac{(I_{+1}I_{-1})^{1/2}}{I_{+1} + I_{-1}}\cos(2\theta p) \ ,$$

where $p = 1$ for polarization directions within the plane of incidence and $p = 0$ for light polarized perpendicular to the plane of incidence. In the following we will assume that the hologram is so thin that the interaction between the two recording beams or interference between the read-out beam and the diffracted beam within the hologram (i.e., beam coupling effects) can be neglected. For thin holograms with $\alpha d \ll 1$ the attenuation factor in (2.12) can be neglected, the main features of the formation of the space-charge field can be described by a one-dimensional model (along the z-axis), the generation of photoelectrons can be assumed to be uniform over the crystal length x, and the generation rate can be written

$$g(z) = g_0(1 + m\cos Kz) \ ; \quad g_0 = \Phi\alpha I_0/h\nu \ . \tag{2.13}$$

In the following treatment all three processes described by (2.8) are allowed to contribute to the photocurrents, and the characteristic charge transport lengths, i.e., the diffusion length $L_D = (D\tau)^{1/2}$ where D is the diffusion coefficient, the drift length $L_E = \mu\tau E$, where μ is the mobility, and the photovoltaic drift length L_{ph}, can be arbitrary, and in particular comparable or larger than the fringe spacing of the grating. Using the results to be derived, one is able to predict and describe the photorefractive characteristics of electro-optic crystals, to understand the primary differences between short and long transport length recording, and to show that long migration lengths can produce a phase shift of the refractive index pattern with respect to the light interference pattern, thus causing transfer of optical energy between the recording beams. Such beam coupling effects are very important in dynamic holography, image amplification and self-pumped phase conjugation, as will be shown later.

The constant applied voltage configuration is analyzed in the following. The short circuit situation is a special case of this configuration (the constant applied voltage being zero). The alternative to the constant applied voltage case is the open circuit (zero current). However, in practice the open circuit case is difficult to achieve.

The accumulation rate of the space-charge density ϱ_{sc} at any point and time is given by the one-dimensional continuity equation

$$\frac{\partial \varrho_{sc}(z,t)}{\partial t} = \frac{\partial j(z,t)}{\partial z} \ . \tag{2.14}$$

Combining this with Poisson's equation gives

$$\frac{\partial E_{sc}}{\partial z} = \frac{\varrho_{sc}(z,t)}{\varepsilon \varepsilon_0} = -\frac{1}{\varepsilon \varepsilon_0} \int_0^t \frac{\partial j(z,t')}{\partial z} \, dt' + G(t) \ , \tag{2.15}$$

where ε is the static dielectric constant of the material and $G(t)$ is determined from the boundary conditions. A constant applied voltage requires

$$-\int_0^L E_T(z,t) \, dz = V \ , \tag{2.16}$$

where $E_T(z,t) = E_{sc}(z,t) - (V/L)$ is the total electrostatic field. This means that

$$\int_0^L E_{sc}(z,t) \, dz = 0 \ . \tag{2.17}$$

If the electron transport length is very small $n_1(z) = \tau g(z)$, the carrier concentration is simply proportional to the generation rate. Examination of (2.9) and (2.11) shows that if $n_1(z) \simeq \tau g(z)$, then $\partial j/\partial z \simeq 0$. That is, there is very slow spatial change in the charge distribution. In the present analysis, the continuity equation is solved directly for $n(z,t)$ rather than assuming $n_1(z) = \tau g(z)$. Since the time scale of the change in $n_1(z,t)$ (starting a few carrier lifetimes after illumination) is very slow compared to τ, then with $dn_d/dt = 0$

$$\frac{\partial n_1(z,t)}{\partial t} \ll g(z) - \frac{n_1(z,t)}{\tau} \quad \text{and} \tag{2.18}$$

$$-\frac{1}{e}\frac{\partial j}{\partial z} \simeq g(z) - \frac{n_1(z,t)}{\tau} \ . \tag{2.19}$$

Thus, as was first pointed out by *Young* et al. [2.29], the difference between g and n_1/τ represents the rate of change of the space charge and of the grating formation. At saturation, when the change in the space-charge field stops $(\partial j/\partial z = 0)$, $n_1 = \tau g$.

The coupled system of differential equations (2.8, 9, 15) subject to the constraint given by (2.17) cannot be solved analytically in general. However, analytical solutions for $n(z, t)$ and $E_{sc}(z, t)$ with the initial condition $E_{sc}(z, 0) = 0$ can be obtained in the initial short writing time approximation and in the steady-state saturation time limit.

2.3.1 Short Time Limit

In the short writing time approximation, the carrier concentration is assumed to be unchanging $(\partial n/\partial t) = 0$ and the feedback effect of the space-charge field is neglected [$E_{sc}(z, t)$ is neglected in the current density equation]. Solving (2.8, 9 and 11) subject to these approximations gives [2.30]

$$n(z) = n_d + \tau g_0 \left(1 + \frac{m'}{[1 + (KL_{ph})^2]^{1/2}} \cos(Kz - \Phi_{ph} + \Phi_1)\right) \tag{2.20}$$

and

$$E_{sc}(z, t) = \frac{eg_0 t}{\varepsilon\varepsilon_0 [1 + (KL_{ph})^2]^{1/2}} [mL_{ph}\cos(Kz - \Phi_{ph})$$

$$-m'L_E\cos(Kz - \Phi_{ph} + \Phi_1) + m'KL_D^2\sin(Kz - \Phi_{ph} + \Phi_1)] , \tag{2.21}$$

where L_D, $L_E = \mu\tau(V/L)$, and L_{ph} are the transport lengths associated with diffusion, drift, and the bulk photovoltaic effect, respectively, the phase angles $\Phi_{ph} = \tan^{-1} KL_{ph}$ and $\Phi_1 = \tan^{-1}\{KL_E/[1 + (KL_D)^2]\}$, and $m' = m/\{[1 + (KL_D)^2]^2 + (KL_E)^2\}^{1/2}$.

For the case of diffusion alone $(L_E, L_{ph} = 0)$, (2.21) reduces to

$$E_{sc}(z, t) = \frac{eg_0 tm}{\varepsilon\varepsilon_0} \frac{KL_D^2}{[1 + (KL_D)^2]} \sin(Kz) . \tag{2.22}$$

For the case of drift alone $(L_D = L_{ph} = 0)$, (2.21) yields

$$E_{sc}(z, t) = -\frac{eg_0 mt}{\varepsilon\varepsilon_0} \frac{L_E}{[1 + (KL_E)^2]^{1/2}} \cos(Kz + \Phi_E) , \tag{2.23}$$

where $\Phi_E = \tan^{-1} KL_E$. For the bulk photovoltaic effect only, (2.19) reduces to

$$E_{sc}(z, t) = \frac{eg_0 mt}{\varepsilon\varepsilon_0} \frac{L_{ph}}{[1 + (KL_{ph})^2]^{1/2}} \cos(Kz - \Phi_{ph}) . \tag{2.24}$$

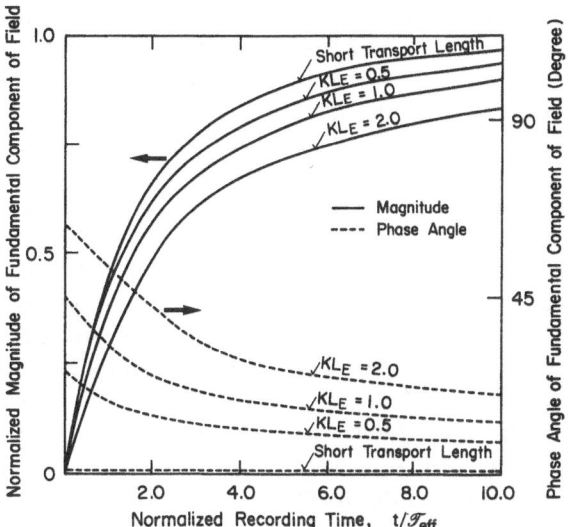

Fig. 2.4. Time development of the magnitude and spatial phase of the fundamental Fourier component of the space-charge field due to drift in an applied or photoinduced field for different values of KL_E including the short drift length approximation where $KL_E < 1$ for $m = 0.9$ $n_d/(g_0\tau) = 10^{-3}$ and $\mu/D = 40/V$ [2.30]

Equation (2.24) appears to be similar to (2.23) (constant applied field case) if L_{ph} is replaced by $-L_E$. However, the similarity is superficial and exists only for the short writing time approximation, since the effective drift length is proportional to the total drift field, which may include applied and photo-induced space-charge fields. That is, it changes during hologram formation, whereas the photovoltaic transport length is (presumed) independent of the real fields whether applied or from space charge, being characteristic of the bulk photovoltaic process.

In Fig. 2.4 the time development of the amplitude and phase of the fundamental Fourier component of the field has been plotted for different values of KL_E assuming $(L_{ph} = L_D = 0)$ and using (2.23 and 25). The fundamental component is normalized with respect to its saturation value. As a result of the normalization, these curves do not depend on the individual values of $\mu\tau$, V/L, and K (for constant $\mathcal{T}_d$) but on their product KL_E. Figure 2.4 clearly shows that for a constant value of $\mathcal{T}_d = \varepsilon\varepsilon_0/(e\mu\tau g_0)$, smaller values of the product KL_E produce larger space-field amplitudes. However, the dependence is not strong; for relatively large values of KL_E (up to 0.5) the field amplitude is within 5% of that obtained from the short drift length approximation $(KL_E \ll 1)$. However, the difference in the phase shift of the fields for these two cases is quite significant. For example, $\Phi_E = 0$ for $KL_E \ll 1$ and $\Phi_E = 26.6°$ at $t = 0$ for $KL_E = 0.5$. Thus the phase shift, which affects the coupling between the two writing beams, is strongly dependent on KL_E whereas the amplitude of the field is not. It should be pointed out that the phase shift of the field will always decrease asymptotically to zero.

Equations (2.23 and 25) and Fig. 2.4 show that the phase shift, and thus the beam coupling, is strongly influenced by the transport length, whereas the amplitude of the space-charge field is not. The saturation value of the space-charge field is found to be independent of the drift and diffusion transport lengths. The rate of development of the space-charge field, on the other hand, varies strongly with migration length.

For a given value of the transport length, larger grating spacings (smaller values of KL) produce larger amplitudes of the space-charge field. Since the short transport length approximation means that $KL \ll 1$ and L has a nonzero value (which may be large or small depending on the value of K), this approximation results in larger space-charge fields.

In most electro-optic materials the characteristic transport lengths L_D, L_E and L_{ph} are short compared with the grating spacing. Therefore, many cycles of photoexcitation, charge transport and trapping are necessary before the charges are finally trapped in dark areas of the hologram. An increase in the operative transport length (up to a certain limit comparable to the fringe spacing of the hologram), as observed in a series of electro-optic photoconductors (see next section), leads to very efficient hologram writing.

2.3.2 Saturation Time Limit

At saturation, the total current density is constant and $\partial n/\partial t = 0$. Therefore, from (2.9 and 11), $n(z) = n_d + \tau g(z)$. Substituting $n(z)$ into (2.8) and using the constraint of constant applied voltage given by (2.17), the space-charge field is [2.30]

$$E_{sc}(z) = \frac{bE'_D \sin Kz}{1 + b \cos Kz} + \left(E'_v - \frac{V}{L} \right) \left(1 - \frac{(1-b^2)^{1/2}}{1 + b \cos Kz} \right), \qquad (2.25)$$

where the effective modulation ratio is $b = n/(1+\Sigma)$ with $\Sigma = \sigma_d/\sigma_p = n_d h\nu/(\Phi\tau\alpha I_0) = \Sigma_1/I_0$ which takes the dark conductivity into account, $E'_D = K(k_B T/e) + KL_{ph}E'_v$, and $E'_v = L_{ph}/\mu\tau[1 + (KL_{ph})^2]$.

The steady-state electric field distribution for oxidized KNbO$_3$ with 650 ppm Fe and $m = 0.9$, calculated from photoconductivity data (Table 2.1), is shown for one grating period in Fig. 2.5. It illustrates the difference in the space-charge fields due to diffusion and drift [2.5]. Since in this material $KL_i \ll 1$ (where $L_i = L_{ph}$ or $L_i = L_E$) for all transport processes involved, $E'_D \simeq Kk_B T/e$ and $E'_v \simeq L_{ph}/\mu\tau$ and the first and second terms in (2.25) are due to diffusion and drift, respectively.

Since the resulting field distribution is not sinusoidal for large intensities, the fundamental components $E_{sc}^{D_1}$ and $E_{sc}^{d_1}$ of the Fourier series expansion of both

Table 2.1. Absorption constants $\alpha(\lambda = 500$ nm), dark conductivities σ_d and transport parameters, derived from photoconductivity measurements and relevant for the photosensitivity, for a series of electro-optic materials. (Room temperature values)

Material	$\alpha\,[\text{cm}^{-1}]$	$\sigma_d\,[\Omega\,\text{cm}]^{-1}$	$\Phi L_{ph}\,[\text{pm}]$	$\Phi\mu\tau\,[\text{cm}^2\,\text{V}^{-1}]$	$E_{ph} = \dfrac{L_{ph}}{\mu\tau}$ $[\text{V cm}^{-1}]$	$\mu\left[\dfrac{\text{cm}^2}{\text{Vs}}\right]$	References
LiNbO$_3$	$10^{-2} - >10^2$	$10^{-8} - 10^{-19}$	$2.4-7.2$	$0.03-0.6\times10^{-12}$	$400-15000$	0.8	[2.31–33]
LiTaO$_3$	–	$\lesseqgtr 10^{-20}$	$6-(60)$	2.6×10^{-12}	$1100-2600$	–	[2.34]
KNbO$_3$	$0.8-5$	$10^{-9} - 10^{-14}$	$5-500$	$1.9\times10^{-8}-4.9\times10^{-12}$	$2-1500$	0.5	[2.5, 35, 36]
K(TaNb)O$_3$ ($T_c=40\,^\circ$C)	–	10^{-17}	<25	$\lesssim 10^{-9}$	2.5	3	[2.37]
K(TaNb)O$_3$ (cubic)	–	9×10^{-16}	0	$\lesssim 7.6\times10^{-9}$	0	3	[2.37]
BaTiO$_3$	2.4	1.3×10^{-12}	–	$10^{-9}-10^{-11}$	–	0.5	[2.38]
Ba$_{0.4}$Sr$_{0.6}$Nb$_2$O$_6$:Ce	$0.3-11.5$	1.7×10^{-11} to 3.5×10^{-13}	8.7	$1.4\times10^{-8}-2.2\times10^{-10}$	$\lesssim 4$	–	[2.14, 39] ($\lambda=440$ nm)
Ba$_{0.25}$Sr$_{0.75}$Nb$_2$O$_6$:Ce	8.6	–	9.6	2.9×10^{-12}	3.3	–	[2.39] ($\lambda=440$ nm)
Ba$_{0.25}$Sr$_{0.75}$Nb$_2$O$_6$	1	7.5×10^{-12}	–	8.4×10^{-8}	–	–	[2.40] ($\lambda=436$ nm)
Ba$_{2.09}$Na$_{0.72}$Nb$_{5.02}$O$_{15}$	$0.25-10$	10^{-14}	9	$0.76-3.6\times10^{-10}$	$2.5-10$	–	[2.40] ($\lambda=436$ nm)
Bi$_{12}$SiO$_{20}$	$1.3-3$	10^{-15}	0	1×10^{-7}	0	0.03	[2.41]
Bi$_{12}$GeO$_{20}$	$1.3-3$	10^{-14}	0	0.84×10^{-7}	0	–	[2.41]
KH$_2$PO$_4$	0.58	10^{-17}	20	–	–	–	[2.42] ($\lambda=300$ nm)
InP	$0.2-2^{\,a}$	–	–	10^{-7}	–	4600	Chap. 8
GaAs	$1-5^{\,a}$	3×10^{-8}	–	10^{-7}	–	8500	Chap. 8

^{a} $\lambda = 1.1$ μm

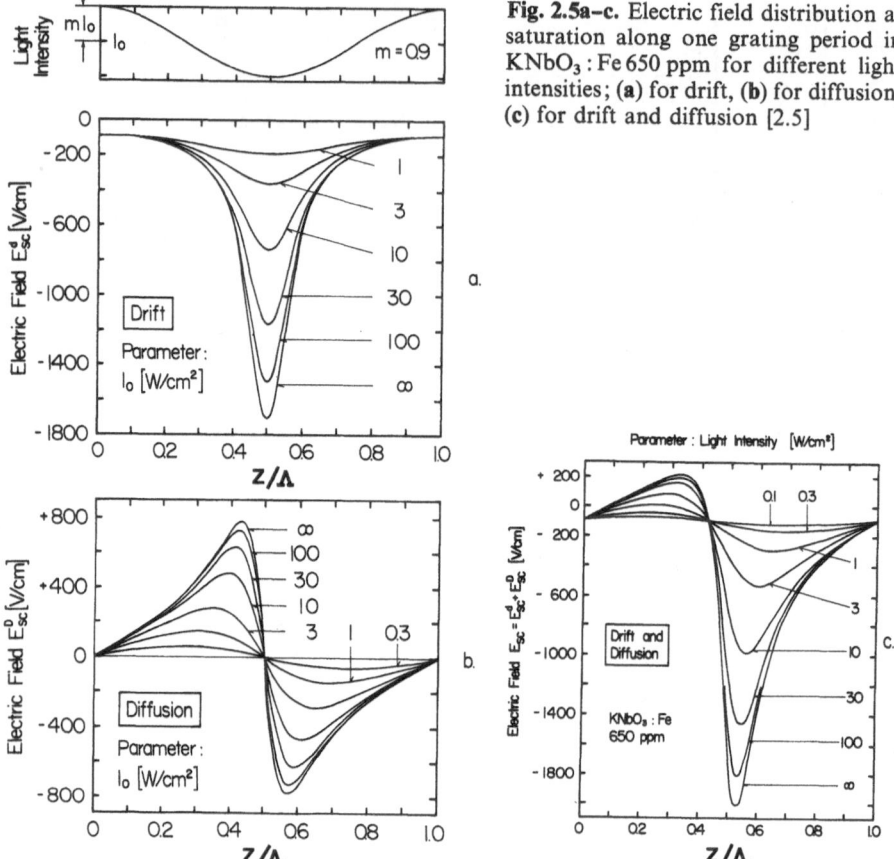

Fig. 2.5a–c. Electric field distribution at saturation along one grating period in $KNbO_3$: Fe 650 ppm for different light intensities; (**a**) for drift, (**b**) for diffusion, (**c**) for drift and diffusion [2.5]

terms in (2.25) were calculated:

$$E_{sc}^{D_1} = 2 E_D' h(b) \quad \text{and} \tag{2.26}$$

$$E_{sc}^{d_1} = \left(E_v' - \frac{V}{L} \right) g(b) , \tag{2.27}$$

where $h(b)$ and $g(b)$ are shown in Fig. 2.6. These components can be used to calculate the intensities of the first diffraction order of the written hologram.

The intensity dependence of $E_{sc}^{D_1}$ and $E_{sc}^{d_1}$ is determined by the relative influence of dark and photoconductivities through $b = m/(1 + \Sigma_1/I_0)$ and has been plotted in Fig. 2.7. For large light intensities the dark conductivity can be neglected and $E_{sc}^{d_1}$ and $E_{sc}^{D_1}$ become intensity independent, but this is not the case for small intensities where dark conductivity is dominant. For low light

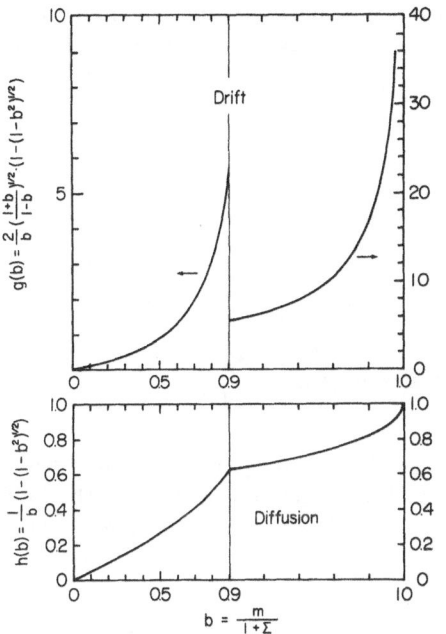

Fig. 2.6. Functions $g(b)$ and $h(b)$ for calculating the amplitudes of the drift and diffusion fields $E_{sc}^{d_1}$ and $E_{sc}^{D_1}$ (amplitudes of first Fourier coefficients of space-charge field distribution (2.25) [2.5])

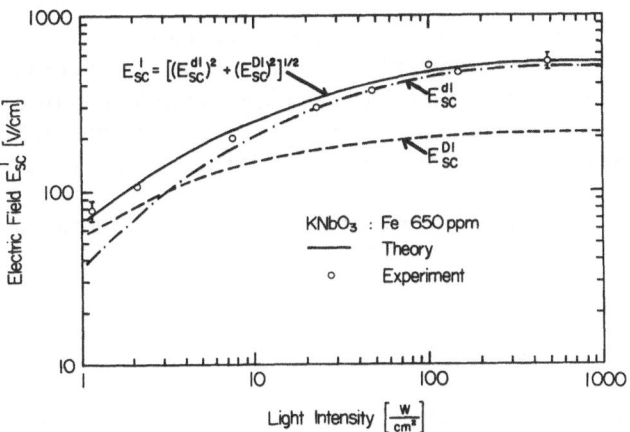

Fig. 2.7. Intensity dependence of the first Fourier component of the space-charge field in $KNbO_3$: Fe 650 ppm [2.5]

intensities the diffusion component dominates over drift. The intensity dependence of

$$E_{sc}^1 = [(E_{sc}^{d_1})^2 + (E_{sc}^{D_1})^2]^{1/2} \tag{2.28}$$

for both drift and diffusion, which can be determined by measuring the light intensity diffracted from the photoinduced grating, is in agreement with the theoretical result (2.1) and photoconductivity data from Table 2.1. The amplitudes and phases of the complete Fourier series expansion (2.25), with arbitrary transport lengths, is given by

$$E_{sc}(z) = -2E_e \sum_{h=1}^{\infty} \left[\left(\frac{1}{b^2} - 1 \right)^{1/2} - \frac{1}{b} \right]^h \cos(hKz - \Phi_g) , \qquad (2.29)$$

where

$$E_e = \left[E_D'^2 + \left(E_v' - \frac{V}{L} \right)^2 \right]^{1/2} \quad \text{and} \quad \tan \Phi_g = E_D' / \left(E_v' - \frac{V}{L} \right) .$$

Clearly, the saturation value of all the Fourier components of the field is independent of the transport length (for constant E_v') and of the way the field develops with time. Also, all the harmonic components have the same spatial phase shift.

2.3.3 Trap Density Limited Space-Charge Field

Under the assumptions of the above analysis there are no apparent limitations on the spatial frequency of the electric field which induces the phase grating. However, it should be clear that the magnitude of the field can be limited by the available trap density N_A of the material. This can be seen by calculating the maximum field amplitude produced by the unneutralized charge density ϱ_{sc} for a given trap density N_A.

For simplicity we can assume that the recorded field pattern is sinusoidal:

$$E = E_q b \cos(Kz + \Phi_g) , \qquad (2.30)$$

corresponding to a space-charge density

$$\varrho_{sc} = -\varepsilon\varepsilon_0 bKE_q \sin(Kz + \Phi_g) = eN_A \sin(Kz + \Phi_g) . \qquad (2.31)$$

From (2.15) the maximum field E_q for a given trap density N_A and $b=1$ is given by

$$E_q = \frac{eN_A \Lambda}{2\pi\varepsilon\varepsilon_0} . \qquad (2.32)$$

For a peak field $E_q = 10^4$ V cm^{-1}, a fringe spacing $\Lambda = 1$ μm, and a dielectric constant $\varepsilon = 50$, the required trap density would be $N_A = 1.7 \times 10^{16}$ cm^{-3}. If the trap density is substantially lower, the photoinduced space-charge fields will be limited by E_q, rather than the field $E_0 = (V/L) - E_v'$ and $E_D = K(k_B T/e)$ derived from the transport equations (2.8, 9), assuming the density of empty traps to be sufficiently large. The influence of trap filling on the photorefractive effect can be described by introducing a parameter ξ, which expresses the photoinduced fields in terms of the maximum possible field E_q given by the trap density N_A. In this approximation ξ_D and ξ_E are smaller than unity so that we get

$$E_D = \xi_D E_q = \left(\frac{l_D}{\Lambda}\right)^2 E_q \tag{2.33}$$

for the diffusion field, and

$$E_d = \xi_E E_q = \left(\frac{l_E}{\Lambda}\right) E_q \tag{2.34}$$

for the drift field, where l_D is called the space-charge screening length of electron tightening by the electric field E_0 and, using (2.32),

$$\xi_D = \frac{4\pi^2 \varepsilon\varepsilon_0 k_B T}{e^2 N_A \Lambda^2} \quad \text{and} \tag{2.35}$$

$$\xi_E = \frac{2\pi\varepsilon\varepsilon_0 E_0}{e\Lambda N_A} \; . \tag{2.36}$$

For ξ_E or $\xi_D \ll 1$ the slight deformations of the free carrier distribution with respect to the intensity distribution, due to diffusion and electric tightening of the carriers near the fringe pattern minima, can be taken into account by replacing the field modulation factor b in all the above relations by the more general expression

$$b_1 = b(1 - \xi_D - \xi_E) \; . \tag{2.37}$$

Note that for the most commonly used approximation $\xi_D = \xi_E = 0$ ($E_q \gg E_0, E_D$), $b_1 = b$.

For ξ_E or $\xi_D \ll 1$ where (2.33 and 34) do not hold, the photoinduced space-charge field reaches the limiting value given by (2.32).

2.4 Photoinduced Refractive Index Changes

In electro-optic crystals the electric fields described above will cause a spatial variation of the refractive indices. Depending on the form of the electro-optic tensor, which depends on the point group symmetry of the crystal, and the direction of the induced space-charge field with respect to the crystallographic axes, the magnitude or orientation of the index ellipsoid will be changed. In the following we will describe the most important configurations used for obtaining photoinduced refractive index changes.

Electro-optic effects are most commonly described by considering field-induced changes of the index ellipsoid:

$$\Delta\left(\frac{1}{n^2}\right)_{ij} = r_{ijk}E_k + R_{ijkl}E_kE_l \ , \tag{2.38}$$

where r_{ijk} is the linear electro-optic tensor (Pockels effect) and R_{ijkl} the quadratic electro-optic tensor (Kerr effect). Since the impermeability tensor $(1/n^2)_{ij}$ is symmetric in i and j, it follows that $r_{ijk}=r_{jik}$ and $R_{ijkl}=R_{jikl}$. This leads to a reduced notation where the interchangeable indices i and j, as well as k and l, are contracted into new indices m and n with $m, n = 1, 2, \ldots, 6$. Table 2.2 shows the nonvanishing elements of the linear and quadratic electro-optic tensor elements. Table 2.2 enables us to discuss different photorefractive configurations. We can distinguish between grating configurations, where a probe beam diffracted from the photoinduced grating has the same polarization direction as the incident wave, and configurations which lead to a rotation of the polarization direction of the probe beams.

2.4.1 Transmission Gratings Without Rotation of Index Ellipsoids

The basic configuration for recording transmission grating is shown in Fig. 2.2.c. The two recording beams with incidence angles $\pm\theta$ (measured within the electro-optic crystal) generate a photoinduced space-charge field $E_{\text{sc},3}$ along the c-axis with a periodicity $\Lambda = \lambda/2\cos\theta$, where λ is the wavelength within the crystal. The changes in the index ellipsoid induced by this field in a ferroelectric material (e.g., point group symmetries $mm2$, $4mm$, $3m$ or $6mm$) is then given by (see Table 2.2)

$$\Delta\left(\frac{1}{n^2}\right)_{ij}(z, t) = r_{113}E_{\text{sc},3}(z, t) + r_{223}E_{\text{sc},3}(z, t) + r_{333}E_{\text{sc},3}(z, t)$$

$$+ 2r_{233}E_{\text{sc},3}(z, t) + 2r_{133}E_{\text{sc},3}(z, t) + 2r_{123}E_{\text{sc},3}(z, t) \ , \quad (2.39)$$

which means that all three main indices of the index ellipsoid are changed. From (2.39) the magnitude of the refractive index changes can be calculated, e.g., Δn_3 is in good approximation given by

$$\Delta n_3(z, t) \simeq -\tfrac{1}{2}n_3^3 r_{33} E_{\text{sc}}(z, t) \ . \tag{2.40}$$

Table 2.2. Tabulation, in contracted notation, of allowed linear and quadratic electro-optic tensor elements for the 32 point groups[a] (from [2.43])

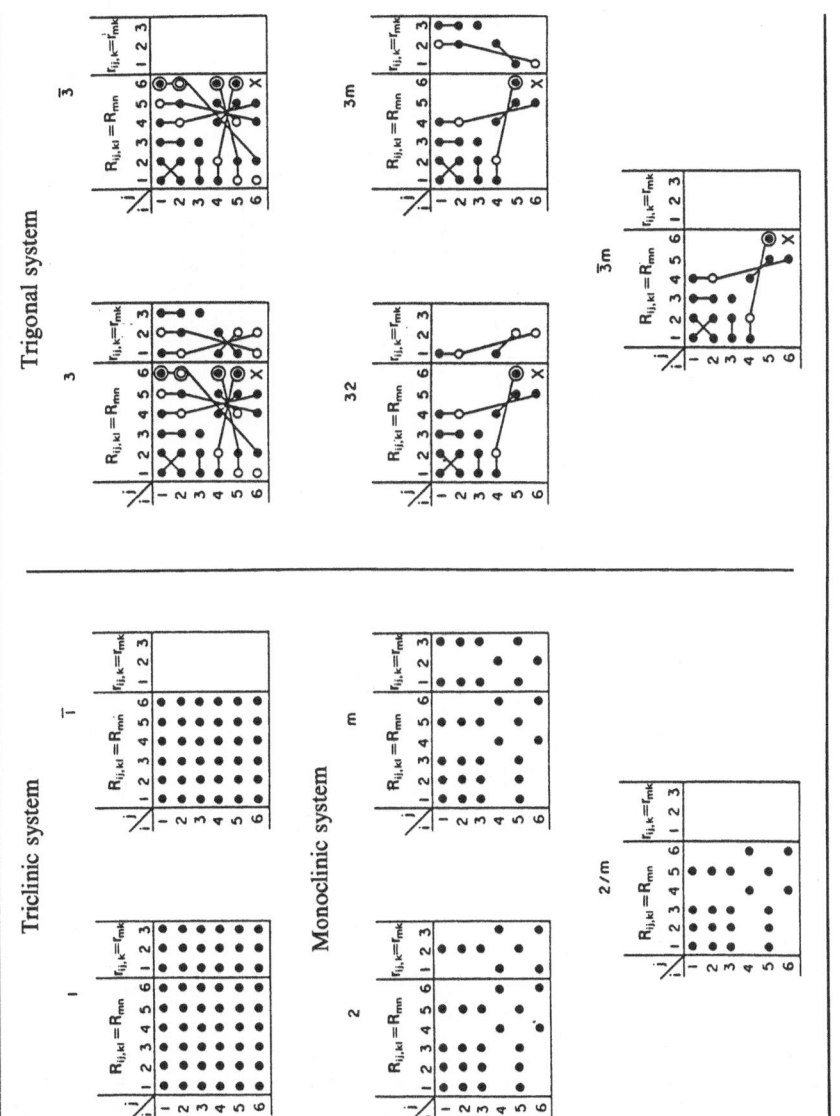

Table 2.2 (continued)

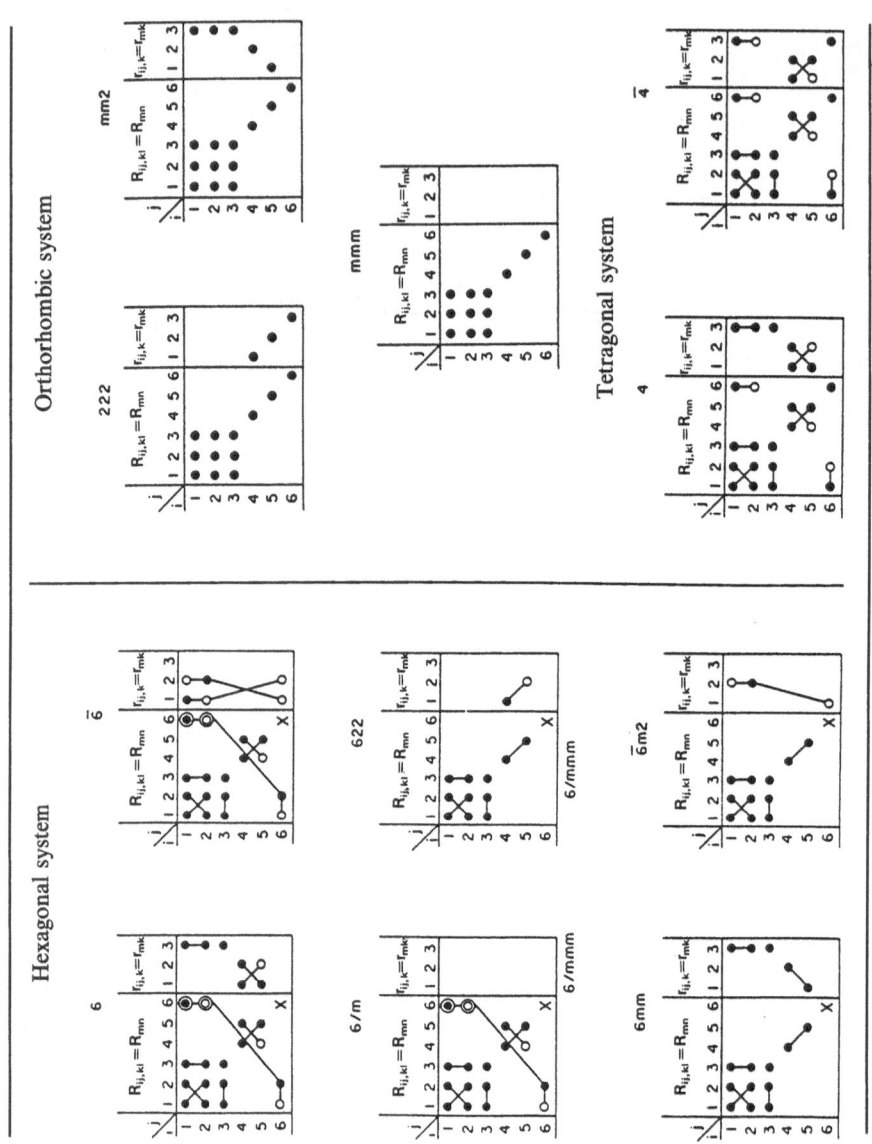

Table 2.2 (continued)

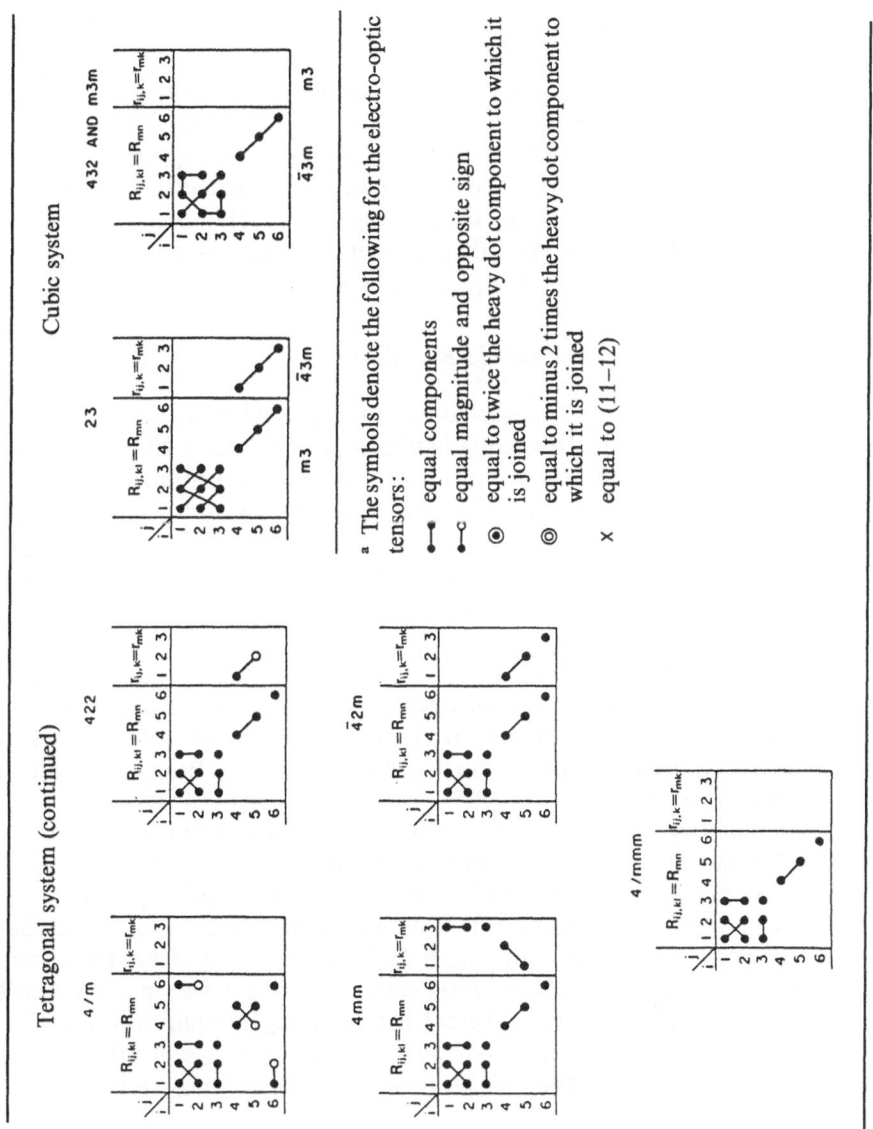

Tetragonal system (continued)

Cubic system

^a The symbols denote the following for the electro-optic tensors:

⟝⟝ equal components

⟝⟜ equal magnitude and opposite sign

◉ equal to twice the heavy dot component to which it is joined

◎ equal to minus 2 times the heavy dot component to which it is joined

× equal to (11–12)

2.4.2 Reflection Gratings

The basic configuration for recording reflection gratings is shown in Fig. 2.2.d. The direction of photoinduced space-charge fields in this configuration is (almost) normal to the polarization directions of recording or probe beams. Therefore mainly the refractive index changes perpendicular to the direction of the space-charge field are seen by the probing or recording beams. Examples of refractive index changes useful for reflection gratings for space-charge fields along the c-axis [see (2.39)] are r_{113}, r_{223} and r_{123}.

2.4.3 Photoinduced Index Ellipsoid Rotations

Nondiagonal elements of the electro-optic tensor such as $r_{42} = r_{232}$, $r_{51} = r_{131}$, can lead to a photoinduced grating having a periodic index ellipsoid rotation. Diffraction at such gratings leads to a change in the polarization direction of the probe beam since off-diagonal elements of the optical susceptibility $\Delta\varepsilon_{ij}$ are induced by the optical wave $E_j(z, t)$, and since the induced polarization at optical frequencies P_i which generates the diffracted wave is orthogonal to the polarization direction of the incident wave,

$$P_i(z, t) = \varepsilon_0 \cdot \Delta\varepsilon_{ij}(z, t) E_j(z, t) , \quad (i \neq j) . \tag{2.41}$$

If the recording material is birefringent, the incident and diffracted waves will propagate with different velocities. Therefore a generalized configuration allowing wave vector conservation has to be fulfilled (Bragg condition for anisotropic diffraction, see Sect. 2.5).

The electro-optically formed phase grating has a spatial index variation given by (2.39), related to the original light distribution in the crystal through (2.21 and 25). It is seen from Fig. 2.5b and (2.25) that the refractive index pattern is phase shifted with respect to the optical interference pattern if diffusion is the charge transport process or if the drift length L_{ph} becomes comparable to the fringe spacing ($KL_{ph} \simeq 1$). Such phase shifts cause coupling between the recording beams which can be used for coherent light or image amplification [2.1].

The phase shifted gratings can also lead to optical self-diffraction and photoinduced light scattering, which is discussed in the next section.

2.5 Isotropic and Anisotropic Bragg Diffraction from Photoinduced Gratings

Optimum light diffraction by sufficiently thick refractive index gratings occurs if the Bragg condition is satisfied. This means that the grating wave vector K_g is exactly equal to the difference between the incident and diffracted optical wave vectors in the medium. Figures 2.8a and 2.9a show the assumed diffraction geometries.

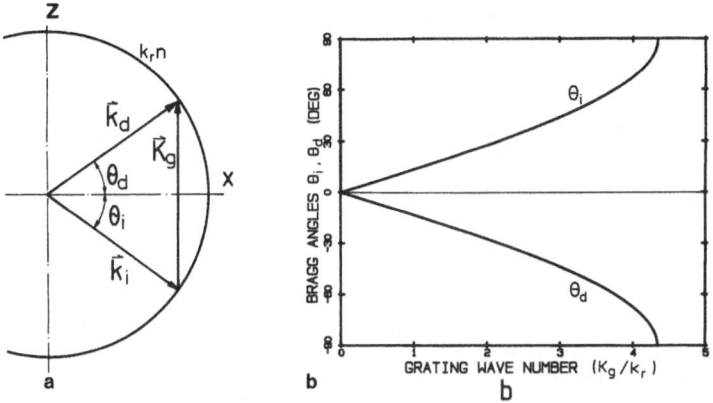

Fig. 2.8. (a) Wave vector diagram for isotropic Bragg diffraction. (b) Wave vector dependence of angles of incidence and diffraction θ_i and θ_d for isotropic Bragg diffraction (for $n=2.2$)

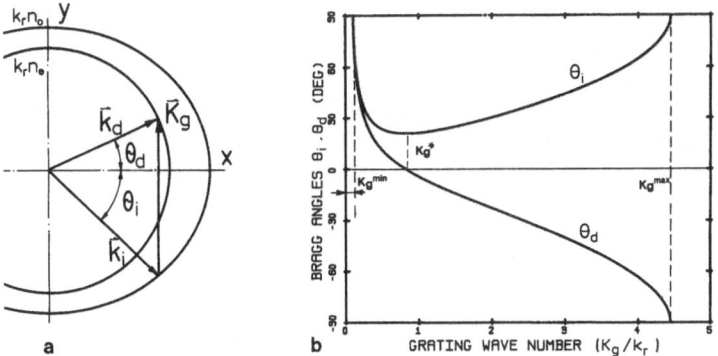

Fig. 2.9. (a) Wave vector diagram for anisotropic Bragg diffraction. (b) Wave vector dependence of angles of incidence and diffraction θ_i and θ_d for anisotropic Bragg diffraction (for $n_o=2.287$ and $n_e=2.192$ as in LiNbO$_3$ for $\lambda_r=633$ nm)

2.5.1 Isotropic Bragg Diffraction

First we shall discuss ordinary isotropic diffraction where only the direction of incident and diffracted waves are changed and not the magnitude of the wavevector. In an isotropic medium, where the light velocities are independent of the propagation directions, the Bragg condition is determined easily from the laws of conservation of energy and momentum (Figs. 2.8a and 2.9a). This leads to

$$\omega_d = \omega_i \tag{2.42}$$

$$k_d = k_i \pm K_g \ . \tag{2.43}$$

Here ω_i, ω_d are the frequencies and k_i, k_d are the wave vectors of the incident and diffracted waves, respectively ($k_{i,d} = n_{i,d} 2\pi/\lambda_r$; n_i and n_d stand for the effective refractive indices of the incident and diffracted waves; λ_r is the vacuum wavelength of the read-out beam).

Conservation of momentum (2.43) leads to the condition for the diffraction angle (Fig. 2.8a)

$$\sin\theta_i = \sin\theta_d = \frac{\lambda_r}{2\Lambda n} , \qquad (2.44)$$

where θ_i, θ_d are the angles of incidence, and diffraction, respectively, measured inside the crystal, $\Lambda = \lambda/2\sin\theta$ is the grating spacing and n is the refractive index. In optically isotropic materials $k_i = k_d$ leads to $\theta_i = \theta_d$.

Figure 2.8b shows the Bragg angles as a function of the relative grating wave number K_g/k_r ($k_r = 2\pi/\lambda_r$) for a material with a refractive index $n = 2.2$. In the case of maximal K_g the two wavevectors k_i and k_d are antiparallel ($K_g = 2k_i = 2k_d$), the incident wave is retroreflected by the phase grating (reflection grating, distributed feedback configuration).

The intensity diffracted by a photoinduced refractive index grating for the Bragg configuration has been derived by *Kogelnik* [2.44]. For a phase grating with a sinusoidal index variation of amplitude Δn_3 and noninteracting beams, the diffraction efficiency η, defined as the ratio of the intensities of diffracted and incident waves, is given by [2.44]

$$\eta = \exp\left(-\frac{\alpha d}{\cos\theta_0}\right)\sin^2\left(\frac{\pi\Delta n_3 d}{\lambda\cos\theta_0}\right) , \qquad (2.45)$$

where d is the grating thickness and θ_0 the Bragg angle corresponding to the incident wavelength λ. The factor $\exp(-\alpha d/\cos\theta_0)$ takes into account a possible uniform background absorption. Equation (2.45) has been generalized to include beam coupling effects [2.45] and the more general case of anisotropic Bragg diffraction [2.46] to be explained in the next section.

2.5.2 Anisotropic Bragg Diffraction in Uniaxial Crystals

The first experiments on anisotropic Bragg diffraction in optically uniaxial crystals were performed by *Stepanov* et al. [2.47] in $LiNbO_3$. They also showed that this type of diffraction can improve the holographic resolution when the read-out wavelength is different from the recording wavelength [2.48].

In an anisotropic medium with $n_d \neq n_i$ the length of incident and diffracted wave vectors is changed, i.e., $k_d = k_r n_d \neq k_i = k_r n_i$ ($k_r = 2\pi/\lambda_r$, where λ_r is the vacuum wavelength of the incident beam). To derive a graphical method for

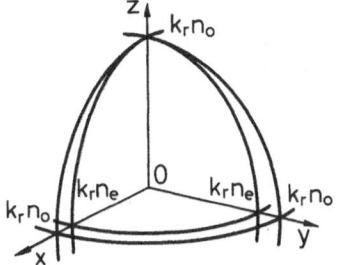

Fig. 2.10. Wave vector surface for an uniaxial crystal

determining the Bragg condition, let us rewrite (2.43) in the form

$$k_r n_d d = k_r n_i i \pm K_g \ , \tag{2.46}$$

where i and d are the unit vectors in the direction of the propagation vectors of the incident and diffracted waves. The refractive indices n_d, n_i of the two waves are obtained as a function of the direction of the wave vectors in the following way. For a given wave vector there are two propagation modes which are subjected to different refractive indices. We can construct a three-dimensional surface with two shells, where when any wave vector is considered, the points of intersection of this wave vector direction with the shells have distances from the origin that correspond to the two wave numbers $k_r n$. Figure 2.10 shows such a "wave vector surface" for an optically uniaxial crystal. (e. g., LiNbO$_3$ or BaTiO$_3$ at room temperature).

With the help of these surfaces one can now determine the geometry which fulfils the Bragg condition (2.46). Any two points lying on different shells form a triangle with the center point 0. Thus, for arbitrary diffraction geometries we can easily find the necessary diffraction parameters.

Let us suppose, that the photoinduced grating vector K_g is parallel to the y-axis (Fig. 2.10) and the light propagation and polarization direction of the incident wave lies in the (x, y)-plane. In this case a nonvanishing electro-optic tensor element of the type r_{yzy} gives rise to a diffracted wave polarized along the z-direction (e. g., r_{42} in LiNbO$_3$ or BaTiO$_3$).

Figure 2.9a shows a center section of the wave vector surface perpendicular to the z-axis. It consists of two circles, the radii of which are $n_o k_r$ and $n_e k_r$, respectively (n_o and n_e are the ordinary and extraordinary refractive indices of an optically uniaxial crystal). From Fig. 2.9b we see that there exists a lower limit for the grating wave number K_g^{min} below which no diffraction is possible. Then k_i and k_d are in the same direction and K_g^{min} is antiparallel to them (Fig. 2.11); the diffracted wave travels in the same direction as the incident wave but the polarizations of both waves are different:

$$K_g^{min} = |n_o - n_e| k_r \ . \tag{2.47}$$

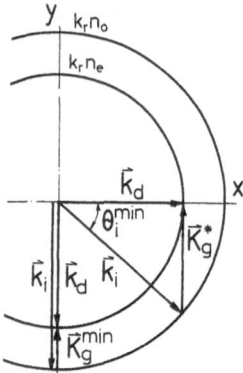

Fig. 2.11. Configurations for anisotropic Bragg diffraction with minimum grating wave vector K_g^{min} and grating wave vector K_g^* where the angle of incidence θ_i^{min} is a minimum (noncritical configuration)

As in isotropic Bragg diffraction, in the case of maximal K_g the two wave vectors k_i and k_d are antiparallel. But, in contrast to the isotropic case, here the retroreflected wave has changed its state of polarization.

Another special configuration of great practical interest is the one where the angle of incidence θ_i^{min} is a minimum and where the diffraction angle is exactly zero. In this case the grating wave number K_g^* is given by

$$K_g^* = \sqrt{|n_o^2 - n_e^2|}\, k_r \ . \tag{2.48}$$

This configuration is also indicated in Fig. 2.11. It is shown in the next section that it can be applied to optically controllable light deflectors. In Table 2.3 we have calculated grating spacings Λ^{max} and Λ^* corresponding to the two grating wave vectors K_g^{min} and K_g^* for several optically uniaxial and biaxial photorefractive materials at room temperature.

Table 2.3. Maximal grating spacing Λ^{max} and Λ^* needed for minimal angle of incidence for anisotropic Bragg diffraction in photorefractive materials [2.49]. (All refractive index data interpolated from [2.50] except KNbO$_3$ from [2.51])

Material	Refractive indices for $\lambda = 633$ nm	Λ^{max}[µm]	Λ^*[µm]
LiNbO$_3$	$n_1 = n_2 = n_o = 2.287$, $n_3 = n_e = 2.192$	6.7	0.97
LiTaO$_3$	$n_1 = n_2 = n_o = 2.177$, $n_3 = n_e = 2.182$	144	5.57
BaTiO$_3$	$n_1 = n_2 = n_o = 2.404$, $n_3 = n_e = 2.316$	14.7	1.40
KNbO$_3$	$n_1 = 2.279$, $n_2 = 2.329$, $n_3 = 2.168$	5.7[a]	0.76[a]
		3.9[b]	0.88[b]

[a] For grating wave direction parallel to x_1 (effective electro-optic coefficient r_{42})

[b] For grating wave direction parallel to x_2 (effective electro-optic coefficient r_{51})

2.5.3 Light Deflection by Anisotropic Bragg Diffraction in Photorefractive Materials

In this section we will discuss anisotropic Bragg diffraction occurring near the minimal angle of incidence θ_i^* (Fig. 2.11) where the angle of diffraction θ_d is around zero degrees.

The broad minimum in the angle of incidence around K_g^* (Fig. 2.9b) is very useful in light deflectors because the light incident at a fixed angle can be effectively deflected by a wide range of grating wave numbers K_g (or recording wavelength λ_0), resulting in a large range of deflection angles. For a small variation of the grating wave number K_g the incidence Bragg angle does not have to be changed whereas the deflection angle can vary considerably. Such a use of anisotropic diffraction for noncritical light beam deflection with photorefractive materials has recently been proposed [2.52] and will be discussed in the following.

The change of the diffraction angle θ_d as a function of the recording wavelength change can be obtained by using the configuration shown in Fig. 2.9a, which yields

$$\frac{K_g}{k_r} = \sin\theta_i' - \sin\theta_d' , \tag{2.49}$$

where θ_i' and θ_d' are the angles in air.

Differentiation of (2.49) near $\theta_d' = 0$ for a fixed angle of incidence leads to

$$\Delta\theta_d' = -\frac{\Delta K_g}{k_r} . \tag{2.50}$$

Using $K_g = 4\pi/\lambda_0 \sin\theta$ we get for the variation of the diffraction angle

$$\Delta\theta_d' = \left(\frac{K_g}{k_r}\right)\left(\frac{\Delta\lambda_0}{\lambda_0}\right) . \tag{2.51}$$

Experimental data for the relative diffraction efficiency as a function of the deflection angle θ_d induced by changing the interaction angle between the writing beams at fixed wavelength of $\lambda_0 = 488$ nm are shown in Fig. 2.12. The measurements where performed in a nominally pure $KNbO_3$ crystal with a thickness of 2.55 mm (plate oriented perpendicular to the a-axis).

In another experiment a He-Ne laser beam incident at the Bragg angle was deflected in several directions by changing the wavelength of the writing beams between 457.9 nm and 514.5 nm (Ar^+ laser lines). Figure 2.12d is a photograph of the diffracted spots for the different wavelengths. The variation of the deflection angle $\Delta\theta_d$ was 5.7° for $\Delta\lambda_0 = 57.7$ nm. No adjustment of all angles of incidence was required in these experiments.

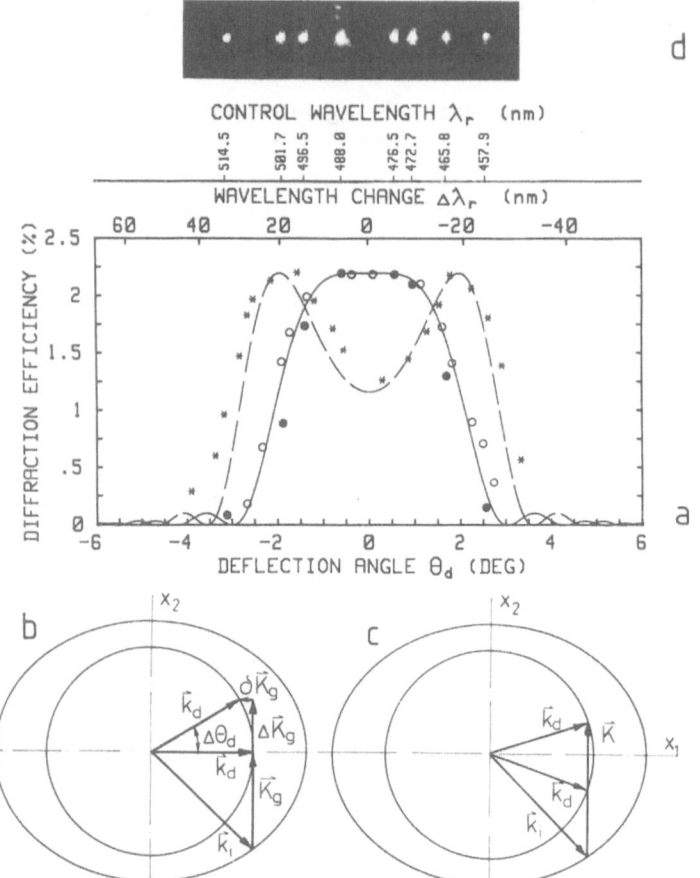

Fig. 2.12. (a) Diffraction efficiency and deflection angle for anisotropic Bragg diffraction at photorefractive gratings in KNbO$_3$. O, *: Recording at fixed wavelength $\lambda_i = 488$ nm and fixed angle of incidence $\theta_i = 56.23°$ (O) or $\theta_i = 56.30°$ (*); tuning of the deflection angle by changing the angle $2\theta_0$ between the recording beams. ●: Recording at fixed angle between the recording beams $2\theta_0 = 36.9°$ and fixed angle of incidence $\theta_i = 56.23°$; tuning of the deflection by changing the recording wavelength. (b) Wave vector diagram for the experimental configuration of the (O, ●) point. (c) Wave vector diagram for the experimental configuration of the (*) points. (d) Photograph of the diffracted spots for six different Ar$^+$ ion laser wavelengths

To stay within a 50% range of the maximum diffraction efficiency for all wavelengths used, a slight increase of the Bragg angle (about 0.07°) can be employed. The diffraction efficiency for this configuration is shown in Fig. 2.12a. The dip at $\theta_d = 0$ occurs because of the slight off-Bragg adjustment for $\Delta K_g = 0$.

2.5.4 Anisotropic Self-diffraction in Photorefractive Materials

In the isotropic diffraction case the two beams recording the photoinduced grating naturally fulfil the Bragg condition for the grating they have created. Therefore the two beams will be partly diffracted into each other's directions. This can lead to the well-known beam coupling effect, where energy is exchanged between the two beams [2.35].

In the case of anisotropic self-diffraction no coupling of the writing beams occurs. They cannot interact with the phase grating they have produced because they do not fulfil the Bragg condition for their own grating (they do not "see" their own grating). Only for a well-defined angle of incidence θ and for a specific write light polarization is the photoinduced grating capable of diffracting one or both of the writing beams. Figure 2.13 illustrates the phase-matching condition needed for anisotropic self-diffraction in a negative uniaxial crystal (e.g., BaTiO$_3$). The two writing beams with wave vectors k_{-1} and k_{+1} have extraordinary polarization and enter the crystal at the incidence angle $\pm\theta$. They create a phase grating with a wave vector K. If one of the vector $(k_{-1}\pm K)$ or $(k_{+1}\pm K)$ ends on the outer shell of the wave vector surface, then the corresponding wave with wave vector k_{+1} or k_{-1} will be self-diffracted from this grating.

Anisotropic self-diffraction was first proposed by *Kukhtarev* [2.53] and experimentally observed in BaTiO$_3$ crystals by *Kukhtarev* et al. [2.54]. They have also shown that self-diffraction is accompanied by optical phase conjugation.

For a negative uniaxial crystal (e.g., BaTiO$_3$) the critical angles are given by [2.54]

$$\sin\theta' = \tfrac{1}{8}\sqrt{(n_o - n_e)(n_o + n_e)} \cdot , \qquad (2.52)$$

$$\sin\psi' = 3\sin\theta' . \qquad (2.53)$$

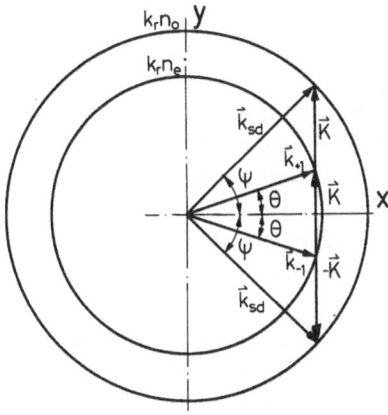

Fig. 2.13. Wave vector and grating configuration for anisotropic self-diffraction in an optical uniaxial crystal (k_{sd}: wave vector of self-diffracted wave, K: grating vector)

The angles θ, ψ are propagation angles inside and θ', ψ' outside the crystal. For optical biaxial crystals (e.g., KNbO$_3$) the critical angles are

$$\sin\theta' = \sqrt{\frac{(n_y - n_z)\cdot(n_y + n_z)}{9n_y^2/n_x^2 - 1}} \qquad (2.54)$$

$$\sin\psi' = 3\sin\theta' \ , \qquad (2.55)$$

where n_x, n_y, n_z are the main refractive indices.

The measurements are in very good agreement with these formulas for negative uniaxial BaTiO$_3$ [2.54] and for the case of optically biaxial KNbO$_3$ [2.46].

2.5.5 Anisotropic Light Scattering in Photorefractive Materials

Anisotropic scattering of a linearly polarized laser beam into a conical ring of different polarization has been observed in LiTaO$_3$:Cu [2.55] and BaTiO$_3$ [2.56, 57] and KNbO$_3$ [2.46]. The intensity in the ring is maximal in the direction perpendicular to the *c*-axis and zero along the *c*-axis, as shown in the photographs of Fig. 2.14. The explanation of this phenomenon is as follows.

When a single laser beam is propagating in a crystal, part of its intensity will be scattered at inhomogeneities, impurities, etc. These scattered waves, which mostly have the same polarization as the strong incident beam, create noisy photorefractive gratings of small amplitude by interference with the incident beam. Some of these gratings are capable of anisotropically diffracting the incident beam. Figure 2.15 illustrates the wave vector matching for this type of scattering. For simplicity only the (x, y) scattering plane for a negative uniaxial crystal (e. g., BaTiO$_3$) is shown. The incident beam k_i scatters in many directions, but only two of the scattered beams k_{isc} lie at an angle θ_{sd} such that the grating vector K_g allows anisotropic self-diffraction of the incident beam.

The scattering angle for a negative uniaxial crystal is given by [2.55]

$$\sin\psi'_{sc} = \left[n_o^2 - \left(\frac{3n_e^2 - n_o^2}{4n_e}\right)^2 \right]^{1/2} \ ; \qquad (2.56)$$

ψ'_{sc} is the scattering angle measured outside the crystal.

The disappearance of scattered light along the *c*-axis is easy to explain by anisotropic self-diffraction. A noisy intensity grating along the *c*-direction cannot induce an anisotropic phase grating as described before. In the crystals discussed here (see Table 2.2) all electro-optic coefficients of the form r_{ij3} (with $i \neq j$) vanish by symmetry and therefore no diffraction or scattering can take place in this direction.

In the case of biaxial KNbO$_3$ the scattering geometry is more complicated. The scattering angle ψ_{sc} is not constant around the direction of the incident beam and this leads to an oval shape of scattered light as shown in Fig. 2.14.

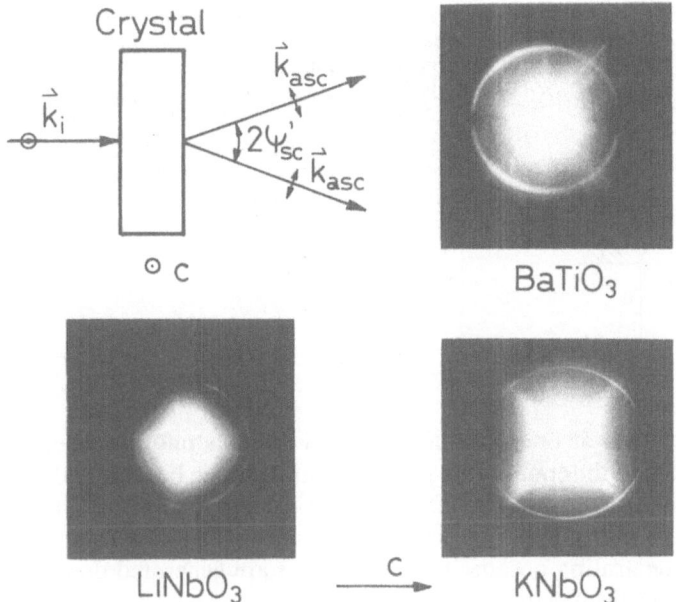

Fig. 2.14. Anisotropic scattering of a laser beam polarized along the c-axis incident normal to the crystal. The photographs illustrate the scattering into a ring of light polarized normal to the c-axis in BaTiO$_3$, LiNbO$_3$ and KNbO$_3$

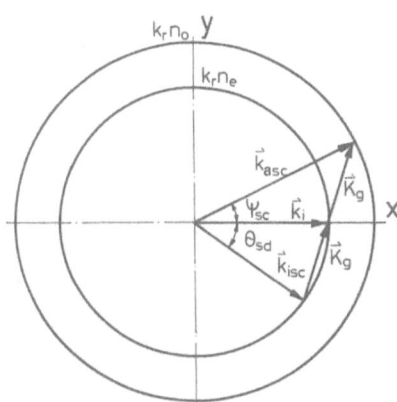

Fig. 2.15. Wave vector matching condition for anisotropic light scattering (diffraction from photoinduced "noisy" gratings)

2.6 Multiphoton Photorefractive Effect

Photorefractive gratings in electro-optic crystals can also be recorded by multiphoton excitation. The principle of multiphoton photorefraction is shown in Fig. 2.16. The crystal is transparent at frequencies ω_1 and ω_2, yet by

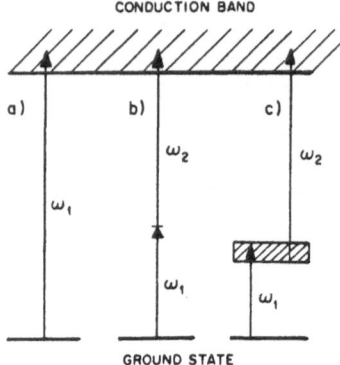

CONDUCTION BAND

a)   b)   c)

GROUND STATE

Fig. 2.16a–c. One- and two-photon excitation mechanisms for photorefractive recording. (a) Single-photon absorption: read-in, read-out and erasure with the same frequency ω_1. (b) Two-photon absorption via a virtual intermediate level; read-in with frequencies ω_1 and ω_2, read-out at ω_1 (or ω_2). (c) Two-step absorption via a real intermediate level: read-in and erasure with ω_1 and ω_2, read-out at ω_2 [2.61]

simultaneous absorption of one photon at ω_1 and another photon at ω_2 free carriers can be excited to the conduction band. The grating can be recorded at ω_1 (or ω_2) and only during recording is the formation process "sensitized" by the second frequency ω_2 (or ω_1). Reconstruction of the grating by light of frequency ω_1 does not erase the grating because no free carriers are generated due to the absence of the second frequency.

This can be illustrated by the two-phonon recording experiments in LiTaO$_3$:Fe performed by *Vorman* and *Krätzig* [2.58]. They recorded gratings via the intermediate ^5E Fe^{2+} state excited by the 1.06 μm emission line of a Nd:YAG laser (Fig. 2.17a). Simultaneous irradiation with the doubled laser frequency (0.53 μm) caused electron transitions from the ^5E state to the conduction band. The grating may be erased by homogeneous illumination at both frequencies simultaneously. Reading the grating with the recording wavelength ($\lambda = 1.06$ μm) does not erase it (Fig. 2.17b).

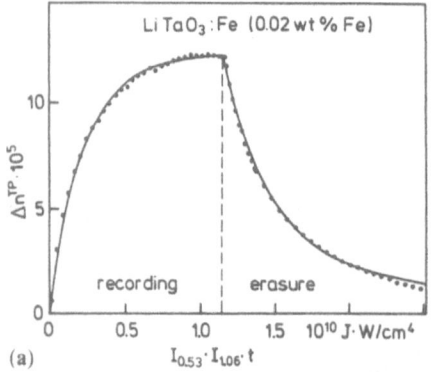

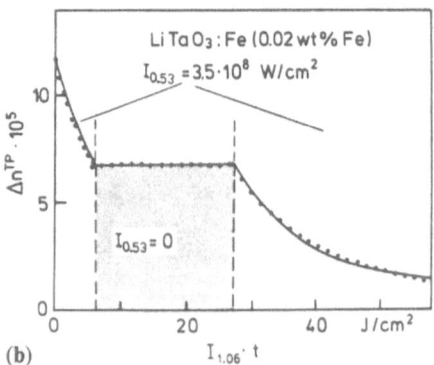

Fig. 2.17. (a) A full two-photon record-erase cycle. The refractive index amplitude Δn^{TP} is plotted versus the product of green and IR intensities and time: $I_{0.53}I_{1.06}t$. (b) The same experiment as in (a), but now Δn^{TP} is plotted versus the product of time and IR intensity only. Thus it is possible to show that in the absence of a green beam ($I_{0.53}=0$) no erasure occurs

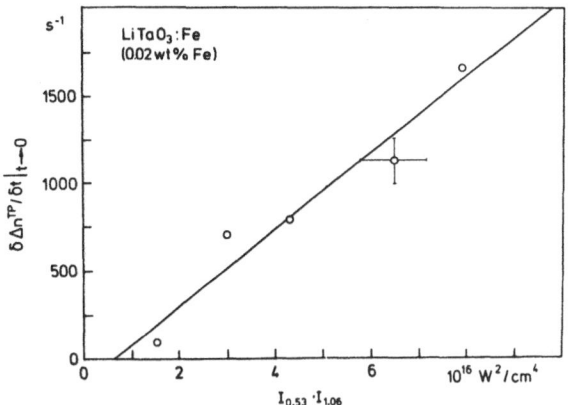

Fig. 2.18. Change of refractive index amplitude $\delta \Delta n^{TP}/\delta t|_{t\to 0}$ versus product of green and IR intensities: $I_{0.53}I_{1.06}$. The slope of the solid line is the two-photon sensitivity S^{TP}

For two-photon photovoltaic grating recording with pulse duration t_p much smaller than the excited state lifetime the refractive index change at the beginning of the recording process ($t\to 0$) for small drift length and modulation ratio is

$$\Delta n^{TP}(t) = \frac{n^3 r \kappa \alpha_{1.06} \sigma^* t_p}{\varepsilon\varepsilon_0 h\nu_{1.06}} I_{1.06}I_{0.53}t \ , \tag{2.57}$$

where κ is the two-photon absorption constant, σ^* is the excited state absorption cross section [2.58] and t the recording time of the mode-locked laser pulse train. Figure 2.18 illustrates that $(\Delta n^{TP}/t)_{t\to 0}$ depends linearly on the product at the two recording intensities, as expected for two-photon recording.

The photosensitivity for two-photon recording has been shown to be much larger than for conventional one-photon processes in several materials including $K(Nb,Ta)O_3$ [2.59] and undoped $LiNbO_3$ [2.60], especially if long-lived real intermediate states such as Cr^{3+} in $LiNbO_3$ and $LiTaO_3$ are used [2.61].

2.7 Fixing of Phase Gratings

2.7.1 Fixing by Domain Reversal

Fixing of photoinduced gratings by means of polarization reversal in regions of high space-charge field has been accomplished in $BaTiO_3$ [2.62] and $Sr_{0.75}Ba_{0.25}Nb_2O_6$ [2.63] by using an external poling field below the the coercive field for bulk ferroelectric switching, as shown in Fig. 2.19. The stored grating can be erased by applying an external field larger than the coercitive field of the entire crystal volume during uniform illumination.

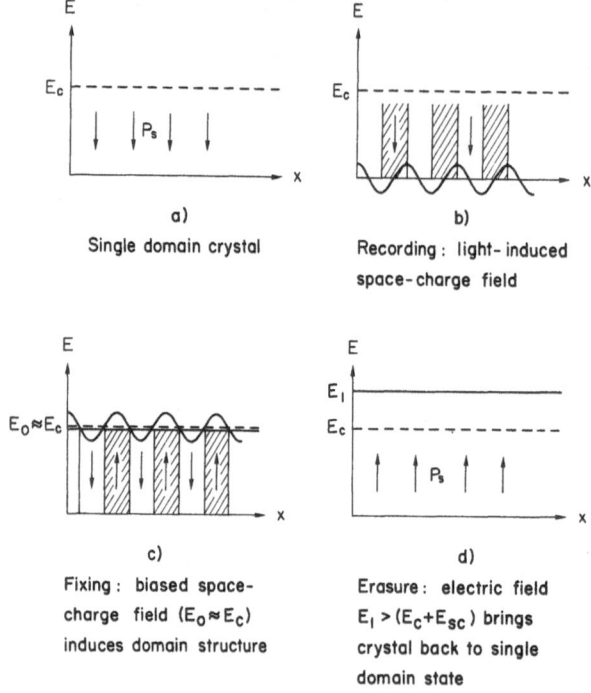

Fig. 2.19a–d. Electrical fixing and erasure of holograms stored in ferroelectric crystals (schematic)

2.7.2 Fixing by Ionic Space-Charge Fields

Complete fixing of an image can be accomplished by transforming the electronic space-charge pattern into a stable ionic charge pattern. In $LiNbO_3$ at 120 °C, the proton conductivity [2.64] exceeds the electronic conductivity and proton motion relaxes the space-charge fields [2.65]. When the crystal is returned to room temperature and uniformly illuminated, the electronic pattern is at least partially erased, leaving the ionic pattern which gives rise to the space-charge field. This process requires a large amount of energy and must be repeated at intervals as the space-charge fields are relaxed by electronic motion. Furthermore, to erase the memory the crystal volume must now be heated above 120 °C. The theory for the ionic relaxation process closely follows that given for optical erasure.

The field pattern decays exponentially with a time constant given by a dielectric relaxation time T_i. The difference is that ionic conductivity, instead of photoconductivity, is used to calculate it, i.e.,

$$T_i = \frac{\varepsilon\varepsilon_0}{\sigma_i} \ . \tag{2.58}$$

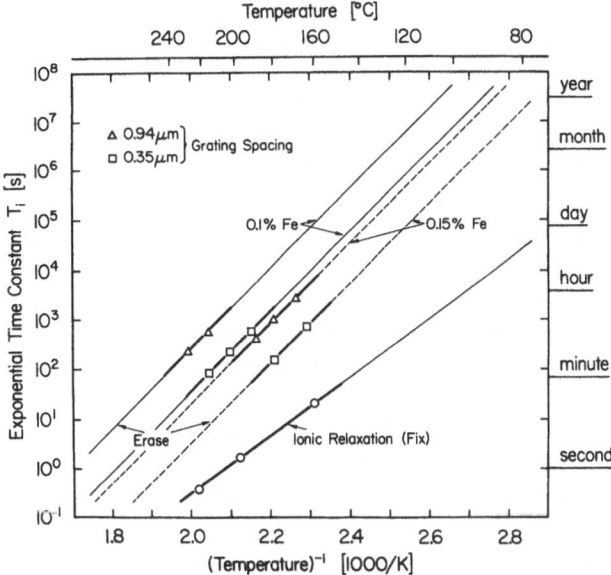

Fig. 2.20. Time constants for thermal fixing *(lower line)* and thermal erasure *(upper lines)* of holograms recorded in Fe-doped LiNbO₃. Erasure time depends on the grating spacing and doping level, the fixing time does not [2.66]

Values of T_i as a function of temperature are shown in Fig. 2.20. The dependence has the form

$$T_i = T_1 e^{(E_a/k_B T)} , \tag{2.59}$$

where T_1 is a constant and E_a is the activation energy for ionic transport ($E_a \simeq 1.1$ eV for the LiNbO₃ used in Fig. 2.20).

Using Fig. 2.20 the maximum shelf life of the recorded and fixed holograms is estimated to be $\simeq 10^5$ yr at room temperature. Indeed, holograms recorded and simultaneously fixed in LiNbO₃ have shown no degradation in either diffraction efficiency or image quality [2.67].

2.8 Connections with Nonlinear Optics

In a formal analogy between holography and nonlinear optics, the photoinduced refractive index changes can be related to the third-order nonlinear optical susceptibility $\chi_{NL}^{(3)}$. Whereas in "ordinary" nonlinear optics the response is instantaneous, the response of the "nonlinearity" in photorefractive materials shows some inertia, which depends on the build-up time of the photoinduced space-charge fields. Most real-time nonlinear-optics experiments (four-wave mixing, optical phase conjugation, etc.) can also be performed in "quasi real time" with dynamic holography in photorefractive media, but at much lower optical power densities than in ordinary nonlinear-optical mixing experiments.

The general approach to the description of nonlinear optical interactions, which has now become the classical one, was introduced by Bloembergen and his co-workers in 1962 [2.68]. They started with the local, time-dependent polarization $P(r, t)$ which, in general, is a functional $f(E)$ of the local, time-dependent electric field $E(r, t)$. The functional character (involving temporal and spatial operators) is of relevance mostly for transient effects; in the steady state and assuming local response of the medium, $f(E)$ becomes a function of E.

The steady-state function $f(E)$ can be expanded into a power series in E. The right-hand side of the linear relation $P_i = \varepsilon_0 \chi_{ij} A_j$ is the first term of this series; at high fields, higher-order terms in A_j have to be considered, giving rise to nonlinear polarization terms P_i^{NL}. (χ_{ij} is the linear susceptibility).

To find the interaction between a certain number of light waves, Maxwell's equations have to be solved using the new expression for the polarization. This cannot be done in a general way; however, since the higher-order contributions to $P(r, t)$ are normally small, it is possible to consider separately the different light beams with frequencies $\omega_1, \omega_2, \omega_3, \ldots$ and wave vectors $k_1, k_2, k_3, \ldots$. The respective higher-order terms can then be introduced into the wave equation as source terms. Moreover, it is convenient to go from the instantaneous and local field E and polarization P to the amplitudes $A(\omega_1, k_1)$, $A(\omega_2, k_2)$, $A(\ldots)$ and $P^{NL}(\ldots)$ of the different waves involved.

In order to describe a certain source term, only those combinations of amplitudes are allowed for which the frequencies and wave vectors add up to the desired values of the source term. Careful bookkeeping of vector and tensor components with indices $i, j, k, \ldots$ (each index can have the values x, y, z) is required to properly account for anisotropic interactions. The resulting general, basic expression for a source term at a frequency combination $\pm \omega_1 \pm \omega_2 \pm \omega_3$ is

$$P_i^{NL}(\pm \omega_1 \pm \omega_2 \pm \omega_3 \pm \ldots; \pm k_1 \pm k_2 \pm k_3 \pm \ldots)$$

$$= \varepsilon_0 [\chi_{ijk}^{(2)}(\ldots) A_j(\ldots) A_k(\ldots) + \chi_{ijkl}^{(3)} A_j(\ldots) A_k(\ldots) A_l(\ldots) + \ldots] , \qquad (2.60)$$

where $\chi_{ijk}^{(2)}$ and $\chi_{ijkl}^{(3)}$ are nonlinear susceptibility tensors of rank two and three, respectively.

Negative values of the frequencies in the combination $\pm \omega_1 \pm \omega_2 \pm \omega_3 \ldots$ are obtained using $A_1(-\omega_1, k_1) = A^*(\omega_1, k_1)$, etc. All the coefficients are functionals (transient effects) or functions (stationary conditions) of the frequencies and, in principle, of the k-vectors involved [abbreviated (...) in (2.60)]. It is assumed in this derivation that the material is spatially homogeneous.

We shall now apply the general expression (2.60) to the case of photoinduced gratings. We are interested in the source term for the scattered beam $P_d(\omega_d, k_d)$ in a four-wave mixing scheme as shown in Fig. 2.21. Two plane pump waves $E_1(r, t)$ and $E_2(r, t)$ with frequency $\omega_1 = \omega_2 = \omega$ and antiparallel propagation direction

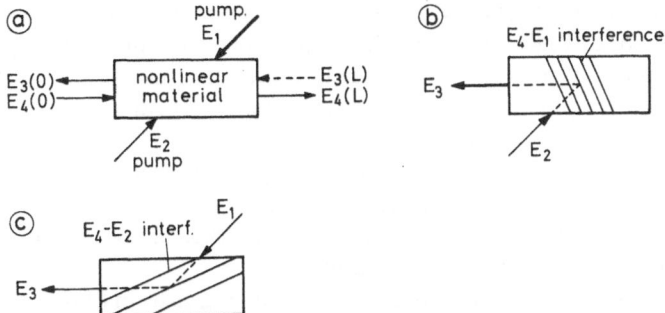

Fig. 2.21. Four-wave mixing configuration (**a**) and grating read-out interpretation for photorefractive gratings produced by E_4–E_1 interference (**b**) and E_4–E_2 interference (**c**) [2.69]

($k = -k_1$), interact with a signal wave $E_4(r, t)$ and generate through nonlinear mixing in the photorefractive material a new wave $E_3(r, t)$.

We assume all interacting waves to have the form

$$E_j(r, t) = \tfrac{1}{2} A_j(r) e^{(i\omega t - k_j z)} + \text{c.c.} \tag{2.61}$$

The second-order term in (2.60) involves the combination of two field amplitudes and hence the sum or difference of the available frequencies: $2\omega, 0$. Thus, this term cannot contribute at the frequency ω. The same is true with the other even-order terms. Thus we are left, to lowest order, with the third-order term, which provides the required combination $\omega_d = \omega_1 - \omega_2 + \omega_4 = \omega$. Similarly, possible combinations exist at all odd-order terms, but higher-order terms are usually smaller. The source term with ω radiates a new wave A_3 which builds up monotonically if the respective wave vectors are equal:

$$k_3 = k_1 + k_2 - k_4 = -k_4 \ . \tag{2.62}$$

Equation (2.62) is called the phase-matching condition in nonlinear optics. It corresponds to the first-order Bragg condition, which connects the directions of the probe wave k_4 and the wave k_3 diffracted at a grating with a grating vector $K_g = k_1 - k_4$.

If the phase-matching condition (2.62) is not fulfilled, a scattered wave can still be observed if the sample is sufficiently thin. This situation corresponds to diffraction at a thin grating (Raman-Nath diffraction).

Considering only terms radiating at frequency ω the basic equation (2.60) reduces to

$$P_{3,i} = \varepsilon_0 (\chi_{ij}^{(1)} A_{3,j} + \chi_{ijkl}^{(3)} A_{1,k} A_{4,l}^*) \ . \tag{2.63}$$

All the nonlinear interactions of the grating generation and detection process are now expressed in one single quantity, the third-order susceptibility $\chi_{ijkl}^{(3)}$. Considering the different physical mechanisms leading to this quantity, it is not surprising that $\chi_{ijkl}^{(3)}$ is a complicated parameter: it is a tensor of rank four with complex components.

Introducing the source term (2.63) into the wave equation, the diffracted field strength A_3 can be calculated. Thus, it is possible to describe the directions and intensities of the diffracted waves by nonlinear optical interactions (degenerate four-wave mixing) of the pump and probe beams.

We shall now show that the general nonlinear-optical-source term (2.63) can be reduced to a change $\Delta\chi_{ij}$ of the linear susceptibility for the probe beam. In our case of degenerate four-wave mixing, the polarization of the diffracted beam can be written in analogy to (2.63) as

$$P_{3,i} = \varepsilon_0(\chi_{ij}A_{3,j} + \Delta\chi_{ij}A_j) \quad \text{with} \tag{2.64}$$

$$\Delta\chi_{ij} \equiv \chi_{ijkl}^{(3)} A_{1,k} A_{4,l}^* . \tag{2.65}$$

In the case of an isotropic medium and $A_1 \| A_2$, $\Delta\chi_{ij}$ becomes a scalar $\Delta\chi$ which can be expressed as a change Δn of the refractive index according to

$$\Delta n \simeq \frac{\Delta\chi}{2n} = \frac{\chi^{(3)}}{2n} |A_1| |A_4^*| . \tag{2.66}$$

Similarly, the nonlinear refractive index n_3 can be introduced, converting (2.66) into

$$\Delta n \equiv n_3 |A_1| |A_4^*| \quad \text{with} \tag{2.67}$$

$$n_3 = \frac{\chi^{(3)}}{2n} . \tag{2.68}$$

In summary, the general approach of nonlinear optics includes the photo-induced grating situation as a special case. The formal description by nonlinear susceptibilities does not provide new information on the material excitations involved, but is an elegant shorthand notation for the evaluation of electromagnetic beam interactions.

2.9 Required Properties of Photorefractive Materials

Photorefractive materials are defined as electro-optic materials in which refractive indices are changed by photo-induced space-charge fields via the electro-optic effect. We have already pointed out that the photoinduced space-charge fields are due to generation and transport of charge carriers. The main three charge transport processes are: diffusion, the volume photovoltaic effect and photoconductivity (drift in an applied electric field). Since diffusion occurs in any electro-optic material, all such materials which have a photoexcitation mechanism are in principle also photorefractive. The dominant charge transport process depends on photoconductivity parameters and on the experimental conditions (i.e., fringe spacing, external electric field, etc.). Other charge transport processes have been observed not only with visible or near ultraviolet light, but, depending on the material, also at ultraviolet and infrared wavelengths. In order to find out whether an electro-optic material is useful for photorefractive applications, the charge transport parameters for the wavelength to be used have to be considered. In this part we will summarize the main properties of the photoconductive electro-optic materials. In order to evaluate materials for applications in hologram storage, coherent light amplification or optical phase conjugation, we will also list and compare the most important fundamental and derived materials requirements.

The main qualities to be considered in choosing electro-optic materials for photorefractive applications are:

1. Photorefractive sensitivity
2. Dynamic range (maximum refractive index change)
3. Phase shift between refractive index and light intensity distribution
4. Photorefractive recording and erasure time
5. Spatial frequency dependence
6. Electric field dependence
7. Laser wavelength for inducing refractive index change
8. Resolution
9. Signal-to-noise ratio
10. Room temperature operation.

In the following each of these attributes and their importance in applications such as hologram storage, light amplification or optical phase conjugation will be discussed separately.

2.9.1 Photorefractive Sensitivity

There are two figures of merit useful for characterizing the photorefractive sensitivity of a given material. The first one describes how much optical energy is needed to produce a given refractive index change, whereas the second one, important in hologram storage, quotes how much energy is needed to give 1% diffraction efficiency for a 1 mm thick storage material. In the first case the photorefractive sensitivity $S_{n_i} = dn_i/dw$ is defined as the refractive index change per unit absorbed energy and volume ($w = \alpha W_0$, where W_0 is the incident optical energy) at the initial stage of photorefractive recording. In the framework of our one-dimensional model we can find an expression for S_{n_1} using (2.21) and (2.40):

$$S_{n_1} = \left(\frac{n_3^3}{2} \frac{r_{33}}{\varepsilon_{33}\varepsilon_0}\right)\left(\frac{\Phi}{h\nu}\right)(eL_{\text{eff}})m \simeq \left(\frac{n_3^3}{2} f_{33}\right)\left(\frac{\Phi}{h\nu}\right)^8 (eL_{\text{eff}})m \ . \qquad (2.69)$$

In (2.69) we have introduced the polarization-optic coefficient $f_{33} = r_{33}/[\varepsilon_0(\varepsilon_{33} - 1)]$ [2.70]. By writing S_{n_1} in this form, the significance of the different terms is emphasized. The first term reflects the electro-optic effect. Although the electro-optic r coefficients can vary appreciably, the polarization-optic f coefficients are similar in different electro-optic materials [2.70]. For oxides it has been shown that $f_{33} \simeq (1/4)P$ in $m^2 C^{-1}$, where P is the total polarization of the crystal (spontaneous and field-induced along the polar axis in Cm^{-2}). With the maximum value of P being $1\ Cm^{-2}$, which is greater than but of the same order of magnitude as the spontaneous polarization of many ferroelectric oxides [2.70] one gets $f_{33} \simeq 0.25\ m^2 C^{-1}$.

The third term in (2.69) is the change of electric dipole moment due to the photoinduced charge separation. The effective transport length L_{eff} depends on the charge transport properties of the material as well as on experimental parameters such as the applied electric field or the fringe spacing Λ of the grating.

For homogeneous illumination, or small drift or diffusion lengths ($KL \ll 1$), $L_{\text{eff}} = L_{\text{ph}} + L_E$ [2.71]. Expressions for L_{eff} are more complicated if $KL \geq 1$ because the different terms for E_{sc} in (2.21) have different phases. For most photorefractive materials, one of the following two cases applies:

(1) Photovoltaic effect only ($L_E, L_D \ll L_{\text{ph}}$):

$$L_{\text{eff}} = \frac{L_{\text{ph}}}{[1 + (KL_{\text{ph}})^2]^{1/2}} \ . \qquad (2.70)$$

(2) No photovoltaic effect ($L_{\text{ph}} = 0$):

$$L_{\text{eff}} = \frac{[L_E^2 + K^2 L_D^4]^{1/2}}{\{[1 + (KL_D)^2]^2 + (KL_E)^2\}^{1/2}} \ . \qquad (2.71)$$

The equations (2.69–71) show that the photorefractive sensitivity depends mainly on the transport lengths L_{ph}, L_E and L_D and increases with L_{eff}. There is however a penalty for optimizing the sensitivity by increasing the transport lengths: if KL_i is larger than unity $L_{\text{eff}} K^{-1} = \Lambda/2\pi$ for any of the above-mentioned charge transport processes.

This is shown also in Fig. 2.22 where we have plotted L_{eff}/L_i as a function of (KL_i). For the drift case (L_E or $L_{\text{ph}} \neq 0$, $L_D = 0$) $L_{\text{eff}}/L_{E,\text{ph}}$ becomes maximum for $KL_i \to 0$. For the case of diffusion only, a maximum value of $L_{\text{eff}} = 1/2\ L_D$ is observed for $KL_D = 1$ ($\Lambda_{\text{max}} = 2\pi L_D$). A measurement of the fringe spacing

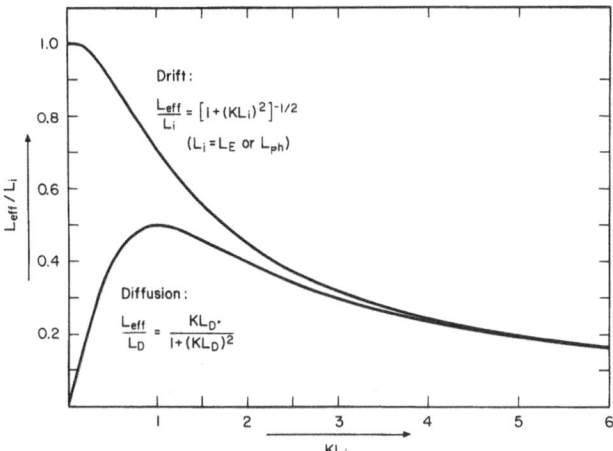

Fig. 2.22. Effective transport length L_{eff}/L_i vs KL_i for diffusion ($L_i = L_D$), drift ($L_i = L_E$) or photovoltaic effect ($L_i = L_{ph}$)

dependence of the photorefractive sensitivity can thus be used to evaluate the diffusion length L_D.

For the case of drift alone ($L_{ph} \neq 0$ or $L_E \neq 0$, $L_D = 0$) a simple relationship for L_{eff} can be derived from (2.70) and (2.71):

$$\frac{1}{L_{eff}^2} = \frac{1}{L_{ph}^2} + \frac{1}{K^2} = \frac{1}{L_{ph}^2} + \frac{(2\pi)^2}{\Lambda^2} , \qquad (2.72)$$

i.e., the analogue to the relationship used for calculating the shunted resistance.

In Fig. 2.23 we have plotted L_{eff} as a function of L_{ph}^{-1} or L_E^{-1} and $2\pi/\Lambda$. Using this scale the curves of constant L_{eff} are circles around the origin. It is also evident from (2.72) and Fig. 2.23, that the limiting value of L_{eff} is $\Lambda/2\pi$ (for L_E or $L_{ph} = \infty$), and $L_{eff} = L_E$ or L_{ph} for homogeneous illumination ($\Lambda = \infty$). Dependences $L_{eff}(\Lambda)$ or $L_{eff}(L_E)$ can be obtained from the intersection of the L_{eff} curves and constant L_E or Λ lines in Fig. 2.23, respectively.

From the two materials properties mentioned above, the change of electric dipole moment per excited electron, described by the second and third terms in (2.69), differs much more from material to material than the electro-optic effect, and can be influenced by experimental parameters. Tables 2.1 and 2.4 list some experimentally obtained values for the most important materials.

By using (2.69–71) and assuming $L_{eff} = 1 \, \mu m$, $f_{33} = 0.25 \, m^2 C^{-1}$, $\Phi = 1$, $h\nu = 2.5 \, eV$ and $n_3^3 = 10$, we get for the ultimate sensitivity of the photorefractive process

$$S_{n_1} = 0.5 \, cm^3 J^{-1} .$$

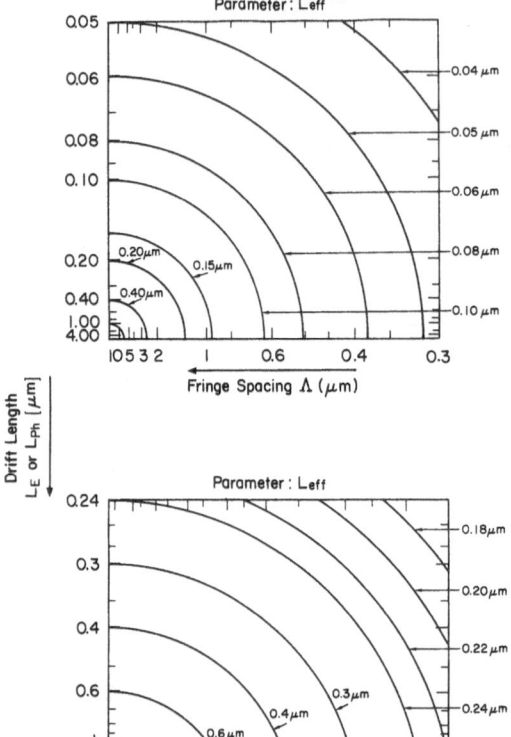

Fig. 2.23. Relationship between the effective transport length L_{eff}, the fringe spacing Λ and the drift length L_E or L_{ph}

Another definition of the photorefractive sensitivity besides

$$S_{n_1} = \frac{dn_i}{d(\alpha W_0)} \tag{2.73}$$

is the refractive index change per unit incident (not absorbed) energy density:

$$S_{n_2} = \frac{dn_i}{dW_0} = \alpha S_{n_1} \ . \tag{2.74}$$

In hologram storage applications one is primarily interested in the intensity diffracted by the photoinduced refractive index change. For a phase hologram with a sinusoidal index variation of amplitude Δn_3 and noninteracting beams, the diffraction efficiency η, defined as the ratio of the intensities of diffracted and

Table 2.4. Electro-optical coefficients and other relevant materials parameters characterizing photoinduced refractive index changes. (Most data have been taken from [2.49]; data for λ = 488 nm if no other wavelength is indicated)

Material	n_i	ε_{ij}	r_{ij} [pm V⁻¹]	f_{ij} [m² C⁻¹]	B_1 [10³ cm² J⁻¹]	Notes
LiNbO₃	2.259	29	31[a]	0.12[a]	2.74	[a] $\lambda = 633$ nm
LiTaO₃	2.227	43	31[a]	0.09[a]	1.77	
KNbO₃	2.227	55	64	0.13[a]	2.85	
		1000	380[a]	0.043[a]	0.93	r_{42}
KNb₀.₆₅Ta₃₅O₃	2.362	10000	1100[b]	0.013[b]	3.22	[b] $E_0 = 5$ kV/cm
K(NbTa)O₃ ($T_c = 88$ °C)	2.35	270	400	0.16	4.27	
BaTiO₃	2.365[c]	168	80	0.054	1.4	[c] $\lambda = 546$ nm
		4300	1640	0.043	1.12	$r_{51} = r_{42}$
Ba₀.₇₅Sr₀.₂₅Nb₂O₆	2.260[a]	180	45[a]	0.028[a]	0.64	
Ba₀.₃₉Sr₀.₆₁Nb₂O₆	2.3	750	216	0.033	0.7	[2.39]
Bi₁₂SiO₂₀	2.540[d]	56	5[a]	0.01[a]	0.32	[d] $\lambda = 620$ nm
Bi₁₂GeO₂₀	2.55[e]	47	3.4[a]	0.03	0.27	[e] $\lambda = 510$ nm
KH₂PO₄	1.474	21	10	0.06	0.34	
GaAs	3.5	13.2	1.2	0.01	2.2	$\lambda \simeq 1.06$ μm
CdTe	2.82	9.4	6.8[f]	0.09	13.9	[f] $\lambda \simeq 3.39$ μm

$B_1 = n^3 \cdot r \cdot e \cdot m/(2\varepsilon_0 \varepsilon h \nu)$, figure of merit for photosensitivity in drift recording

incident waves, is given by [2.44]

$$\eta = e^{(-\alpha d/\cos\theta_0)} \sin^2\left(\pi \Delta n_3 \frac{d}{\lambda \cos\theta_0}\right) , \tag{2.75}$$

where d is the hologram thickness and θ_0 the Bragg angle corresponding to the incident wavelength λ. One advantage of phase hologram storage in contrast to storage based on changes in the absorption coefficient $\Delta\alpha$ (amplitude holograms) lies in the fact that, in the former, maximum values of η can approach 100%. This occurs, however, only in weakly absorbing media and correspondingly the recording sensitivity is very low. For maximum index change per exposure, however, (2.13, 21 and 40) show that a high absorption is desirable, but at the same time the maximum diffraction efficiency decreases, being limited by the material transmission $T_c = \exp(-\alpha d/\cos\theta_0)$. The optimum value is $T_c = 1/3$ corresponding to a maximum diffraction efficiency of 33%. This value compares favorably with the maximum diffraction efficiency of 3.7% achievable with absorption holograms. The optimum crystal length d_{opt} for near normal incidence ($\cos\theta_0 \simeq 1$) is thus given by

$$d_{opt} = \frac{\ln 3}{\alpha} \simeq \frac{1.1}{\alpha} . \tag{2.76}$$

More practical definitions of the photorefractive sensitivity for characterizing the holographic recording potential of a material have thus been defined as the changes of diffraction efficiencies per unit absorbed energy and volume or per incident energy density for unit crystal length, respectively, at the initial stage of hologram formation:

$$S_{\eta_1} = \frac{d(\eta^{1/2})}{d(\alpha W_0)} \frac{1}{d} \quad \text{and} \tag{2.77}$$

$$S_{\eta_2} = \frac{d(\eta^{1/2})}{dW_0} \frac{1}{d} \ . \tag{2.78}$$

Again $S_{\eta_2} = \alpha S_{\eta_1}$.

Since at the initial state of hologram formation the diffraction efficiency is small, the sine in (2.75) can be approximated by its argument and, for sufficiently thin crystals, the absorption term can be neglected:

$$\eta \simeq \left(\frac{\pi \Delta n_3 d}{\lambda \cos \theta_0} \right)^2 , \quad \text{and} \tag{2.79}$$

$$S_{\eta_1} = \frac{\pi}{\lambda \cos \theta_0} S_{n_1} \ . \tag{2.80}$$

In Tables 2.5 and 2.6 the inverse of the sensitivities, $S_{n_1}^{-1}$ and $S_{\eta_1}^{-1}$, are given for a number of electro-optic materials. The origin of the different values is discussed in Sect. 2.10 which treats the materials and their properties.

2.9.2 Dynamic Range (Maximum Refractive Index Change)

The dynamic range of a phase storage medium is the maximum possible photoinduced change of its refractive index. The dynamic range determines the largest diffraction efficiency that can be recorded in a crystal of a given thickness, and the number of different holograms that can be recorded in a given volume. Measured values are shown in Tables 2.5 and 2.6.

There are two theoretical limits to the dynamic range in electro-optic crystals. The first is the density of empty or occupied traps, whichever is lower. This determines the largest space-charge density that can be built up during exposure to light. From (2.31), and assuming sinusoidal gratings, one can see that the charge density required for a given index change [given E_q in (2.31)] is inversely proportional to the grating spacing. For a 1 µm grating spacing, an index change of 2×10^{-5} is produced with a trap density of only 10^{15} cm^{-3}. Impurity-doped crystals do not normally present a problem in this respect, because the trap concentrations are in excess of 10^{18} cm^{-3}. The only case where the traps can limit

Table 2.5. Photorefractive recording sensitivity in photovoltaic recording [2.1]

Materials	α [cm^{-1}]	$S_{n_1}^{-1}$ [10^3 J cm^{-3}]	$S_{n_2}^{-1}$ [10^3 J cm^{-2}]	$S_{n_1}^{-1}$ [mJ cm^{-2}]	$S_{n_2}^{-1}$ [mJ cm^{-1}]	Δn_{max}	Storage time	Comment
LiNbO$_3$	10^{-1} ->10^2	20–200	—	(1000)	300	10^{-5}–10^{-3}	100 h–1 yr	—
LiTaO$_3$	1 –10	—	—	(50)	10	10^{-4}	10 yr	—
KNbO$_3$	1 – 5	6– 60	0.1	—	—	5×10^{-5}	1 s–1 d	—
K(NbTa)O$_3$	—	—	0.1	—	—	—	7 months	T_c=40 °C
BaTiO$_3$	—	—	—	—	0.24	—	15 h	—
Ba$_{0.4}$Sr$_{0.6}$Nb$_2$O$_6$: Ce	10	—	—	50 –1000	—	—	1 h–1 month	—
Ba$_{0.6}$Sr$_{0.4}$Nb$_2$O$_6$: Ce	0.29–11.5	12–75	7.2–30	15	1.5	2.2 × 10^{-5}	1 h–17 months	λ=633 nm
K(D$_{0.7}$H$_{0.3}$)$_2$PO$_4$	0.58	5	9	2.5–15	1.6–6	—	7 d	λ=300 nm, T=113 K

Table 2.6. Photorefractive recording sensitivity in photoconductive recoding [2.1]

Material	α [cm^{-1}]	$B_1 \mu\tau$ $\left[\frac{cm^2}{JV}\right]$ calc.	exp.	$S_{n_1}^{-1}$ $\left[10^3 \frac{J}{cm^3}\right]$	$S_{n_2}^{-1}$ $\left[10^3 \frac{J}{cm^2}\right]$	$S_{n_1}^{-1}$ $\left[\frac{mJ}{cm^2}\right]$	$S_{n_2}^{-1}$ $\left[\frac{mJ}{cm}\right]$	Δn_{max}	E_0 [kV/cm]	Λ [μm]
LiNbO$_3$ [a]	0.1–100	0.2 × 10^{-8}	(0.2–0.4)	—	6	—	—	10^{-4}	50	—
LiTaO$_3$ [a]	3 – 4	0.5 × 10^{-8}	– × 10^{-8}	—	—	66	22	10^{-4}	15	—
KNbO$_3$	3.8	11.6 × 10^{-5}	0.6 × 10^{-5}	0.08	0.02	—	—	10^{-5}	7	10
KTa$_{0.65}$Nb$_{0.35}$O$_3$	—	2.4 × 10^{-5}	—	0.03	—	0.13	0.05	—	10	—
BaTiO$_3$	—	1.8 × 10^{-7}	—	—	—	0.1–10^{b}	—	5×10^{-5}	10	—
Ba$_2$NaNb$_5$O$_{15}$	—	1.1 × 10^{-6}	—	—	—	—	3	10^{-5}	3	—
Bi$_{12}$SiO$_{20}$	2.3	7.6 × 10^{-5}	—	0.014	0.006	0.7	0.3	—	6	5
Bi$_{12}$GeO$_{20}$	2.1	4.6 × 10^{-5}	—	0.05	0.024	2.6	1.7	—	6	5
PLZT ceramic (9/65/35)	—	—	—	—	—	—	100–600	10^{-3}	10	—

[a] λ = 531 nm   [b] Depending on I_0

the diffraction efficiency in a doped crystal is for those prepared so as to have a very low density of empty traps, i.e., reduced crystals.

The second limitation on the dynamic range is the equilibrium space-charge field. Since the steady-state refractive index changes can be derived from (2.25, 29 and 40), the limiting quantities are given by these relations. The maximum refractive index change $(h = b = 1) \Delta n_3^0$ can be derived from (2.29) and (2.40) by using $E_e = [E_D'^2 + (E_v' - V/L)^2]^{1/2}$

$$\Delta n_3^0 = n_3^3 r_{33} E_e . \tag{2.81}$$

Using $b = 1$ means $m = 1$ and $\Sigma = \sigma_d / \sigma_p = 0$ (2.25), i.e. $\sigma_p \gg \sigma_d$. For the diffusion-only case, $\Delta n_3^0 = n_3^3 r_{33} K_g k_B T / e$, i.e., Δn_3^0 depends on the electro-optic coefficient[1] and refractive index as the only materials quantities, and increases with increasing spatial frequency K_g. For photoconductivity to be dominant, $\Delta n_3^0 = n^3 V / L$, and it is seen that again only the materials quantities involved in the electro-optic effect are important. If photovoltaic effects are dominant, Δn_3^0 depends also on the charge transport parameters describing the photovoltaic currents (2.5, 72). This is illustrated in Fig. 2.24 where the intensity dependences of the saturation refractive index changes for a series of $KNbO_3$ crystals have been plotted.

2.9.3 Phase Shift Between Refractive Index and Light Intensity Distribution

a) Steady-State $\pi/2$ Phase Shifted Component of Space-Charge Field

It has been shown in Sect. 2.3 that the refractive index grating can be phase-shifted with respect to the light intensity distribution [see e.g. Fig. 2.4 and (2.20, 25)]. In thick crystals this phase shift leads to an intensity transfer between recording beams and to interference of an incident light beam with its own diffracted beam inside the recording material. This effect causes the continuous recording of a new grating that, depending on the photoinduced refractive index change, may be phase-shifted with respect to the initial grating, and need not be uniform throughout the thickness of the material. The stationary or transient spatial mismatch between the fringes and the grating leads to a series of applications in the field of dynamic holography, such as coherent light amplification and optical phase conjugation. The same space-charge fields,

---

[1] In most of the early publications dealing with the photorefractive effect the authors assume the crystals to be mechanically clamped and thus use clamped electro-optic coefficients. However, in [2.1] we proposed that even if the whole illuminated area (see, e.g., Fig. 2.2a) remains mechanically clamped by the surroundings, positive and negative elastic strains are produced by the piezo-electric effect in areas with respectively positive and negative space-charge fields (Fig. 2.2a). Therefore, sufficiently far away from the boundary of illuminated and dark areas the free electro-optic coefficients have to be used for calculating refractive index changes. This view has found experimental support in [2.5] and [2.72]

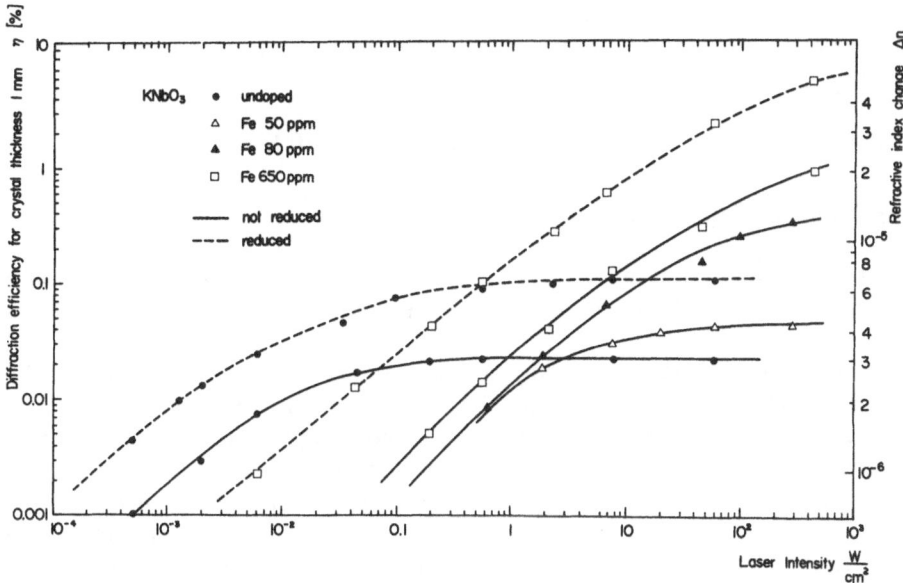

Fig. 2.24. The interrelated quantities saturation diffraction efficiency and refractive index changes vs laser intensity for $\lambda = 488$ nm

diffraction efficiencies, and energy transfer for volume holograms stored in electro-optic crystals can be derived using a "dynamic" theory which also takes into account any possible changes of the fringe-pattern contrast along the crystal length due to the intensity distribution between the two writing beams [2.45].

We can separate the space-charge field into two components (2.25): one which remains unshifted ($A = E'_v - V/L$) and another which is offset by $\pi/2$ with respect to the fringe pattern ($B = E'_D$). The phase mismatch can then be expressed in terms of these two components as

$$\tan \Phi_g = \frac{E'_D}{E'_v - V/L} \ . \tag{2.82}$$

The energy transfer of coupled recording beams is determined by the $\pi/2$-shifted component of the fringe pattern [2.45]. This phase shift remains constant along the hologram depth in the steady-state stage of photorefractive recording.

The exponential gain Γ resulting from this interaction of coupled beams defined as $I_{-1} \simeq I_{-10} \exp[(\Gamma - \alpha)l]$ for $\beta_0 \ll 1$ expressed in terms of materials parameters is given by [2.45]

$$\Gamma = -\frac{2\pi}{\lambda} n_3^3 r_{33} E'_D \cos \theta \ . \tag{2.83}$$

The gain Γ is thus again determined by the materials quantities used to describe the electro-optic response of a material (n_3 and r_{33}) and by the $\pi/2$-shifted component of the electric field E_D'.

In KNbO$_3$ ($n_3^3 = 10.6$, $r_{33} = 64$ pm V^{-1}) a maximum gain $\Gamma = 42$ cm^{-1} can be derived from (2.83) for $\lambda = 488$ nm and $E_D' \cos\theta = 10$ kV cm^{-1}.

b) Transient Phase Shift

A transient phase shift and energy transfer appears in media with local response when the recording time is either less than or comparable with the grating decay time.

At the initial stage of recording for step-like light pulses the light fringes in the symmetrical recording scheme are the planes normal to the incidence crystal surface, as are the isophase surfaces of the holographic grating. Then due to diffraction in the grating the phases of two beams vary and these changes of phase depend on the intensities of the beams. As a result light fringes become tilted and curves and spatial mismatch of the holographic grating and fringes occurs. This gives rise to energy redistribution of the output beams determined by (i) the phase shift Φ_g derived fom (2.82) and (ii) the parameters of the transport equation determining the rise- and decay-time of the grating (see Sect. 2.9.4).

2.9.4 Response Time of the Photorefractive Effect

The time constant to build up a grating is also a specific property of the photorefractive effect. The refractive index changes are due to eletro-optic effects driven by space-charge fields, and the time required to record a grating depends on the efficiency of the charge generation and transport process. The inertia in the nonlinear response of photorefractive media makes an important difference with other nonlinear media where the refractive index change is of electronic origin and thus occurs instantaneously. A complete analysis of the time evolution of the grating formation is presented by *Kukhtarev* in [2.73, 74] for continuous wave illumination and by *Valley* [2.75] for high irradiance nanosecond pulses. We will give in the following the expression for the grating time constant $\mathcal{T}_{eff}$ valid for cw illumination and for charge transport lengths which may be larger or comparable with the grating. In such conditions the crystal response has an overdamped oscillatory behavior with a response time given by [2.1, 76, 77]

$$\mathcal{T}_{eff} = \mathcal{T}_d \frac{\left(1 + \dfrac{\tau_R}{\tau_D}\right)^2 + \left(\dfrac{\tau_R}{\tau_E}\right)^2}{\left(1 + \dfrac{\tau_R \mathcal{T}_d}{\tau_D \tau_I}\right)\left(1 + \dfrac{\tau_R}{\tau_D}\right) + \left(\dfrac{\tau_R}{\tau_E}\right)^2 \left(\dfrac{\mathcal{T}_d}{\tau_I}\right)} , \tag{2.84}$$

where $\mathcal{T}_d$ is the dielectric relaxation of the crystal:

$$\mathcal{T}_d = \frac{\varepsilon\varepsilon_0}{n\mu e} = \frac{\varepsilon\varepsilon_0}{(n_d + n_L)\mu e} \ ,$$ (2.85)

n_L is the free carrier concentration due to the incident illumination I_0, n_d the thermally excited free carrier concentration, and μ the mobility of the photo-carriers

$$n_L = \tau_R \frac{\alpha\Phi}{h\nu} I_0 \ .$$ (2.86)

The charge recombination time τ_R may be written as

$$\tau_R = \frac{1}{\gamma_R N_A} \ ,$$ (2.87)

where γ_R is the recombination coefficient, γ_E and τ_D are the drift and diffusion times, respectively, of the charges, given by

$$\tau_E = \frac{1}{K_g \mu E_0} \ ,$$ (2.88)

$$\tau_D = \frac{1}{\mu k_B T K_g^2} \ ,$$ (2.89)

τ_I is the inverse of the sum of photogeneration rate sI_0 and ion recombination rate $\gamma_R n_0$:

$$\tau_I = \frac{1}{sI_0 + \gamma_R n_0} \ .$$ (2.90)

A simple expression for the time dependence of the space-charge field during grating recording is

$$\Delta E_{sc} = mE_{sc}[1 - e^{-t/\mathcal{T}'_{eff}}] \ .$$ (2.91)

During erasure by uniform illumination the photoinduced space-charge field decreases according to the relation

$$\Delta E_{sc} = mE_{sc} e^{-t/\mathcal{T}'_{eff}} \ ,$$ (2.92)

where E_{sc} is the initial amplitude of the field and $\mathcal{T}'_{eff}$ a time constant similar to that for writing.

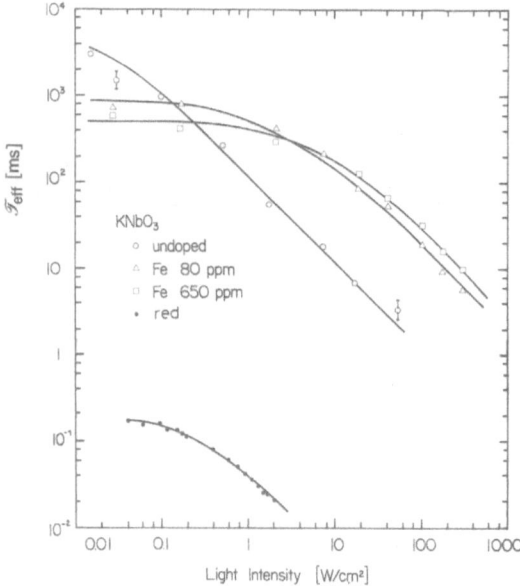

In Fig. 2.25 $\mathscr{T}_{\mathrm{eff}}$ has been plotted as a function of the light intensity for crystals with different Fe concentrations. A behavior described by (2.48) can be observed. At low light intensities $\mathscr{T}_{\mathrm{eff}}$ is constant and given by the dark carrier concentration. As the writing intensity increases, photoconductivities dominate, leading to $\mathscr{T}_{\mathrm{eff}}$ inversely proportional to the absorbed light intensity. The intensity dependence of the saturation diffraction efficiency and the corresponding changes, which also show the relative influence of dark conductivity and photoconductivity, have been plotted in Fig. 2.24 for the same crystals as in Fig. 2.25. Saturation on $\eta(I)$ occurs if photoconductivity dominates.

The time for which the original distribution remains recorded after illumination is also given by (2.85). It is determined mainly by the dark conductivity σ_{d}. It can be seen from Table 2.5 that electro-optic crystals typically have large dielectric constants and they can be highly insulating (Table 2.1). Dielectric relaxation times of up to 2 yr can be reached.

Erasure, however, can occur during read-out of holograms. The light uniformly excites electrons out of the traps so that they can redistribute themselves uniformly, thus bringing the sample back to its original state. With the use of multiphoton absorption schemes (Sect. 2.8) or fixing methods (Sect. 2.7) erasure can be avoided.

In real-time holographic interferometry, signal processing, coherent light amplification or optical phase conjugation, long storage times are undesirable. The main requirement in these applications is a large charge displacement to give a high photorefractive sensitivity. In photoconductive materials the charge displacement can be controlled by an external field, which makes these materials especially interesting for real-time signal processing and for the other applications mentioned above.

2.9.5 Angular and Wavelength Selectivity

The Bragg condition for constructive interference of light scattered from different parts of a grating limits the range of angles of incidence of the read-out light beam over which reconstruction of the original wave front can occur, and the range of wavelengths over which the wave front can be constructed, permitting wavelength selectivity to be used to access different holograms. These two conditions are related through the Bragg relation

$$\lambda = 2 \Lambda \sin \theta_0 \ . \tag{2.93}$$

Angular selectivity is most commonly used for multiple storage, since the variation in angle can be easily obtained by rotation of the storage medium or by deflection of the read-out laser beam. Variation of wavelength is not a practical approach, because of the broad range of wavelengths of coherent light required for the storage of a large number of holograms (319 Å/degree of angular spacing of 488 nm). This approach has been used, however, for the direct recording of color objects using the blue, green and red output from a mixed Ar^+/Kr^+ ion laser in Fe-doped $LiNbO_3$ [2.67].

A measure of the angular selectivity of phase holograms is the angular deviation of first zeros in diffraction efficiency with respect to the Bragg angle $\Delta\theta - \theta$, which can be approximated by [2.44]

$$\Delta\theta = \frac{\lambda}{2d} \frac{1}{\sin \theta_0} \ . \tag{2.94}$$

In an approximation similar to that used to compute the angular selectivity (2.94), one gets for the wavelength selectivity

$$\Delta\lambda \simeq \frac{\lambda^2 \cos \theta_0}{2 dn \sin^2 \theta_0} \ . \tag{2.95}$$

The first zeros for a hologram recorded in a 1 cm $LiNbO_3$ crystal at 0.488 μm with a Bragg angle $\theta_0 = 15°$ occur at ± 0.22 mrad. To obtain a read-out efficiency greater than 50% of the peak diffraction efficiency ($\Delta\theta = 0$), the repeatability of the accessing device must be ± 0.11 mrad or less for the 1.0 cm thick crystal.

The wavelength selectivity for these parameters is $\Delta\lambda \simeq 2$ Å. Consequently, faithful read-out of a volume grating with a broad range of spatial frequencies can only be achieved with the same wavelength as used for writing.

While these low values pose some problems of alignment for read-out of holograms. it has the advantage that individual holograms can be superimposed in the same volume of material by varying either the wavelength or the angle of the writing beam outside the ranges given in (2.94 and 95).

Up to 1000 holograms can in principle be superimposed with a total angular variation of only 0.1 rad in a 1 cm thich crystal. Experimentally, up to 500 holograms with a maximum capacity of up to 5 Mbits/mm^2 surface area have already been reversibly stored in LiNbO$_3$ [2.31], corresponding to storage densities of the order of 10^{11} bits/cm^3.

2.9.6 Other Requirements

a) Resolution

Since, for the most part, these materials are single crystals, the spatial resolution is in principle limited only by the distance between traps. Consider a trap density of 10^{15}, the minimum value required for an index change of 2×10^{-5} at a grating spacing $\simeq 1$ μm. The distance between traps is on the average only 100 nm, an order of magnitude smaller than the grating spacing. Thus, the hologram can clearly by recorded at this spatial frequency, but it may have some scattering noise due to statistical fluctuations of the trapped electrons. The noise would be even more severe at smaller grating spacings. To compound the problem, a smaller grating spacing would lead to a lower refractive index change, assuming a trap-limited situation. For these reasons, the use of impurity doping is quite important. Concentrations in excess of 10^{18} cm^{-3} can be easily achieved, giving a distance between traps of only 10 nm. This should be sufficient for any holographic application. PLZT ceramics, which are of particular interest for hologram storage because they can be prepared in quite large dimensions, contain pores as small as 0.5 μm [2.78]. The resolution limits due to these defects have not been studied.

b) High Signal-to-Noise Ratio

Another problem associated with the lack of latency, i.e., the immediate appearance of a refractive index change, is the buildup of optical scattering [2.79, 80]. This occurs when any photosensitive electro-optic crystal is exposed to a coherent beam of light. A defect that scatters light produces a spherical wave that then interferes with the original beam. The resulting interference patterns are recorded as index changes that, in turn, lead to more scattering. The result is a rapid buildup of complicated scattering patterns that produce noise in the read-out image. This process occurs during storage and thus limits the total exposure (and thus the diffraction efficiency) that can be used. It also occurs during read-out and can be a particularly severe problem for holograms fixed in highly sensitive materials. Other scattering sources are discussed in [2.81]. Much better signal-to-noise (S/N) ratios than the minimum of $\simeq 20$ dB corresponding to a theoretical error rate of 10^{-11} [2.82] have been achieved in high quality electro-optic crystals at spatial frequencies allowing the recording of more than 500 holograms.

c) Wavelength

The electro-optic materials considered for photorefractive recording should be photosensitive at a convenient laser wavelength. Materials and suitable dopants have already been found which allow recording in a broad range of wavelengths extending from the near UV to the near infrared range.

2.10 Materials and Properties

Photorefractive effects have been studied in many electro-optic materials, and a great deal of research has been devoted to understanding the effect and optimizing the materials. Since the oxygen-octahedra ferroelectrics show the largest electro-optic effects, as well as other promising properties, most of the work concerning the photorefractive effect has been concentrated on this group of materials. The primary source of photocarriers in undoped crystals was found to be small concentrations of Fe impurities [2.83] ($\simeq 10$ ppm) incorporated into the lattice during the high temperature Czochralsky growth. While the electro-optical properties seem not to depend on these impurities, the charge transport parameters and optical absorption, and thus the photorefractive sensitivity, strongly depend on the concentration of impurities and on their valence states, determined by chemical treatment (oxidation, reduction).

2.10.1 Oxygen-Octahedra Ferroelectrics

a) Transition Metal Dopants as Photorefractive Centers

Transition metal elements are useful activators for photorefractive oxides, because they are capable of giving up and recapturing d electrons. These elements can enter the ABO_3 lattice substitutionally in either the A or B sites, both of which are surrrounded by a roughly octahedral array of oxygen atoms. The valency of the dopants, usually different from that of the A or B metals, can be compensated by oxygen vacancies [2.32, 84].

Iron and other transition metal dopants are added to the ABO_3 melt during growth in the form of oxides. About 5%–90% of the amount added is incorporated in the solid. Apart from iron, other impurities such as chromium [2.79], nickel [2.79], copper [2.79], manganese [2.79, 85], rhodium [2.85], uranium [2.87], cobalt [2.79, 88] and cerium [2.89] have been used as dopants, but the sensitivity S_{η_1} appears to be greatest for Fe^{2+}. In the case of $LiNbO_3$ it has been shown [2.83] that the photorefractive sensitivity is improved only for

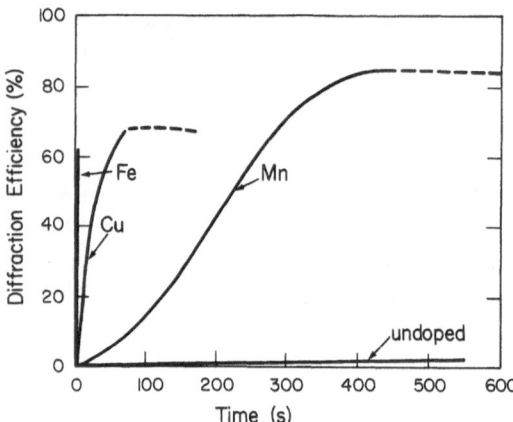

Fig. 2.26. Time dependence of the diffraction efficiency for undoped $LiNbO_3$ and manganese-, copper- and iron-doped $LiNbO_3$ [2.79]

Mn, Fe and Cu, and probably also for Ce [2.89] and U [2.87], where the excited electrons are coupled to the conduction band and only loosely bound to the transition metals, whereas in the case of Cr, Co and Ni the optical absorption arises from localized electronic states. Figure 2.26 illustrates the differences of the time dependence of the diffraction efficiency for undoped $LiNbO_3$ and for $LiNbO_3$ doped with Mn, Cu and Fe.

In the Fe-doped crystals the sensitivity is correlated with the Fe^{2+}/Fe^{3+} ratio [2.90]. The ratio can be controlled by suitable heat treatment in an oxidizing or reducing environment [2.91] or by electrochemically reducing the crystals [2.25]. It has been shown that in $LiNbO_3$ the photovoltaic current density depends linearly on the concentration of donors (Cu^+ or Fe^{2+}) [2.92]. The photoconductivity, however, depends on the donor and acceptor concentration. In Fig. 2.27 the electron and hole contributions to the photoconductivity of Fe-doped $LiNbO_3$ have been reproduced. For large Fe^{2+}/Fe^{3+} ratios, so that Fe^{2+} centers are excited into the conduction bands, electron transport predominates [2.93]. For small Fe^{2+}/Fe^{3+} ratios, however, holes in the valence band are generated by excitation of electrons from oxygen π orbitals to Fe^{3+} ions, and hole conductivity predominates [2.93]. If both electrons and holes contribute (Chap. 5), the conductivity in the above equations has to be modified to become [2.92]

$$\sigma_p = \frac{\sigma_{pe} - \sigma_{ph}}{\sigma_{pe} + \sigma_{ph}} \, , \qquad (2.96)$$

where σ_{pe} and σ_{ph} are the electron and hole conductivity, respectively. The concentration of Fe^{2+} determines the absorption coefficient near the Fe^{2+} peak at 2.55 eV (Fig. 2.1.a). By controlling this concentration, the optimum optical density of 0.3 for maximum overall recording–read-out efficiency (2.75) at a

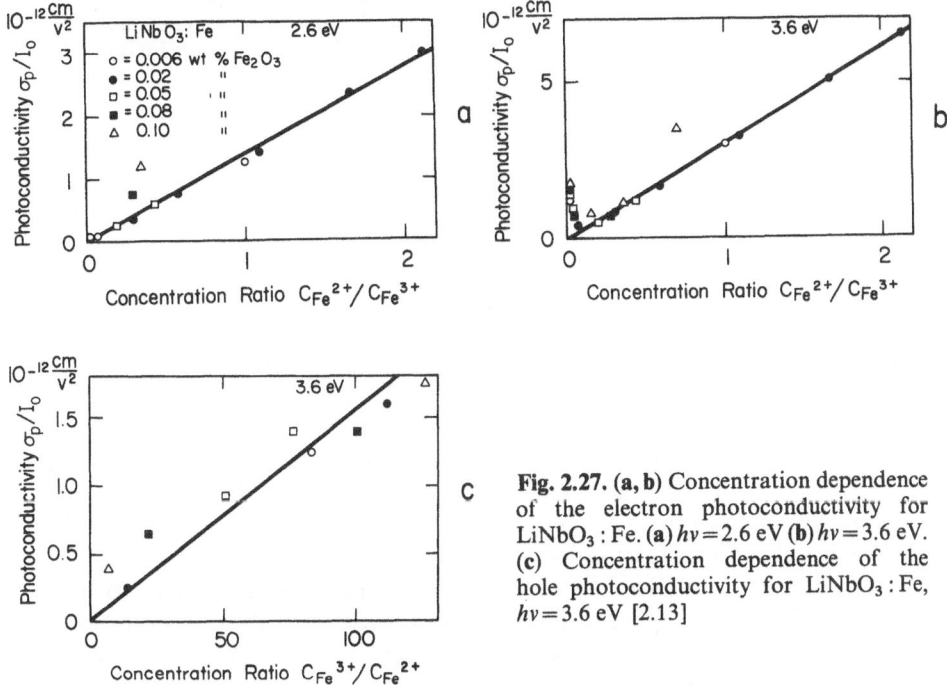

Fig. 2.27. (a, b) Concentration dependence of the electron photoconductivity for LiNbO₃ : Fe. (a) $h\nu = 2.6$ eV (b) $h\nu = 3.6$ eV. (c) Concentration dependence of the hole photoconductivity for LiNbO₃ : Fe, $h\nu = 3.6$ eV [2.13]

given laser wavelength may be obtained. In Tables 2.5 and 2.6 we give values for the sensitivity and some other parameters. Data for the charge transport parameters can also be found in Table 2.1 for different dopants.

b) Lithium Niobate and Lithium Tantalate

Among all the electro-optic materials the photorefractive effects have been investigated most extensively in lithium niobate (LiNbO₃). For LiNbO₃ and LiTaO₃ (lithium tantalate), from room temperature to the Curie temperature ($T_c = 1210\,°C$ and $665\,°C$, respectively) the point group symmetry is $3\,m$. High quality crystals of up to 6 cm diameter can be Czochralsky grown and subsequently or simultaneously poled, to yield large single domain crystals suitable for high resolution hologram storage. The main optical and electro-optical properties of these crystals are summarized in Table 2.2. More data on the preparation and physical properties of LiNbO₃ can be found in [2.94]. The most commonly used dopant is Fe. It enters the lattice as either Fe^{2+} or Fe^{3+}. In the photorefractive process, the Fe^{2+} ions are occupied traps, and the Fe^{3+} ions are empty traps. The concentrations of these two species determine the photorefractive behavior of the crystals. The dependence of the saturation refractive index change on the Fe concentration is shown in Fig. 2.28. Depending on the Fe^{2+} concentration $C_{Fe^{2+}}$, absorption constants in the range of 0.1–100

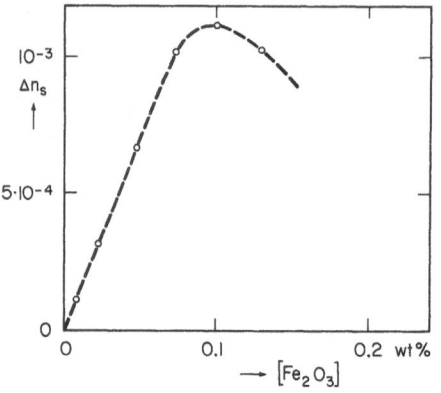

Fig. 2.28. Saturation value of refractive index change as a function of the concentration of Fe_2O_3 doping for $LiNbO_3$ [2.31]

cm^{-1} have been measured at $\lambda = 488$ nm. For a thickness of 2 mm, a total absorption of 67% at 488 nm requires an Fe^{2+} concentration of $\simeq 10^{17}$ cm^{-3} [2.3]. As can be seen from Table 2.5, the sensitivity for photovoltaic recording also increases with increasing Fe^{2+} concentration.

On the other hand, the dark conductivity and photoconductivity depend on the carrier lifetime τ, which is inversely proportional to the concentration of traps (Fe^{3+} mainly). For Fe concentrations smaller than 0.19%, the photoconductivity $\sigma_p = g_0 \tau e \mu \simeq C_{Fe^{3+}}/C_{Fe^{3+}}$, whereas for higher concentrations σ_p grows more rapidly, indicating additional contributions [2.90]. The dark conductivity can be considerably enhanced by increasing $C_{Fe^{2+}}$ (reduction treatment). By this method crystals with dark conductivities in the range of 10^{-8}–10^{-19} $(\Omega cm)^{-1}$ can be obtained. The largest values of σ_d, however, presumably originate from the formation of new Nb^{4+} conduction bands as discussed in [2.95].

c) Potassium Niobate, Potassium Niobate-Tantalate and Barium Titanate

Potassium niobate ($KNbO_3$) and barium titanate ($BaTiO_3$) are ferroelectrics with the perovskite structure. Both materials have the same sequence of structural phase transitions: cubic→tetragonal→orthorhombic→rhombohedral with decreasing temperature. The transition temperatures T_c are higher in $KNbO_3$, and therefore $KNbO_3$ is orthorhombic (point group symmetry $mm2$) at room temperature, whereas $BaTiO_3$ is tetragonal (point group symmetry 4 mm).

Potassium tantalate ($KTaO_3$) remains cubic at all temperatures. Since $KTaO_3$ and $KNbO_3$ have similar lattice parameters mixed crystals [$K(NbTa)O_3$] can be prepared and the transition temperature can be composition controlled. At room temperature cubic $KNb_{0.37}Ta_{0.63}O_3$ [2.96], with $T_c = 20\,°C$, and tetragonal $KNb_{0.4}Ta_{0.06}O_3$ [2.96], with $T_c = 40\,°C$, show interesting electro-optic properties [2.97], due to the large values of the dielectric susceptibilities close to the transition temperatures. Since the transition temperatures in $KNbO_3$ and $BaTiO_3$ are not too different from room temperature, interesting electro-optic properties have been measured [2.98–101]. The most important results for photorefractive recording have been included in Table 2.4. The extremely large electro-optic coefficients r_{42} and r_{51} can give very efficient photorefractive effects. However, since the space-charge field has to be along the b- or a-axis (perpendicular to the polar axis), respectively, photoconductivity or diffusion has to be used as the charge transport process.

The diverse, strongly first order phase transitions between melting and room temperatures in $KNbO_3$ and $BaTiO_3$ make it much more difficult to prepare large single domain samples than for $LiNbO_3$. The maximum single domain volume is therefore presently limited to about 1–2 cm^3. Both crystals have been Fe doped in order to give larger photorefractive effects. The charge transport parameters for these crystals are given in Table 2.1, and the resulting sensitivities in Tables 2.5 and 2.6.

In the case of $KNbO_3$ the sensitivity of photovoltaic recording is several times larger than in a similarly doped crystal of $LiNbO_3$. As can be seen in Tables 2.1 and 2.6, reduced $KNbO_3$ yields an extremely large sensitivity for photoconductive recording; i.e., recording with an applied electric field. The dark conductivities σ_d are relatively large $10^{-9}-<10^{-13}$ $(\Omega cm)^{-1}$. These large values probably cannot not be explained by the Fe^{2+} concentration alone. The presence of ionic conductivity may also contribute in oxidized samples, as suggested in [2.36]. The large values of the dark conductivity and photoconductivity limit the maximum refractive index changes (Fig. 2.24). It has also been shown [2.5] that in slightly reduced crystals diffusion, photoconductivity and the photovoltaic effect contribute similarly to the photorefractive effect in $KNbO_3$ and that their relative importance can be controlled by the fringe spacing or the applied electric field. This is illustrated in Fig. 2.29 where we have plotted the steady-diffraction efficiency as a function of the external electric field (Fig. 2.29a) and the fringe spacing (Fig. 2.29b) for $KNbO_3$: Fe 300 ppm.

Photorefractive effects in $BaTiO_3$ have been reported for melt grown crystals [2.102–105], for flux grown crystals [2.106] and for Fe-doped flux grown crystals [2.107]. Since the physical parameters for the two crystal growth methods can vary considerably and since the melt grown samples show better qualities, we have included in Table 2.1 mainly data from [2.102, 103]. The crystals investigated [2.102, 103] show photoconductivity without the photovoltaic effect. The photoconductivity has been reported to be mainly p-type and smaller than in reduced $KNbO_3$.

The main interest of $KNbO_3$ and $BaTiO_3$ comes from their large electro-optic effects, which seem promising for optical phase conjugation and coherent light amplification [2.108].

In $K(NbTa)O_3$ as in $KNbO_3$ very large photoconductivities have been measured, making these materials also very sensitive for photoconductive recording. Photovoltaic recording is also possible at room temperature for ferroelectric $K(NbTa)O_3$ with higher transition temperatures. Since in the paraelectric material the quadratic electro-optic effect is the lowest-order one, we have used by biasing electric field of $E = 5$ kV/cm in Table 2.4 in order to compare the "linearized" electro-optic response with the linear effect observed in

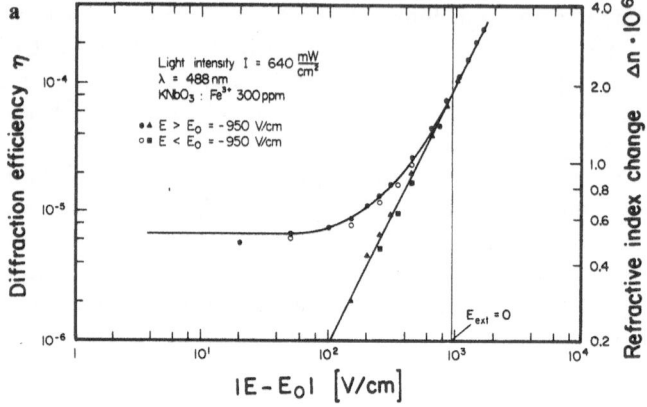

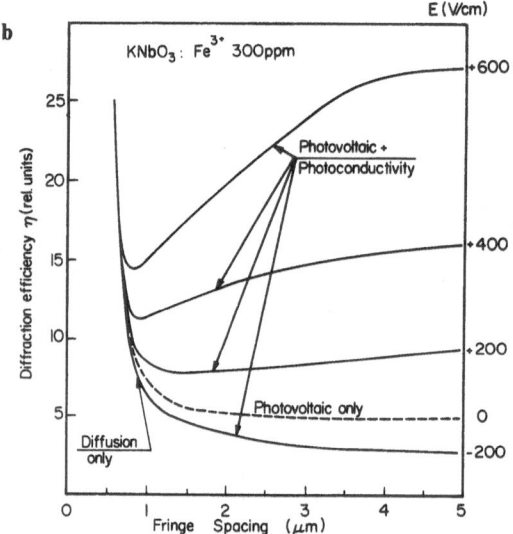

Fig. 2.29. (a) Electric field dependence of diffraction efficiency for $KNbO_3 : Fe^{3+}$ 300 ppm. **(b)** Fringe spacing dependence of the diffraction efficiency for different external fields for $KNbO_3 : Fe^{3+}$ 300 ppm (smoothed curves, with relative errors of $\pm 5\%$; $\lambda = 488$ nm

other materials. The ability to control the electro-optic properties of $K(NbTa)O_3$ by controlling the Nb/Ta ratio and the large photoconductivity [2.109] make the KTN system very interesting for photorefractive applications. However, the growth of homogeneous crystals is very difficult.

d) Barium Sodium Niobate and Barium Strontium Niobate

Barium sodium niobate ($Ba_2NaNb_5O_{15}$) is a ferroelectric with promising nonlinear optical and electro-optical properties [2.110]. It is ferroelectric (point group $4mm$) below $T_c = 560\,°C$. At $T_1 = 300\,°C$ a ferroelastic phase transition ($4mm \rightarrow mm2$) occurs and therefore samples have to be poled and detwinned. Photorefractive effects in $Ba_2NaNb_5O_{15}$ have been reported by *Amodei* et al.

[2.111] and *Voronov* et al. [2.40, 112]. The observed sensitivity is rather low and the material seems therefore to be more suitable for nonlinear optical applications where low photoinduced refractive index changes are required.

Barium strontium niobate [(BaSr)Nb$_2$O$_6$] is a ferroelectric with the tungsten bronze structure (point group symmetry 4*mm*). Similarly to KTN, the transition temperature can be composition controlled from $T_c \simeq 205\,°C$ for Ba/Sr = 3 to $T_c \simeq 60\,°C$ for Ba/Sr = 1/3. Crystals of good optical quality as large as 1 cm in diameter and 7 cm in length can be grown.

Thaxter and *Kestigan* [2.113] and *Micheron* and *Bismuth* [2.114] have observed that in (BaSr)Nb$_2$O$_6$ photoconductive recording can enhance the effect and thus the electrical field dependence can be used to control the refractive index change. *Megumi* et al. [2.115] have shown that very high sensitivities in photovoltaic recording can be reached by 0.1 wt % CeO$_2$ doping. The mobility in this material is comparable to the one in LiNbO$_3$, so a high quantum efficiency seems to be responsible for the increased sensitivity.

2.10.2 Sillenites

The materials Bi$_{12}$SiO$_{20}$(BSO), Bi$_{12}$GeO$_{20}$(BGO), and Bi$_{12}$TiO$_{20}$(BTO) are paraelectric electro-optic and photoconductive. Very large, high quality samples can be grown from the melt [2.116]. Having the cubic point group symmetry $\overline{4}3m$ they are optically isotropic if no electric field is applied. With an applied electric field they become birefringent due to the nonvanishing Pockels coefficient $r_{41} = 3.4$ pm/V (BSO). The crystals are also strongly optically active ($\varrho_0 = 45°\ \text{mm}^{-1}$ at $\lambda = 514.5$ nm for Bi$_{12}$SiO$_{20}$). Therefore the polarization states of the diffracted, incident and scattered light are all different [2.117]. This has the advantage that coherently scattered noise can be suppressed by a properly oriented Polaroid in the diffracted image plane, improving the signal-to-noise ratio. On the other hand, Bragg diffraction of a volume phase grating recorded in such an optically active medium can be smaller than in Kogelnik's classical theory [2.118] for storage media without optical activity. Two different geometries have been found: the first one gives a maximum diffraction efficiency, while the second one gives better beam coupling of recording beams and is used in light amplification or optical phase conjugation [2.119, 120].

The energy band diagram for Bi$_{12}$SiO$_{20}$ (and Bi$_{12}$GeO$_{20}$) is shown in Fig. 2.30. The energy gap in BSO is $E_g = 3.25$ eV [2.121]. A broad shoulder in the absorption curve is attributed to a silicon (or germanium) vacancy. Its level is situated 2.60 eV below the conduction band, when occupied by an electron. A luminiscence center at 1.30 eV has been identified, direct excitation of which results in a broad band centered at 2.2 eV. This center is believed to participate in the charge transfer process; its energy level is 2.25 or 1.30 eV below the conduction band, depending on whether it is occupied by an electron or not. Photoconductivity is n-type and dark conductivity is p-type. It is seen from Table 2.1 that the photoconductivity parameter is extremely large and that the

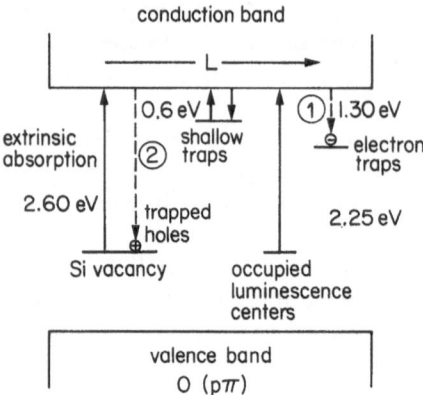

Fig. 2.30. Energy-band diagram of $Bi_{12}SiO_{20}$ at room temperature

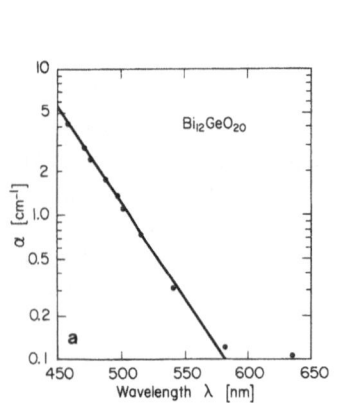

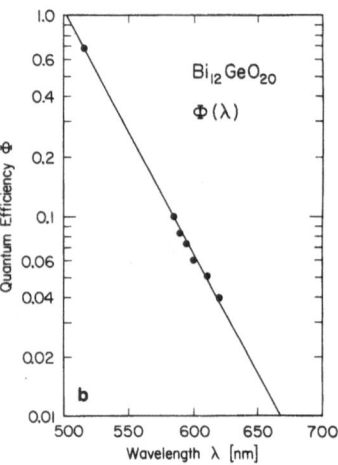

Fig. 2.31a, b. Wavelength dependence of the absorption constant α (**a**) and of the quantum efficiency Φ for photoexcitation of an electron (**b**) for $Bi_{12}GeO_{20}$ [2.1]

dark conductivity is small enough to allow sufficient storage times for many applications. The absorption constant α and the quantum efficiency Φ, which determine the photo-excitation efficiency, are plotted for $Bi_{12}GeO_{20}$ in Fig. 2.31 as a function of the wavelength [2.1]. It is shown in Fig. 2.31 that the quantum efficiency can be as large as 0.7 for $\lambda = 514.5$ nm. The drift length for relatively small electric fields, i.e., the optimum length for holographic recording, is on the order of micrometers. This leads to the very large photoconductive photorefractive sensitivity (see Table 2.6).

2.10.3 Semiconductors (GaAs, InP, CdFe,...)

Photorefractive effects have been investigated in the following III-V semiconductor materials: pure and Cr-doped GaAs [2.122], Fe-doped InP and CdFe
[2.123]. The materials have good photoconductivities due to large carrier
mobilities, are very photosensitive, and show fast recording speed in the near
infrared wavelength range from 0.95–1.35 μm [2.122–125]. However, the
electro-optic coefficients are rather small (Table 2.4) and therefore the photoinduced refractive index changes are relatively small. For more details on the
photorefractive effects and applications of semiconductors we refer to Chap. 8.

2.10.4 Electro-optic Ceramics

Highly transparent ferroelectric ceramics of the lead zirconate titanate family
with perovskite structure show interesting electro-optical effects and can be
cheaply produced in large pieces. The lead ions are substituted partly by
lanthanum ions and the chemical compositions is $(Pb_{1-x}, La_x)(Zr_y, Ti_z)O_3$,
abbreviated to $PLZT_{x/y/z}$. Different electro-optic effects can be optimized by
adjusting the Zr/Ti ratio or the La content. At high La contents one obtains the
optically isotropic cubic paraelectric perovskite structure and the electro-optic
effect is quadratic (Kerr effect). At low La contents, two ferroelectric phases are
observed, one with a tetragonally distorted perovskite cell for the Ti-rich
compositions, the other rhombohedrally distorted in Zr-rich material. At electric
field strengths lower than the switching field, these compositions exhibit the
linear electro-optic Pockels effect.

Photorefractive effects have been observed in both linear [2.126] and
quadratic [2.127] electro-optic materials. Transition metal substitutions [2.128]
only slightly increase the relatively low sensitivity. The low values of dark
conductivity lead to long storage times.

Up to 10^4-fold increase in the photosensitivity can be achieved in ion-
implanted rhombohedral phase of PLZT [2.129, 130] excited in the near
ultraviolet. Ions of hydrogen, helium or argon, or coimplantation of argon and
neon ions induce disorder into the near surface structure, decreasing the dark
conductivity and increasing the absorption constant, the efficiency of the
photoexcitation process and the density of trapping sites. The disorder increases
with increasing mass of the ion, hence, greater damage, which is more uniform
with depth from the surface, is produced by argon than by helium or protons.
Ultraviolet exposure energies required to store an image were about $10 \, mJ \, cm^{-2}$
for hydrogen implants, about $3 \, mJ \, cm^{-2}$ for helium implants, about $100 \, \mu J \, cm^{-2}$
for argon implants, and about $10 \, \mu J \, cm^{-2}$ for argon-neon implants. Some
interesting electro-optic properties and large photoconductivity have also been
observed in $K(NbTa)O_3$ ceramics [2.131]. However, the optical transparency is
still not sufficient for optical applications.

2.10.5 Ultraviolet Transparent Materials

Photorefractive effects in the ultraviolet wavelength range have been observed in the ferroelectric phases of KH_2PO_4 [2.41, 42] and Rb_2ZnBr_4 [2.132]. Since these crystals become ferroelectric at $T_0 = 123$ K and 187 K, respectively, both crystals have to be cooled below T_0 temperature for photovoltaic recording. Photovoltaic drift lengths and sensitivities for illumination in the region of the intrinsic optical absorption edge are comparable to the ones observed in $LiNbO_3$. Other room temperature ultraviolet transparent ferroelectrics or other photoconductive electro-optic crystals seem to be very promising for holography with ultraviolet lasers since the resolution of high quality crystals should be well above the requirement for short wavelengths. However, no such measurements on other ultraviolet transparent photorefractive crystals have been reported yet.

2.10.6 Infrared Transparent Materials

Oxygen-octahedra ferroelectrics are transparent up to wavelengths of about 5 μm and can thus be used in principle for hologram storage. However, the quantum efficiency for the infrared wavelengths becomes very small in these large bandgap materials, leading to small values for the photosensitivity. Therefore lower band gap materials which offer a better quantum efficiency for photoexciting charge carriers, and which can have an extended transmission range in the infrared, have been used for photorefractive recording. One disadvantage of these materials, however, is the larger dark conductivity in low band gap materials. For this reason these materials are useful only for short-time photorefractive applications, such as holographic interferometry, coherent light amplification or phase conjugation. Transient photoinduced refractive index gratings have been recorded in SbSI [2.133], CdS [2.134], CdSe [2.135] and CdTe [2.136], using Nd^{3+} laser radiation at $\lambda = 1.06$ μm for the latter two compounds.

Of the above-mentioned materials, the ones with large carrier drift length, i.e., the photoconductive electro-optic materials, are the most interesting ones, because they are the most sensitive. In addition to the improved sensitivity in photoconductive recording, we have seen that for large drift lengths the photo-induced refractive index grating receives a further phase shift with respect to the intensity grating in addition to the one due to diffusion. This makes these materials also very interesting for dynamical holography and two- and four-beam mixing experiments.

References

2.1   P. Günter: Phys. Rep. **93**, 199 (1982)
2.2   M. Clarke: J. Chem. Phys. **59**, 6209 (1973)
2.3   D. L. Staebler, W. Philips: Appl. Opt. **13**, 788 (1974)
2.4   W. Philips, D. L. Staebler: J. Electron. Mater. **3**, 601 (1974)

2.5   P.Günter, F. Micheron: Ferroelectrics **18**, 27 (1978)
2.6   E.Krätzig: Ferroelectrics **21**, 635 (1978)
2.7   K.L.Sweeney, L.E..Halliburton: Appl. Phys. Lett. **43**, 336 (1983)
2.8   J.L.Ketchum, K.L..Sweeney, L.E.Halliburton, A.F.Armington: Phys. Lett. **94**A, 450 (1983)
2.9   L.E.Halliburton, K.L.Sweeney, C.Y.Chen: Nucl. Instrum. Methods Phys. Res. B1, 344 (1984)
2.10 L.Arizmendi, J.M.Cabrera, F. Agullo-Lopez: J. Phys. C. **17**, 515 (1984)
2.11 O.F.Schirmer: J. Appl. Phys. **50**, 3404 (1979)
2.12 O.F.Schirmer, D.von der Linde: Appl. Phys. **33**, 35 (1978)
2.13 E.Krätzig: Ferroelectrics **21**, 635 (1978)
2.14 K.Megumi, H.Kozuka, M.Kobayashi, Y.Furukata: Appl. Phys. Lett. **30**, 631 (1977)
2.15 R.A.Sprague: J. Appl. Phys. **46**, 1673 (1975)
2.16 V.F.Belinicher, B.I.Sturman: Sov. Phys. – Usp. **23**, 199 (1980)
2.17 S.G.Odulov: JETP Lett. **35**, 10 (1982)
2.18 M.G.Moharam, T.K.Gaylord, R.Magnusson: J. Appl. Phys. **50**, 5642 (1979)
2.19 A.M.Glass, D. von der Linde, T.J.Negran: Appl. Phys. Lett. **25**, 233 (1974)
2.20 D. von der Linde, A.M.Glass: Appl. Phys. **8**, 85 (1975)
2.21 G.Chanussot, A.M.Glass: Phys. Lett. **59**A, 405 (1976)
2.22 G.Chanussot: Ferroelectrics **20**, 37 (1978)
2.23 P.N.Günter: Opt. Lett. **7**, 10 (1982)
2.24 A.M.Glass, D.H.Auston: Opt. Commun. **5**, 45 (1972)
2.25 A.M.Glass, D.H.Auston: Ferroelctrics **7**, 187 (1974)
2.26 R.Orlowski, E.Krätzig: Solid State Commun. **27**, 1351 (1978)
2.27 G.C.Valley: J. Appl. Phys. **59**, 3363 (1986)
2.28 F.P.Strohkendl, J.M.C.Jonathan, R.W.Hellwarth: Opt. Lett. **11**, 312 (1986)
2.29 L.Young, W.K.Y.Wong, M.L.W. Thewalt, W.D.Cornish: Appl. Phys. Lett. **24**, 264 (1974)
2.30 M.G.Moharam, T.K.Gaylord, R.Magnusson, L.Young: J. Appl. Phys. **50**, 5642 (1979)
2.31 H.Kurz: Philips Tech. Rev. **37**, 109 (1977)
2.32 H.Kurz, E.Krätzig, W.Keune, H.Engelmann, U.Gonser, B.Dischler, A.Räuber: Appl. Phys. **12**, 355 (1977)
2.33 R.Orlowski, E.Krätzig, H.Kurz: Opt. Commun. **20**, 171 (1977)
2.34 E.Krätzig, R.Orlowski: Appl. Phys. **15**, 133 (1978)
2.35 A.Krumins, P.Günter: Appl. Phys. **19**, 153 (1979)
2.36 A.E.Krumins, P.Günter: Phys. Status Solidi A**55**, K185 (1979)
2.37 R.Orlowski, L.A.Boatner, E.Krätzig: Opt. Commun. **35**, 45 (1980); Ferroelectrics **27**, 247 (1980)
2.38 E.Krätzig, F.Weltz, R.Orlowski, V.Dormann, M.Rosenkranz: Solid State Commun. **34**, 817 (1980)
2.39 I.R.Dorosh, Yu.S.Kuzminov, N.M.Polzkov, A.M.Prokhorov: Phys. Status Solidi A**65**, 513 (1981)
2.40 V.V.Voronov, Yu.S.Kuzminov, I.G.Lukina: Sov. Phys. – Solid State **18**, 598 (1976)
2.41 V.M.Fridkin, B.N.Popov, K.A.Verkhovskaya: Sov. Phys. – Solid State **20**, 730 (1978)
2.42 V.M.Fridkin, B.N.Popov, K.A.Verkhovskaya: Appl. Phys. **16**, 313 (1978)
2.43 P.Günter (ed.): *Electro-optic and Photorefractive Materials*, Proc. Phys., Vol. 18 (Springer, Berlin, Heidelberg 1987)
2.44 H.Kogelnik: Bell Syst. Tech. J. **48**, 2909 (1969)
2.45 N.Kukhtarev, V.B.Markov, S.G.Odulov, M.S.Soskin, V.L.Vinetsky: Ferroelectrics **22**, 949, 961 (1979)
2.46 E.Voit: In *Electro-optic and Photorefractive Materials*, ed. by P. Günter, Proc. Phys., Vol. 18 (Springer, Berlin, Heidelberg 1987) p. 246
2.47 S.J.Stepanov, M.P. Petrov, A.A.Kamshilin: Sov. Tech. Phys. Lett. **3**, 345 (1977)
2.48 M.P.Petrov, S.I.Stepanov, A.A.Kamshilin: Opt. Commun. **29**, 44 (1979)

2.49 Landolt-Börnstein, Numerical Data and Functional Relationships in Science and Technology, New Series, ed. by K.-H. Hellwege (Springer, Berlin, Heidelberg 1979) Vol. 11 of Group 3
2.50 J.-C. Baumert, J. Hoffnagle, P. Günter: Proc Soc. Photo-Opt. Instrum. Eng. **492**, 374 (1985)
2.51 P. Günter: Opt. Commun. **11**, 385 (1974)
2.52 E. Voit, C. Zaldo, P. Günter: Opt. Lett. **11**, 309 (1986)
2.53 N. Kukhtarev: Sov. J. Quantum. Electron. **11**, 878 (1981)
2.54 N. V. Kukhtarev, E. Krätzig, H. C. Külich, R. A. Rupp: Appl. Phys. B**35**, 17 (1984)
2.55 S. Odulov, K. Belabaev, I. Kiselva: Opt. Lett. **10**, 31 (1985)
2.56 D. A. Temple, C. Warde: J. Opt. Soc. Am. B. **3**, 337 (1986)
2.57 R. A. Rupp, F. W. Drees: Appl. Phys. B**39**, 223 (1986)
2.58 H. Vorman, E. Krätzig: Solid State Commun. **49**, 843 (1984)
2.59 D. von der Linde, A. M. Glass, K. F. Rodgers: Appl. Phys. Lett. **26**, 22 (1975)
2.60 D. von der Linde, A. M. Glass, K. F. Rodgers: Appl. Phys. Lett. **25**, 155 (1974)
2.61 D. von der Linde, A. M. Glass, K. F. Rodgers: J. Appl. Phys. **47**, 217 (1976)
2.62 F. Micheron, G. Bismuth: Appl. Phys. Lett. **20**, 79 (1972); ibid. **23**, 71 (1973)
2.63 J. B. Thaxter, M. Kestigan: Appl. Opt. **13**, 913 (1974)
2.64 H. Vormann, G. Weber, S. Kapphan, E. Krätzig: Solid State Commun. **40**, 543 (1981)
2.65 D. L. Staebler, J. J. Amodei: Ferroelectrics **3**, 107 (1972)
2.66 D. L. Staebler, W. Burke, W. Phillips, J. J. Amodei: Appl. Phys. Lett. **26**, 182 (1975)
2.67 W. J. Burke, D. L. Staebler, W. Phillips, G. A. Alphonse: Opt. Eng. **17**, 308 (1978)
2.68 J. A. Armstrong, N. Bloembergen, J. Ducuing, P. S. Pershan: Phys. Rev. **127**, 1918 (1962)
2.69 H. J. Eichler, P. Günter, D. W. Pohl: *Laser-Induced Dynamic Gratings*, Springer Ser. Opt. Sci., Vol. 50 (Springer, Berlin, Heidelberg 1986)
2.70 S. H. Wemple, M. Di Domenico, Jr.: In *Applied Solid State Science*, ed. by R. Wolfe (Academic, New York 1972)
2.71 P. Günter: Ferroelectrics **22**, 671 (1978)
2.72 M. B. Klein, G. C. Valley: J. Appl. Phys. **57**, 4901 (1985)
2.73 N. Kukhtarev, V. Markov, S. Odulov: Opt. Commun. **23**, 338 (1977)
2.74 N. V. Kukhtarev: Sov. Tech. Phys. Lett. **2**, 438 (191976)
2.75 G. C. Valley: IEEE J. QE-**19**, 1637 (1983)
2.76 G. C. Valley, M. B. Klein: Opt. Eng. **22**, 704 (1983)
2.77 J. P. Huignard, J. P. Herriau, G. Rivet, P. Günter: Opt. Lett. **5**, 102 (1980)
2.78 G. Haertling, C. E. Land: J. Am. Ceram. Soc. **54**, 1 (1971)
2.79 W. Philips, J. J. Amodei, D. L. Staebler: RCA Rev. **33**, 94 (1972)
2.80 R. Magnusson, T. K. Gaylord: Appl. Opt. **13**, 1545 (1974)
2.81 D. L. Staebler: In *Holographic Recording Materials*, ed. by H. M. Smith, Topics Appl. Phys., Vol. 20 (Springer, Berlin, Heidelberg 1977) Chap. 4
2.82 L. D'Auria, J. P. Huignard, E. Spitz: IEEE Trans. MAG-**9**, 83 (1973)
2.83 G. E. Peterson, A. M. Glass, T. J. Negran: Appl. Phys. Lett. **19**, 130 (1971)
2.84 W. Keune, S. K. Date, I. Dézsi, U. Gonser: J. Appl. Phys. **46**, 3914 (1975)
2.85 M. Z. Zha, P. Günter: Opt. Lett. **10**, 187 (1985)
2.86 A. Ishida, O. Mikami, S. Miyazawa, M. Sumi: Appl. Phys. Lett. **21**, 193 (1972)
2.87 E. Okamoto, H. Ikeo, K. Muto: Appl. Opt. **14**, 2453 (1975)
2.88 P. Günter, U. Flückiger: unpublished
2.89 K. Megumi, H. Kozuka, M. Kobayashi, Y. Furuhata: Appl. Phys. Lett. **30**, 631 (1977)
2.90 E. Krätzig, H. Kurz: Opt. Acta **24**, 475 (1977)
2.91 G. A. Alphonse, W. Phillips: RCA Rev. **37**, 184 (1976)
2.92 E. Krätzig, R. Orlowski: Opt. Quantum Electron. **12**, 495 (1980)
2.93 R. Orlowski, E. Krätzig: Solid State Commun. **27**, 1351 (1978)
2.94 A. Räuber: In *Current Topics in Materials Science*, Vol. 1, ed. by E. Kaldis (North-Holland, Amsterdam 1978) pp. 481–601

2.95 R. Courths, P. Steiner, H. Höchst, S. Hüfner: Appl. Phys. 21, 345 (1980)
2.96 S. Triebwasser: Phys. Rev. 101, 993 (1956)
2.97 F. S. Chen, J. E. Geusic, S. K. Kurtz, J. G. Skinner, S. H. Wemple: J. Appl. Phys. 37, 388 (1966)
2.98 P. Günter: Opt. Commun. 11, 285 (1974)
2.99 P. Günter: In Proc. Electro Optics/Laser Int. 76, ed. by H. G. Jerrard (IPC Science and Technology Press Ltd., Guildford, Surrey, England 1976) p. 121
2.100 A. R. Johnston, J. M. Weingart: J. Opt. Soc. Am. 55, 828 (1965)
2.101 I. R. Kaminov: Appl. Phys. Lett. 7, 123 (1965, erratum 8, 54 (1966)
2.102 J. Feinberg, D. Heimann, A. R. Tanguay, R. L. Hellwarth: J. Appl. Phys. 51, 1297 (1980)
2.103 M. B. Klein: Proc. Soc. Photo-Opt. Instrum. Eng. 519, 136 (1984)
2.104 M. B. Klein, R. N. Schwartz: J. Opt. Soc. Am. B3, 293 (1986)
2.105 M. B. Klein: This volume, Chap. 7
2.106 R. L. Townsend, G. D. Boyt, J. M. Dziedzik, R. G. Smith, A. A. Ballman, K. Nassau: Appl. Phys. Lett. 9, 72 (1966)
2.107 F. Micheron, G. Bismuth: Appl. Phys. Lett. 20, 15 (1972)
2.108 F. Laeri, T. Tschudi: Opt. Commun. 47, 387 (1983)
2.109 E. Krätzig, R. Orlowski: Ferroelectrics 27, 241 (1980)
2.110 S. Singh, D. A. Draegert, J. E. Geusic: Phys. Rev. B2, 2709 (1970)
2.111 J. J. Amodei, D. L. Staebler, A. W. Stephens: Appl. Phys. Lett. 18, 507 (1971)
2.112 V. V. Voronov, Yu. S. Kuzminov, V. V. Osiko, A. M. Prokhorov: Sov. Phys. – Crystallogr. 25, 691 (1980)
2.113 J. B. Thaxter, M. Kestigan: Appl. Opt. 13, 913 (1974)
2.114 F. Micheron, G. Bismuth: Appl. Phys. Lett. 20, 79 (1972); Appl. Phys. Lett. 23, 71 (1973)
2.115 K. Megumi, H. Kozuka, M. Kobayashi, Y. Furukata: Appl. Phys. Lett. 30, 631 (1977)
2.116 J. C. Brice, M. J. Hight, O. F. Hill, P. A. C. Whiffin: Phillips Tech. Rev. 37, 250 (1977)
2.117 J. P. Herriau, J. P. Huignard, P. Aubourg: Appl. Opt. 17, 1851 (1978)
2.118 A. Marrakchi: In Electro-optic and Photorefractive Materials, ed. by P. Günter, Springer Proc. Phys., Vol. 18 (Springer, Berlin, Heidelberg 1987) p. 339
      A. G. Apostolidis: In Electro-optic and Photorefractive Materials, ed. by P. Günter, Springer Proc. Phys., Vol. 18 (Springer, Berlin, Heidelberg 1987) p. 324
2.119 A. Marrakchi, J. P. Huignard, P. Günter: Appl. Phys. 24, 131 (1981)
2.120 J. P. Herriau, J. P. Huignard: Appl. Phys. Lett. 49, 1140 (1986)
2.121 S. L. Hou, R. B. Lauer, R. E. Aldrich: J. Appl. Phys. 44, 2652 (1973)
2.122 M. B. Klein: Opt. Lett. 9, 350 (1984)
2.123 J. Strait, A. M. Glass: Appl. Opt. 25, 338 (1986)
2.124 A. M. Glass, A. M. Johnson, D. H. Olson, W. Simpson, A. A. Ballman: Appl. Phys. Lett. 44, 948 (1984)
2.125 A. M. Glass, J. Strait: This volume, Chap. 8
2.126 A. E. Krumins, U. Y. Ilyin, V. I. Dimza: to be published
2.127 F. Micheron, C. Mayeux, A. Hermosin, J. Nicolas: J. Am. Ceram. Soc. 57, 306 (1974)
2.128 J. M. Rouchon, M. Vergnolle, F. Micheron: Ferroelectrics 12, 239 (1976)
2.129 C. E. Land, P. S. Peercy: Ferroelectrics 27, 131 (1980)
2.130 C. E. Land, P. S. Peercy: Appl. Phys. Lett. 37, 39, 815 (1980)
2.131 P. E. Debely, P. Günter, H. Arend: Am. Ceram. Soc. Bull. 58, 606 (1979)
2.132 T. Nakamura, V. Fridkin, R. Magomadov, M. Takashige, K. Verkhovskaya: J. Phys. Soc. Jpn. 48, 1588 (1980)
2.133 A. Kh. Zeinally, A. M. Madedov, Sh. M. Efendiev: Sov. Phys. – Solid State 18, 1643 (1977)
2.134 H. J. Eichler, Ch. Hartig, J. Knof: Phys. Status Solidi A45, 433 (1978)
2.135 K. Jarasiunas, J. Vaitkus: Phys. Status Solidi A23, K19 (1974)
2.136 V. Kremenistikii, S. Odulov, M. Soskin: Phys. Status Solidi A57, K71 (1980)

3. Theory of Photorefractive Effects in Electro-optic Crystals

George C. Valley and Juan F. Lam

With 8 Figures

This chapter provides a detailed review of theoretical work on the photorefractive effect. The physical model of photorefractivity used here, which now seems well established by many experimental observations, includes photoionization of free carriers (electrons and/or holes), drift or diffusion of the carriers, retrapping of the carriers with a charge separation balanced by an internal space-charge field, and modulation of the refractive index through the electro-optic effect. Other mechanisms, such as pyroelectric effects, absorption gratings and impurity polarization, which have sometimes been mentioned in passing [3.1–4], are not now believed to have a significant role in the basic photorefractive effect. The drift and diffusion of charge carriers may be replaced by charge hopping [3.5] but to date this alternative microscopic model has not given significantly different results from the band transport model.

The following section of this chapter contains a historical review in which we point out significant contributions to our theoretical understanding of the photorefractive effect. Section 3.2 presents general results for the lowest Fourier component of the space-charge field (and hence the refractive index) in a photorefractive material illuminated by a sinusoidal interference pattern. These results parallel most closely the work of *Kukhtarev* [3.6] in 1976 and reduce to a number of special cases that are widely used. We have added to the results of Kukhtarev the effects of simultaneous electron and hole conduction [3.7–9]. The results can be combined with Maxwell's equations to obtain the nonlinear wave equations for two- and four-wave mixing. Solutions appropriate to picosecond to microsecond pulsed illumination are given in Sect. 3.3. Section 3.4 discusses nearly degenerate four-wave mixing and self-pumped phase conjugation via nearly degenerate backward stimulated two-wave mixing.

3.1 Historical Review

Our current ideas on the photorefractive effect originate with the work of *Chen* et al. [3.2, 10, 11]. As early as 1967 *Chen* [3.10] suggested:

> The process of space charge buildup can be explained by postulating that electrons are photoexcited from traps by the laser beam into the conduction band and drift toward the positive electrode leaving behind stationary space charges. These electrons may be retrapped and reexcited out of the traps as they eventually drift out of the laser spot and are retrapped there.

It is also clear that Chen knew that this space-charge field would modulate the index of refraction through the electro-optic effect. In 1969 Chen discussed "an internal field of not well understood origin" [3.2] that can be interpreted as a concept that is a precursor of the photovoltaic field of *Glass* et al. [3.12]. Rate equations for the electron and trap densities were first given in [3.2].

The next major additions to the theoretical work were made by *Amodei* [3.3, 13] in 1971. He recognized that thermal diffusion of free carriers leads to a space-charge field of magnitude $E_D = Kk_B T/e$ (K is the grating wave number, $k_B T$ is Boltzmann's constant times temperature, e is the magnitude of the charge on the electron). Amodei pointed out that diffusion also leads to nonsinusoidal gratings. In [3.13] he evaluated photorefractive effects using a current equation, Poissons's equation, and the equation of continuity. With both an applied field and diffusion he obtained the result that the steady-state space-charge field is given by $E_{sc} = E_0 + iE_D$ and that the turn-on is proportional to time. He also recognized that the largest space-charge field that the crystal could support is $E_q = eN_A \Lambda/(2\pi\varepsilon\varepsilon_0)$ (where N_A is the empty trap density, Λ is the grating period, ε is the static dielectric constant, and ε_0 is the permittivity of free space) although he did not derive a general expression for the space-charge field including E_q. The dielectric relaxation time was given for the grating erasure time without the correction factors for finite grating period. The major difference between the work of Amodei and the later work of *Vinetskii* and *Kukhtarev* [3.6, 14] is in the treatment of the photoproduction of carriers. Amodei assumed that the carrier number density was simply proportional to the generation rate (i.e., the optical intensity) while Vinetskii and Kukhtarev used a rate equation for the trap density and solved for the carrier number density.

Staebler and *Amodei* [3.15] added a coupled wave theory to the work of Amodei and showed that the diffusion process could lead to unidirectional transfer of energy from one beam to a second (now called beam coupling or two-wave mixing). They also noted that the direction of the energy transfer is controlled by the sign of the charge carrier and the direction of the crystal c-axis.

In 1974 *Young* et al. [3.16] added the grating dependence to time-dependent results and noted that in the transient regime there could be a $\pi/2$ phase shift, which could lead to beam coupling, even when drift dominates. *Glass* et al. [3.12] gave the first formula for the photovoltaic current in $LiNbO_3$, $j = \kappa_1 \alpha I$ (κ_1 is now referred to as the Glass constant, α is the absorption coefficient and I is the irradiance). Also in 1974 *Vinetskii* and *Kukhtarev* [3.14] added a rate equation for the ionized traps to the three equations solved previously by Amodei. They solved these equations in two limits (1) $N_A \gg n$ and $N_D \gg 2N_A$; (2) $N_A \gg n$ and $N_D \ll 2N_A$ (n is the carrier density, N_D is the total density of filled and empty traps).

A number of additional papers in 1975 and 1976 [3.17–21] provided analytical and numerical solutions to the set of equations used by Amodei in 1971. Also in 1976, Kukhtarev published an important paper [3.6] in which he derived the fundamental Fourier harmonic of the steady-state space-charge field and the complex response time (buildup or decay and oscillations) by using the "quasi-steady approximation" in which the mean carrier density is assumed to be a constant and the moduation index of the intensity interference pattern is assumed to vary slowly compared to the response time.

Several papers by Kukhtarev and Vinetskii and their co-workers Markov, Odulov, and Soskin followed in 1977–1979 [3.22–25]. In [3.22] beam coupling in the reflection and transmission geometries was compared. In [3.23], some of the previous work was reviewed and new results were given on higher-order Fourier harmonics, on the effects of a photovoltaic effect and an applied field, and on beam coupling. In addition, comparisons were made between the theory and experimental work on LiNbO$_3$. Transient energy transfer during the build-up phase of a hologram was analyzed and compared to experiment in [3.24] while wave front reversal was discussed in [3.25].

In 1980, *Feinberg* et al. [3.5] applied a different physical model, the "hopping model", to the charge transport that leads to the photorefractive effect in BaTiO$_3$. In this model charges, either electrons or holes, are assumed to hop from filled to vacant sites when exposed to optical radiation. Although the physical basis of the hopping model is statistical as opposed to the deterministic nature of the conduction band model, the models lead to similar expressions for the fundamental Fourier component of the steady-state space-charge field and the complex response time in the limit of no applied field. When the hopping length is large compared to the grating period, however, results of the hopping model are very sensitive to the specific form of the hopping probability distribution used in the calculations [3.26]. One particular form of the hopping probability distribution yields results of the same form as those of Kukhtarev, but the physical basis for use of this model is unclear.

Subsequent to 1980, most theoretical work on photorefractive materials has been devoted to obtaining solutions to the equations of *Vinetskii* and *Kukhtarev* [3.14] in new regimes in which experiments have been performed [3.27–31], to obtaining solutions to the nonlinear wave equations that result in photorefractive materials [3.32–35], and to extending the single carrier, single donor-trap system to more complicated systems [3.7–9, 36–39]. Much of this work is reviewed in this chapter.

3.2 Space-Charge Field in the Quasi-Steady Approximation

In this section we review and extend the work of *Kukhtarev* [3.6] to obtain the lowest Fourier component of the space-charge field in a photorefractive material illuminated by a sinusoidal interference pattern. Kukhtarev's analysis used an approximation in which the mean carrier number density and the modulation index of the interference pattern are assumed to be independent of time (quasi-steady approximation) and obtained the steady-state space-charge field and the complex response time in a medium with a single charge carrier. We have extended this analysis to include both electrons and holes since this involves little additional work and since simultaneous electron and hole conductivity appears to be important in photorefractive materials [3.7–9, 38, 40].

3.2.1 Basic Equations

We start from rate equations for electrons and holes, current equations for both, the equation of continuity, Poissons's equation, and an equation for the total number of dopants participating in the photorefractive process:

$$\frac{\partial n_e}{\partial t} - \nabla \cdot \frac{j_e}{e} = (s_e I + \beta_e) N - \gamma_e n_e N^+ , \tag{3.1}$$

$$\frac{\partial n_h}{\partial t} + \nabla \cdot \frac{j_h}{e} = (s_h I + \beta_h) N^+ - \gamma_h n_h N , \tag{3.2}$$

$$j_e = e\mu_e n_e E + k_B T \mu_e \nabla n_e + \kappa_e s_e NI , \tag{3.3}$$

$$j_h = e\mu_h n_h E - k_B T \mu_h \nabla n_h + \kappa_h s_h N^+ I , \tag{3.4}$$

$$\frac{\partial}{\partial t}(n_e - N^+ - n_h) = \nabla \cdot \frac{j_e}{e} + \nabla \cdot \frac{j_h}{e} , \tag{3.5}$$

$$\nabla \cdot E = -\left(\frac{e}{\varepsilon\varepsilon_0}\right)(n_e + N_A - N^+ - n_h) , \tag{3.6}$$

$$N_D = N + N^+ . \tag{3.7}$$

In these equations n_e and n_h are the electron and hole number densities, N and N^+ are the donor and acceptor number densities, N_D is the total number density of the dopants responsible for the photorefractive effect, and N_A is the number density of negative ions that compensate for the charge N^+ in the dark. The signs are set so that e is the magnitude of the charge on the electron and the mobilities of both electrons and holes are positive. j_i, μ_i, κ_i, γ_i, s_i, β_i are the current. mobility, Glass photovoltaic constant, recombination coefficient, cross section of photoionization and dark generation rate (the subscript i refers to electrons or holes). The system modeled by these equations is shown in Fig. 3.1; electrons are photoionized from N and recombine at N^+ and vice versa for holes.

Solution of (3.1–7) proceeds most easily by eliminating the currents and N and N^+ to obtain three lengthy equations in n_h, n_e and E. Solution of these three nonlinear equations for the lowest Fourier component of the grating in the quasi-steady approximation proceeds by setting $n_e = n_0 + n_1$, $n_h = p_0 + p_1$, $I = I_0$

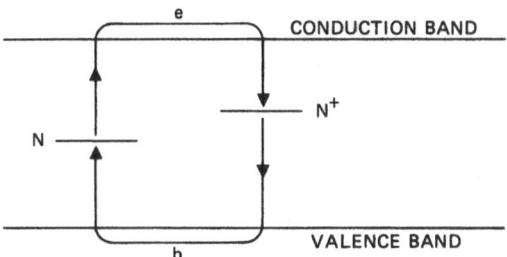

Fig. 3.1. Energy level model for the photorefractive model in which a single set of recombination centers gives rise both to electrons in the conduction band and to holes in the valence band

$+I_1$, and $E = E_0 + E_1$. (Note that E_0 is determined by the external circuit conditions and is equal to the external field only in the absence of the photovoltaic effect.) All subscript zero quantities are assumed to be independent of space and time while the first-order quantities (except I_1) are assumed to be of the form $\exp(iKz + i\delta\omega t)$.

3.2.2 Zeroth-Order Solutions

The zeroth-order equations divide into two sets: one for the number densities and a second for the mean field. The equations for the mean electron and hole densities can be written

$$n_0 = (s_e I_0 + \beta_e) \frac{(N_D - N_A - n_0 + p_0)}{\gamma_e (N_A + n_0 - p_0)} , \tag{3.8}$$

$$p_0 = (s_A I_0 + \beta_h) \frac{(N_A + n_0 - p_0)}{\gamma_h (N_D - N_A - n_0 + p_0)} . \tag{3.9}$$

These equations can be solved numerically for n_0 and p_0, but for typical cw lasers this is unnecessary. Typically, the irradiance of a cw laser is less than 1 W cm^{-2} (2.5×10^{18} photons $\text{cm}^{-2}\text{s}^{-1}$ at a wavelength of 0.5 μm), and in typical photorefractive materials the absorption coefficient is less than 1 cm^{-1} and the recombination time less than 1 μs. This leads to carrier number densities of 10^9–10^{12} cm^{-3} so that $|p_0 - n_0| \ll N_D - N_A$ and $|n_0 - p_0| \ll N_A$ and implies that (3.8, 9) can be reduced to

$$n_0 = \frac{(s_e I_0 + \beta_e)(N_D - N_A)}{\gamma_e N_A} , \tag{3.10}$$

$$p_0 = \frac{(s_h I_0 + \beta_h) N_A}{\gamma_h (N_D - N_A)} , \tag{3.11}$$

since typical values of N_A and $N_D - N_A$ are on the order of 10^{16}–10^{17} cm^{-3}.

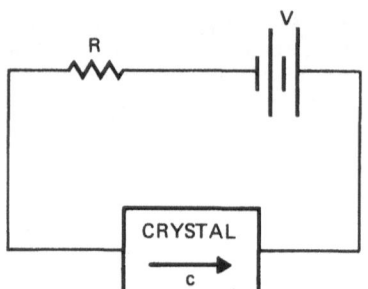

Fig. 3.2. Circuit model containing a photorefractive crystal with a nonzero photovoltaic effect, an external resistor R and an external voltage

After solving for the mean number densities, one can solve for the mean electric field E_0. Adding (3.3) and (3.4) yields

$$j_0 = \sigma_0 (E_0 + E_p) , \tag{3.12}$$

where $j_0 = j_{e0} + j_{h0}$ is the mean current, $\sigma_0 = e\mu_e n_0 + e\mu_h p_0$ is the mean conductivity, and $E_p = (\kappa_e s_e N_0 I_0 - \kappa_h s_h N_0^+ I_0)/\sigma_0$ is the photovoltaic field. (Of course, E_p is not an electric field but the difference between the electron and hole photovoltaic currents divided by the conductivity.) At this point j_0 and E_0 are still unknowns and (3.12) must be supplemented by an equation for the external curcuit around the crystal as shown in Fig. 3.2:

$$j_0 AR + E_0 d = V , \tag{3.13}$$

where A, R, and V are the area, resistance, and voltage of the external circuit and d is the length of the crystal between the electrodes [3.41]. Solution of (3.12 and 13) for E_0 yields

$$E_0 = \left(\frac{V}{d} - \frac{\sigma_0 E_p}{\sigma_{\text{ext}}} \right) \Big/ \left(1 + \frac{\sigma_0}{\sigma_{\text{ext}}} \right) , \tag{3.14}$$

where $\sigma_{\text{ext}} = d/(RA)$. Equation (3.14) has several interesting limits [3.23]:

(1) $R \to 0$ yields $E_0 = V/d$ (short circuit);
(2) $r \to \infty$ yields $E_0 = -E_p$ ($j_0 = 0$) (open curcuit);
(3) $E_0 = 0$ is obtained for $V/d = (\sigma_0/\sigma_{\text{ext}}) E_p$;
(4) $V = 0$ gives $E_0 = -E_p/(1 + \sigma_{\text{ext}}/\sigma_0)$.

In case (1) the effective conductivity of the external circuit is large compared to that of the crystal and the photovoltaic field is negligible. Note that this does not necessarily mean that the photovoltaic effect is unimportant in grating formation. One still has to calculate the first-order terms in (3.3 and 4). Case (2) is an open circuit with no applied field. The mean space-charge field within the crystal just balances the mean photovoltaic field. In case (3) the external voltage exactly compensates the photovoltaic field. Note that the voltage necessary to compensate the photovoltaic field is not equal to E_p, as pointed out by *Spinhirne* et al. [3.42].

3.2.3 General Solution for First-Order Quantities

Before writing down the general solution for the first-order quantities, it is useful to define 6 separate rate constants:

Dielectric relaxation rate, $\Gamma_{die} = e\mu_e n_0/(\varepsilon_0 \varepsilon)$;
Sum of production (photo and dark) and ion recombination rates, $\Gamma_{Ie} = s_e I_0 + \beta_e + \gamma_e n_0$;
Electron recombination rate, $\Gamma_{Re} = \gamma_e(N_A + n_0 - p_0)$;
Mean field drift rate, $\Gamma_{Ee} = K\mu_e E_0$;
Diffusion rate, $\Gamma_{De} = K^2 k_B T \mu_e/e$;
Photovoltaic rate, $\Gamma_{pe} = K\kappa_e s_e I/e$;

and the six corresponding rates for holes (replace subscript e by h, n_0 by p_0 and N_A by $N_D - N_A$). Substitution of these definitions and the approximations discussed above into (3.1–7) yields three simultaneous equations in n_1, p_1 and $A_1 = \varepsilon_0 \varepsilon i K E_1/e$:

$$(i\delta\omega - i\Gamma_{Ee} + \Gamma_{De} + i\Gamma_{pe} + \Gamma_{Ie} + \Gamma_{Re})n_1 + (-i\Gamma_{pe} - \Gamma_{Ie})p_1 + (-\Gamma_{die} + \Gamma_{Ie})A_1$$

$$= \left(1 + \frac{iK\kappa_e}{e}\right) s_e(N_D - N_A - n_0 + p_0)I_1 \ , \tag{3.15}$$

$$(-i\Gamma_{ph} - \Gamma_{Ih})n_1 + (i\delta\omega + i\Gamma_{Eh} + \Gamma_{Dh} + i\Gamma_{ph} + \Gamma_{Ih} + \Gamma_{Rh})p_1 + (\Gamma_{dih} - \Gamma_{Ih})A_1$$

$$= \left(1 + \frac{iK\kappa_h}{e}\right) s_h(N_A + n_0 - p_0)I_1 \ , \tag{3.16}$$

$$(i\delta\omega + \Gamma_{Ie} + \Gamma_{Re} + \Gamma_{Ih})n_1 + (-i\delta\omega - \Gamma_{Ie} - \Gamma_{Ih} - \Gamma_{Rh})p_1 + (i\delta\omega + \Gamma_{Ie} + \Gamma_{Ih})A_1$$

$$= s_e I_1(N_D - N_A - n_0 + p_0) - s_h I_1(N_A + n_0 - p_0) \ . \tag{3.17}$$

The response time $i\delta\omega$ is obtained by setting the determinant of the left-hand side of (3.15–17) equal to zero. The steady-state space-charge field is obtained by Cramer's rule. It does not appear to be that useful to write these expressions down, and we will instead concentrate on a number of special cases that have relatively simple expressions and that can be understood physically.

3.2.4 Special Cases

a) Steady State with No Holes, No Photovoltaic Field, No Applied Field
$(E_0 = \kappa_e = \kappa_h = p_0 = p_1 = \delta\omega = 0)$

Solution of (3.15, 17) yields

$$A_1 = \frac{\Gamma_{De}\alpha_e I_1}{\Gamma_{De}\Gamma_{Ie} + \Gamma_{die}\Gamma_{Ie} + \Gamma_{Re}\Gamma_{die}} \,, \tag{3.18}$$

where $\alpha_e = s_e(N_D - N_A - n_0 + p_0)$ is the absorption coefficient for production of electrons. Consistent with earlier approximation for n_0 and p_0, one can drop the n_0 amd p_0 terms in α_e. In order to cast (3.18) in a more recognizable form it is useful to write $I_1 = mI_0$ where m is the modulation index of the two beams making the interference pattern. Also, one can evaluate the rate constants for parameters appropriate to photorefractive materials using values given in Chaps. 6 and 7 and find that for irradiances typical of cw lasers the dielectric relaxation and photoproduction rates are much smaller than the recombination and diffusion rates. Thus the middle term in the denominator of (3.18) can be dropped. One final approximation, setting the dark production rate to zero in the dielectric relaxation rate, yields the popular formula for the lowest Fourier component of the space charge field E_1 :

$$E_1 = \frac{-iE_D m}{1 + E_D/E_q} \,, \tag{3.19}$$

where $E_D = k_B T K/e$ is the diffusion field and $E_q = eN_A(1 - N_A/N_D)/(\varepsilon_0 \varepsilon K)$ is the limiting space-charge field.

Note the factor i in (3.19), which indicates that the refractive index grating is shifted by 90° from the interference pattern. The diffusion field is that caused by diffusion of charge carriers while E_q is the largest field that can be set up in the crystal with an ionized trap density $N_T = N_A(1 - N_A/N_D)$. The quantity N_T is perhaps more transparent if it is written in the form $N_T = N^+(0)N(0)/[N^+(0) + N(0)]$ where the argument (0) is used to indicate dark number densities $(I = 0)$. Thus N_T reduces to $N^+(0)$ or $N(0)$, whichever is smaller. As will be seen below, E_q takes this form for both electrons and holes – the case where $N^+(0) \ll N(0)$ usually corresponds to electron dominated conductivity and vice versa for holes [3.38].

In the large grating period limit one obtains $E_1 = -imE_D$, while in the small grating period limit $E_1 = -imE_q$. This is because for large periods diffusion is not strong enough to separate enough charge to reach the limiting space-charge field. It is useful to remember that normally one cannot produce space-charge fields due to trapped charges that are larger than E_q in one grating period in photorefractive materials.

b) Response Time with No Holes, Photovoltaic or Mean Field

The equation for the response time for no holes, photovoltaic or mean field is

$$u^2 + (\Gamma_{De} + \Gamma_{Re} + \Gamma_{die} - \Gamma_{Ie})u + (\Gamma_{die} - \Gamma_{Ie})\Gamma_{Re} = 0 \ , \tag{3.20}$$

where $u = i\delta\omega + \Gamma_{Ie}$. This equation has two solutions, one fast and one slow. Again using the approximation that the photoproduction and dielectric relaxation rates are small compared to the recombination and diffusion rates, one obtains for the slow response time

$$\frac{i\delta\omega}{\Gamma_{die}} = -\frac{1 + \Gamma_{Ie}\Gamma_{De}/(\Gamma_{die}\Gamma_{Re})}{1 + \Gamma_{De}\Gamma_{Re}} \ . \tag{3.21a}$$

The fast rate, $i\delta\omega = -(\Gamma_{De} + \Gamma_{Re})$, is not relevant for grating formation with cw beams in typical materials, but for irradiances typical of nanosecond pulses the two rates may become comparable [3.27]. Equation (3.21a) may be cast in a form that is easier to understand physically by defining the Debye length or space-charge screening length as $l_s = [\varepsilon\varepsilon_0 k_B T/(N_T e^2)]^{1/2}$ and the length or "range" that an electron diffuses in one recombination time $r_D = [\mu k_B T/(e\Gamma_{Re})]^{1/2}$:

$$\frac{i\delta\omega}{\Gamma_{die}} = \frac{1 + K^2 l_s^2}{1 + K^2 r_D^2} \ . \tag{3.21b}$$

Note the useful identifies $\Gamma_{Ie}\Gamma_{De}/(\Gamma_{Re}\Gamma_{die}) = E_D/E_q = (Kl_s)^2$. If the grating period is large the response time is given by the dielectric relaxation time in agreement with intuition. If the denominator is negligible in (3.21), as is true for most grating periods in BaTiO$_3$ for example, then the response time becomes longer as the grating period increases, because more "hops" are required to diffuse one grating period. If the numerator is negligible, as in Bi$_{12}$SiO$_{20}$ for most grating periods, then the response time becomes shorter as the grating period increases. This is because the numerator is negligible when the Debye length is small compared to the grating period. When the Debye length is also small compared to the diffusion length, the carriers act collectively and a plasma-like behavior is obtained (see Chap. 6 for a more detailed discussion of this point). This behavior, which was predicted by Kukhtarev in 1976 [3.6], has been experimentally verified for BaTiO$_3$ [3.5] and Bi$_{12}$SiO$_{20}$ [3.26].

c) Applied Field with No Holes or Photovoltaic Effect

The equation for the space-charge field with an applied field reduces, of course, to the two cases discussed above. A compact version of this equation is given by

$$
\Gamma_{\text{die}}\left(1+\frac{\Gamma_D}{\Gamma_{Re}}-\frac{i\Gamma_{Ee}}{\Gamma_{Re}}\right)\frac{\partial E_1}{\partial t}
$$
$$
=-m(E_0+iE_D)+E_1\left(1+\frac{E_D}{E_q}-\frac{iE_0}{E_q}\right). \tag{3.22}
$$

The steady-state solution, which is given by setting the left-hand side equal to zero, shows that application of an external field leads to a space-charge field that is shifted from the interference pattern by an angle

$$
\phi_E=\frac{E_D}{E_0}\left(1+\frac{E_D}{E_q}+\frac{E_0^2}{E_D E_q}\right) \tag{3,23}
$$

with an amplitude of

$$
|E_1|=mE_q\left(\frac{E_0^2+E_D^2}{(1+E_D/E_q)^2+(E_0/E_q)^2}\right)^{1/2}. \tag{3.24}
$$

The complex response time indicates that application of an electric field to a crystal changes its response time. In crystals like $BaTiO_3$ application of an external field causes a modest decrease in response time, while in crystals like $Bi_{12}SiO_{20}$ this causes an increase in the response time [3.43]. The weakness of the effect of an applied field on the response time in $BaTiO_3$ is due to the fact that the diffusion length or hopping range is small compared to the Debye screening length in this material. The charge carriers diffuse without ever noticing the applied field.

The applied field also leads to damped oscillations in the space-charge field [3.6]. In addition, the phase shift between the optical interference pattern and the space-charge field and refractive index varies as a function of time [3.16]. This means that efficient energy exchange between two beams may occur in a transient regime with an applied field when this would not occur under steady-state conditions [3.24]. This effect has also been used to obtain efficient energy exchange under steady-state conditions by applying a small frequency shift to one of the two beams [3.29, 30, 44–46].

d) Space-Charge Field and Response Time with Two Charge Carriers and No Applied or Photovoltaic Field

As is clear from (3.15–17), the formulas for the space-charge field and the response time with two charge carriers are likely to be complicated. The formulas given here are valid for the case $E_0=\kappa_e=\kappa_h=E_p=\beta_e=\beta_h=0$ (no mean fields and no dark conductivity). Also we use the approximation that $n_0 \ll N_A$ and $p_0 \ll N_D - N_A$ and that the photoproduction and dielectric relaxation rates are small compared to the diffusion and recombination rates. All of these approximations can be easily removed with the penalty of lengthier expressions. Next it is useful to note the identity

$$\frac{\Gamma_{Ie}\Gamma_{De}}{\Gamma_{Re}\Gamma_{die}} = \frac{\Gamma_{Ih}\Gamma_{Dh}}{\Gamma_{Rh}\Gamma_{die}} = \frac{E_D}{E_q} \; . \tag{3.25}$$

As this relation indicates, there is one diffusion field and one limiting space-charge field in the crystal and they are given by the same expressions for electron-dominated, hole-dominated or nearly compensated crystals.

The steady-state space-charge field is given by [3.8, 9]

$$E_1 = \frac{-imE_D\left[1 - \dfrac{\sigma_h}{\sigma_e} + \dfrac{E_D}{E_q}\left(\dfrac{\Gamma_{dih}}{\Gamma_{Ih}} - \dfrac{\Gamma_{dih}}{\Gamma_{Ie}}\right)\right]}{\left(1 + \dfrac{E_D}{E_q}\right)\left[1 + \dfrac{\sigma_h}{\sigma_e} + \dfrac{E_D}{E_q}\left(\dfrac{\Gamma_{dih}}{\Gamma_{Ih}} + \dfrac{\Gamma_{dih}}{\Gamma_{Ie}}\right)\right]} \; . \tag{3.26}$$

This expression reduces to the results given previously [3.38, 40] when Γ_{dih}/Γ_{Ih} and Γ_{dih}/Γ_{Ie} are small compared to 1. In this case the one-species space-charge field is simply multiplied by the conductivity factor $\bar{\sigma} = (\sigma_e - \sigma_h)/(\sigma_e + \sigma_h)$ [3.37]. This reflects the fact that when electron and hole conductivities are equal, there is no net diffusion to create a grating. In the opposite case where the third terms in the square brackets of (3.26) dominate (for very small grating periods or in materials like $Bi_{12}SiO_{20}$) one obtains

$$E_1 = -imE_q \frac{\alpha_e - \alpha_h}{\alpha_e + \alpha_h} \; . \tag{3.27}$$

Unfortunately, the rate equations used in [3.38] were incorrect and did not give this solution. Behavior described by both limits has probably been reported in [3.38] where in an anomalous crystal of $BaTiO_3$ measurement of the gain coefficient for beam coupling, which is proportional to the space-charge field when there is no applied field, was observed to change sign as a function of grating period. This could occur if the conductivity and the absorption coefficient factors had opposite signs. The response time for a material in which both carriers are important is obtained from

$$\frac{i\delta\omega}{\Gamma_{die}} = -\frac{\left[1 + \dfrac{\sigma_h}{\sigma_e} + \dfrac{E_D}{E_q}\left(\dfrac{\Gamma_{dih}}{\Gamma_{Ie}} + \dfrac{\Gamma_{dih}}{\Gamma_{Ih}}\right)\right]\left(1 + \dfrac{E_D}{E_q}\right)}{\left(1 + \dfrac{\Gamma_{De}}{\Gamma_{Re}}\right)\left(1 + \dfrac{\Gamma_{Dh}}{\Gamma_{Rh}}\right)} \; . \tag{3.28}$$

Note that the response time is proportional to the intensity as in the case of one charge carrier and that we give only one response time. There are, of course, two other solutions for the response time but these can be omitted on the same physical grounds discussed above for single charge carriers (i.e., if the photoproduction and dielectric relaxation rates are small compared to the diffusion and recombination rates). Two or more response times have been observed in $Bi_{12}SiO_{20}$ [3.47] and $LiNbO_3$ [3.48] and response time proportional to intensity to a fractional power has been observed in $BaTiO_3$ [3.49]. The rate equations given by (3.1–7) do not predict these effects. Alternative and more complicated models such as multiple donor/acceptor systems [3.36, 39] and distributions of levels of the donors or acceptors within the band gap [3,50] appear necessary to explain these effects.

3.3 Grating Formation and Decay with Short Pulses

If gratings are formed or erased in a photorefractive material with lasers whose pulse lengths range from picoseconds to microseconds, many of the approximations used in Sect. 3.2 are not valid and the behavior of the photorefractive material may be substantially changed from that characteristic of cw illumination. One anticipates that the quasi-steady approximation should break down and that the electron and ion number densities may not reach steady state during a laser pulse if the laser pulse is shorter than the recombination time. This will lead to nonexponential time dependence since the mean quantities are not independent of time.

One would also expect that the intensity might become high enough that saturation effects in the carrier and ion densities would be important. This can lead to response times that depend on the square root of intensity or are independent of intensity, and to substantially decreased photorefractive sensitivities. (As the carrier number density saturates, much of the optical energy incident on the material is wasted.) If the carrier number density is sufficiently high, then the approximation that the dielectric relaxation time and the photoproduction time are long compared to the diffusion and recombination time is no longer valid. This can lead to two or more exponential time constants even when there is only one carrier and one photorefractive species. If the pulse length, diffusion or drift time, and read-out time are all small compared to the recombination time, a photorefractive grating can be obtained in the carrier number density as opposed to the trapped charge density. In this case the space-charge field in a photorefractive material can exceed the value of E_q based on the traps and is determined by the maximum carrier number density. Results based on the set of rate equations given by (3.1–7) are available for only a small subset of these problems.

In addition to these relatively straightforward modifications to the cw rate equation theory, one must expect that a number of new photorefractive effects will turn up on illumination with picosecond to microsecond pulses. We will omit any speculation on such effects.

Measurements of the photorefractive effect on nanosecond time scales have been performed in LiNbO$_3$ [3.51, 52] in BaTiO$_3$ [3.53] and in Bi$_{12}$SiO$_{20}$ [3.54]. In LiNbO$_3$ it was observed that less energy per unit area was required to write a grating than in cw experiments. In BaTiO$_3$ the observations were well explained by the cw theory given in the previous section. In Bi$_{12}$SiO$_{20}$ more energy per area was required to write a grating than with cw illumination.

A theoretical development applicable to two cases of short pulse illumination is given in [3.27]. In one case the quasi-steady approximation is extended to include carrier saturation and the approximation that the dielectric relaxation and photoproduction rates are small compared to the diffusion rate is removed. In the second case the pulse length is assumed to be short compared to the diffusion, recombination, dielectric relaxation, and photoproduction rates.

3.3.1 Extension of Quasi-Steady Approximation

First consider the equation for the carrier number density in a material dominated by either electrons or holes

$$\frac{\partial n_0}{\partial t} = sI_0(N_D - N_A - n_0) - \gamma_R n_0(N_A + n_0) \ . \tag{3.29}$$

At low irradiances the steady-state solution is $n_0 = sI_0(N_D - N_A)/(\gamma_R N_A)$; when the "bimolecular" recombination term $\gamma_R n_0^2$ is important, $n_0 = [s(N_0 - N_A)I_0/\gamma_R]^{1/2}$; at high irradiances n_0 saturates at $N_D - N_A$, the total number density of available carriers. These expressions can be substituted as appropriate into the rates Γ_I, Γ_{di}, and Γ_R defined in Sect. 3.2.3 and used in (3.15–17) to extend the results to allow for saturation of the carrier density. Equation (3.18) is valid for this case, but not (3.19). Likewise, solution of the quadratic equation (3.20) for the response times is valid, but (3.21) is valid only when the rates Γ_I and Γ_{di} are small compared to Γ_R and Γ_D. If the carrier number density saturates, this approximation is unlikely to hold, and in the opposite case one obtains two rates, $i\delta\omega \sim \Gamma_I$ and $i\delta\omega \sim \Gamma_{di}$.

3.3.2 Illumination with Delta-Function Pulses

Although no experimental results have yet been published in this regime, theoretical results based on (3.1–7) suggest that gratings may be written or erased in a photorefractive material with high intensity pulses whose duration is shorter than any of the time constants of the material [3.27]. In this case all of the carriers are created instantaneously and then the grating is erased or written by drift and diffusion in the dark. This process takes place until all of the carriers are retrapped, which takes a few recombination times.

If the carrier number density at the end of the pulse is small compared to the trap density N_A, then one can solve for the space-charge field in the single-carrier case of pulse length short compared to photorefractive times:

$$E_1 = E_1(\tau_p) + \left(\frac{\partial E_1}{\partial t}\right)\Bigg|_{t=\tau_p}$$

$$\times \int_0^t dt' \exp\left\{-\left(\frac{\Gamma_{di}}{\Gamma_R}\right)[1 - \exp(-t'/\tau_R)] - (\Gamma_R + i\Gamma_E - \Gamma_D)t'\right\}, \qquad (3.30)$$

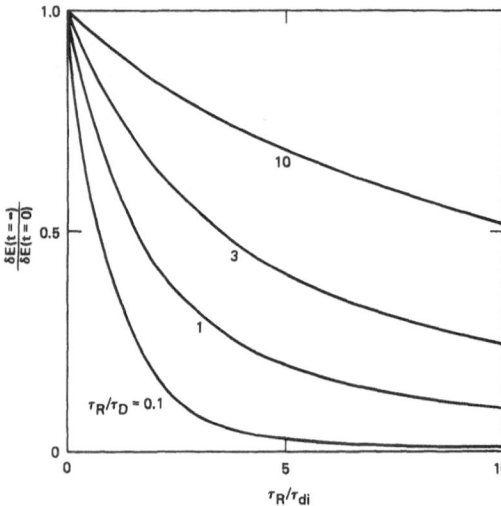

Fig. 3.3. Fractional erasure of a grating $E_1(\infty)/E_1(0)$ at times large compared to the pulse length and recombination time as a function of the ratio of the recombination time to the dielectric relaxation time (τ_R/τ_{di}) with the ratio of recombination time to diffusion time (τ_R/τ_D) as a parameter

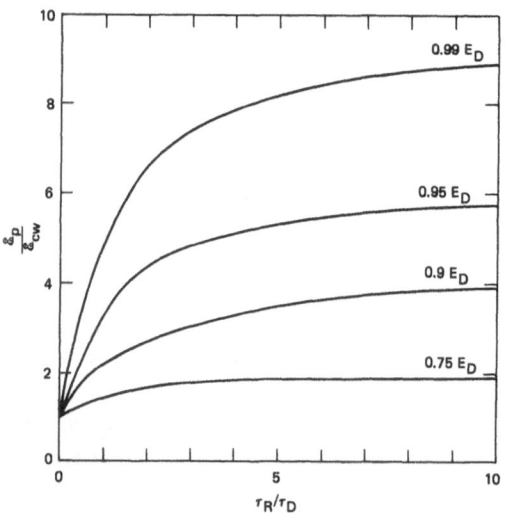

Fig. 3.4. Ratio of pulsed to cw energy per unit area required to write a grating to a fraction of $0.75 E_D$–$0.99 E_D$ as a function of the ratio of recombination time to diffusion time (τ_R/τ_D)

where Γ_{di} is evaluated at $n_0(\tau_p)$ and τ_p is the recombination time. The integral in (3.30) can be expressed in terms of incomplete gamma functions [3.27] or evaluated numerically. The two boundary conditions are $E_1(\tau_p) = 0$ and $(\partial E_1/\partial t)|_{t=\tau_p} = -(E_0 - iE_D)m\Gamma_{di}$ for writing, and $E_1(\tau_p) = E_g$ and $(\partial E_1/\partial t)|_{t=\tau_p} = -\Gamma_{di}E_g$ for erasing, where E_g is the amplitude of the grating being erased. Results for fractional erasure based on (3.30) are shown in Fig. 3.3.

Results for the ratio of pulsed to cw energy required to write a grating of a fraction of E_D are given in the large grating period limit $(E_D/E_q \ll 1)$ in Fig. 3.4. Note that writing with a pulsed laser always requires more energy than with a cw laser (although of course, it will usually be much faster).

3.4 Solutions to the Two- and Four-Wave Mixing Problems

The previous sections of this chapter concern the photorefractive response of electro-optic crystals under the action of external radiation fields and an applied static electric field. The response of the medium is given in terms of a spatially and temporally modulated internal space-charge field. However, the majority of nonlinear optical experiments involve the measurement of optical intensities generated in a nonlinear optical process. Hence, this section is devoted to the formulation and solution of the self-consistent problem of nonlinear optics. Our approach parallels those discussed in [3.23, 28, 29, 31]. Specifically, this section discusses the following cases: nearly degenerate four-wave mixing and self-pumped phase conjugation via nondegenerate backward stimulated two-wave mixing. This section neglects the effect of pump depletion, which is discussed by White et al. in Photorefractive Materials and Their Applications II, Chap. 4.

The interaction of incident and scattered radiation fields in a nonlinear optical material can be described in terms of the coupled scalar wave equations for E_n

$$\hat{k}_n \cdot \nabla E_n = -i\left(\frac{\omega}{2c}\right) n_b^2 \, \hat{e}_n^* \cdot \underline{r} \cdot E_{sc} \cdot \hat{e}_m E_m e^{i(\phi_m - \phi_n)} \; , \tag{3.31}$$

where $\phi_m = k_m \cdot r - \omega_m t$ is the optical phase.

We have assumed that the slowly varying approximation holds for the envelope E_n of the radiation field. The terms in (3.31) have the following interpretation: the term on the left-hand side describes free-space propagation of the optical wave, while the right-hand side describes the nonlinear wave mixing process. The unit vectors $\hat{e}_n$ and $\hat{k}_n$ give the directions of polarization and propagation of E_n; n_b and $\underline{r}$ are the index of refraction and electro-optic tensor of the photorefractive material. The space charge field E_{sc} is determined by the solution of the coupled equations (3.15–17). We remind the reader that the nonlinear optical response of photorefractive materials is a $\chi^{(n)}$ nonlinear optical process.

The analysis provided in the next section involves the following steps. We introduce a Fourier expansion in the nonlinear polarization which represents the interference of all possible combinations of the incident and scattered radiation fields. The coupled wave equation approach leads to a self-consistent formulation for the spatial evolution of each of the Fourier components. In the following, the first two components of the Fourier series are kept in the analysis. The zeroth-order component represents the contribution arising from the dc part of the Fourier expansion; i.e., those terms that are dependent on the average intensity of each radiation field. The first-order component takes into account the sinusoidal grating formation, oscillating at the difference of the two optical phases of the interacting radiation fields. Apart from the details of the photorefractive process itself, the polarization density of the photorefractive response is similar to that obtained in a nearly resonant multilevel system [3.55].

3.4.1 Nearly Degenerate Four-Wave Mixing

Since the first theoretical study of nearly degenerate four-wave mixing (NDFWM) in photorefractive materials [3.28], much of the current research has been devoted to the experimental confirmation as well as applications of the theoretical results. Much progress has been made by Huignard and his group in the practical implementation of NDFWM in optical devices using bismuth silicon oxide and they have extended the theory of NDFWM to the case where an applied static electric field is present [3.30] (see Photorefractive Materials and Their Applications II, Chap. 6).

This section, based on the treatment of *Lam* [3.22], is devoted to a theoretical discussion of the case with no external applied field and with a single species of charge carriers. The geometry of the NDFWM process is shown in Fig. 3.5. The counterpropagating pump fields E_f and E_b oscillate at the same frequency ω, while the probe field E_p oscillates at $\omega + \delta$. Conservation of energy requires that the four-wave mixing signal, generated from the nonlinear process, oscillates at $\omega - \delta$. We assume that the direction of propagation of the probe field makes an angle θ with respect to that of the forward pump field E_f, and the nonlinear interaction length is L. Within the classical regime, the boundary conditions are $E_s(L) = 0$ and $E_p(0)$ is a given quantity. In the undepleted pump regime, the

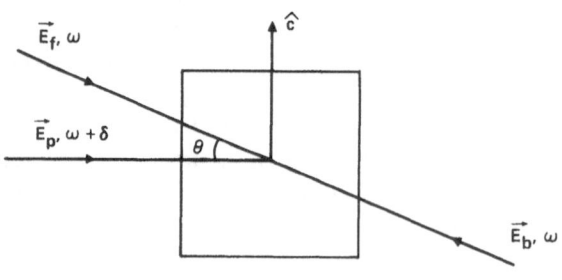

Fig. 3.5. Interaction geometry for the nondegenerate four-wave mixing problem

solution of the coupled wave equations yields [3.28]

$$E_s(0) = \kappa E_p^*(0) \left(\frac{1 - e^{\beta L}}{\lambda_+ - \lambda_- e^{\beta L}} \right)$$ (3.32a)

where

$$\kappa = \frac{\omega}{2c} n_b^2 \left[\hat{e}_n \cdot \underline{r} \cdot (\hat{k}_s - \hat{k}_b) \cdot \hat{e}_n \right] \frac{e}{\varepsilon \varepsilon_0} \left(\frac{\lambda \Delta N}{4\pi \sin(\theta/2)} \right)$$ (3.32b)

is the NDFWM coupling constant and ΔN is the radiation field induced density difference between the electrons and ions

$$\Delta N = \frac{\Gamma_{Ie} \Gamma_{De}}{[(\delta^2 - B)^2 + \delta^2 A^2]^{1/2}} [N_D - N^+(0)] e^{i\Phi_s} ,$$ (3.32c)

where N_D is the total number density of the dopant responsible for the photorefractive effect and $N^+(0)$ is the ion number density in the dark ($I = 0$). The phase shift due to pump-probe detuning is given by

$$\Phi_g = \arctan \left(\frac{\delta A}{\delta^2 - B} \right) .$$ (3.32d)

The parameters A and B are given by

$$A = \Gamma_{die} + \Gamma_{De} + \Gamma_{Re} + \Gamma_{Ie} \quad \text{and}$$

$$B = (\Gamma_{die} + \Gamma_{De}) \Gamma_{Ie} + \Gamma_{Re} \Gamma_{die} .$$

If the k-vector mismatch $\Delta k = -2\delta n_0/c$, then

$$\lambda_\pm = \frac{i\Delta k \pm \beta}{2} \quad \text{and}$$ (3.32e)

$$\beta = [(2\kappa)^2 - (\Delta k)^2]^{1/2} .$$ (3.32f)

Results based on (3.32a–f) show significant novel features that do not appear in the case of degenerate four-wave mixing. First, the four-wave mixing reflectivity $|E_s(0)|^2/|E_p(0)|^2$ depends on the pump-probe detuning as shown in Fig. 3.6a with a full width at half maximum that depends on pump intensity and angle between the beams θ as shown in Fig. 3.6b. The numerical parameters used in these figures are taken to be representative of $Bi_{12}SiO_{20}$ crystals [3.43]: $N_D = 10^{19}$ cm^{-3}, $N_+(0) = 10^{16}$ cm^{-3}, $\mu = 0.03$ cm^2 V^{-1} s^{-1}, $\gamma_r = 2 \times 10^{-11}$ cm^3 s^{-1} and $s_e = 0.415$ cm^2 J^{-1}. (γ_r is the recombination rate

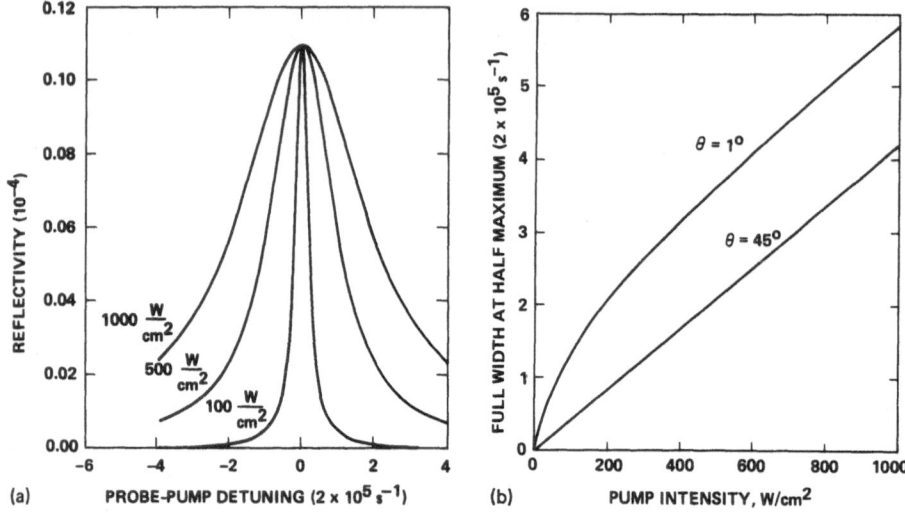

Fig. 3.6. (a) Reflectivity as a function of probe-pump detuning for various values of the pump intensity. **(b)** Full width at half maximum as a function of pump intensity for two different values of the angle θ for the conditions ΔkL and $\kappa L \ll 1$

coefficient for the dominant charge carrier.) The behavior reflects the fact that the response time of photorefractive materials is proportional to the dielectric response time which is inversely proportional to intensity.

Secondly, (3.32a–f) predict that self-oscillation (a pole in the small-signal reflectivity) can occur in photorefractive materials. This phenomenon takes place when the phase shift between the refractive index grating and the interference pattern does not equal 90°; this occurs (1) when there is a nonzero applied or photovoltaic field [3.32] and (2) when the pump-probe detuning δ is nonzero and there is no applied or photovoltaic field. In the second case calculations show that the condition for self-oscillation is given by [3.28]

$$\tan\left[2\sigma L\sin\left(\frac{\phi}{2}\right)\right]=\frac{\sigma\Delta k}{(\sigma^2-\Delta k^2/4)}\cos\left(\frac{\phi}{2}\right), \tag{3.33a}$$

where

$$\sigma=|\kappa|\left\{\left[\cos(2\Phi_g)-\left(\frac{\Delta k}{2\kappa}\right)^2\right]^2+\sin^2(2\Phi_g)\right\}^{1/4}, \tag{3.33b}$$

$$\tan\phi=\frac{\sin(2\Phi_g)}{\cos(2\Phi_g)-(\Delta k/2\kappa)^2}. \tag{3.33c}$$

From these expressions one can calculate the required frequency shift such that a self-oscillation condition can be obtained:

$$\delta = -\frac{A \ln r}{2\pi} \pm \left[\left(\frac{A \ln r}{2\pi} \right)^2 + B \right]^{1/2} \tag{3.34}$$

where r is the ratio of the backward to forward pump intensity. For $r = 1$, one finds that two resonances exist centered around $\delta = 0$.

These effects have been observed by *MacDonald* and *Feinberg* [3.56] who saw enhanced reflectivity for nonzero pump-probe detuning. Their observations are symmetric about $\delta = 0$ and consistent with the existence of a self-oscillation condition in the undepleted pump limit. The observation of self-starting oscillation between an externally pumped phase conjugate mirror and a normal mirror, which has been observed by many authors following the initial work of *Feinberg* and *Hellwarth* [3.57], may also involve self-oscillation due to nonzero pump-probe detuning since the linewidths of the pump lasers used in these experiments greatly exceed the detuning required for self-oscillation. Large reflectivities can also be obtained without detuning, however, so the existence of self-starting oscillators does not necessarily imply the observation of self-oscillation due to nonzero detuning.

Thirdly, unlike the case of degenerate four-wave mixing in the absence of an applied static field, the probe-pump detuning δ introduces an additional phase shift Φ_g which spoils the 90° phase lag between the space charge field E_{sc} and the interference pattern produced in the degenerate mode of operation. This additional phase shift leads to the appearance of a nonlinear phase shift and a simultaneous decrease of the degree of energy transfer between the participating waves. It was shown [3.28] that a detuning of $\delta = \pm B^{1/2}$ gives $\Phi_g = 0°$ and no energy transfer among the waves. For this specific detuning the photorefractive medium behaves in an identical manner to a nonresonant but saturated $\chi^{(3)}$ medium.

3.4.2 Self-pumped Phase Conjugation via Nearly Degenerate Backward Stimulated Two-Wave Mixing

The discovery of self-pumped phase conjugation in an external laser cavity by *White* et al. [3.58] started an exciting period of research in photorefractive materials. This discovery was followed by the demonstration by *Feinberg* [3.59] that total internal reflection inside the electro-optic crystal of barium titanate can also yield phase conjugation with a single input pump beam. At the same time, there has been intense investigation of the physical origin of self-pumped phase conjugation in photorefractive materials. *MacDonald* and *Feinberg* [3.60] proposed that strictly degenerate four-wave mixing is responsible for the generation of phase conjugate waves in the self-pumped phase conjugation process. However, their theory does not appear to explain the origin of the phase conjugation process nor the associated frequency shift that was subsequently observed by a number of researchers [3.61–64]. Heuristic arguments have been given that the counter-propagating waves are complex conjugates in these geometries because this maximizes the spatial overlap integral for the four-wave mixing process, but detailed calculations in three spatial dimensions have not been performed.

An alternative explanation of the self-pumped conjugation process was proposed by *Lam* [3.28] who suggested that stimulated backward two-wave mixing may be responsible. The existence of phase conjugation by stimulated two-wave mixing was confirmed recently in a beautiful experiment by *Chang* and *Hellwarth* [3.65]. But the role of backward two-wave mixing in the other internal and external self-pumped conjugation processes is less clear. Two factors favor backward two-wave mixing: (1) the interaction length is maximized for backscattering as opposed to two-wave or four-wave mixing at large angles; and (2) the photorefractive sensitivity or refractive index change per absorbed energy, which should control start-up conditions, is maximized for the very small grating period of the counterpropagating waves in BaTiO$_3$ [3.43].

The theory of backward stimulated two-wave mixing in photorefractive materials [3.31] follows from (3.1–7) with an important modification of the photovoltaic term. This modification, a term of the form $pnN^+\hat{c}$ in the current equation gives a generalized expression for the scalar photovoltaic effect [3.31]. More importantly, calculations predict that the photovoltaic effect gives rise to a frequency shift in the backscattered wave. Unlike the mean photovoltaic field, this frequency shift cannot be compensated with the application of an external static electric field. The results can be summarized as follows. The coupled wave equations for E_p and E_s are found to be

$$-\frac{\partial E_s}{\partial y} - \frac{i}{2k_s}\, \nabla^2 E_s = -\bar{g}|E_p|^2\, E_s - \alpha_s E_s \ , \tag{3.35a}$$

$$\frac{\partial E_p}{\partial y} - \frac{i}{2k_p}\, \nabla^2 E_p = \bar{g}^*|E_s|^2\, E_p - \alpha_p E_p \ , \tag{3.35b}$$

where α_s and α_p are the linear absorption coefficients of E_s and E_p. The complex gain coefficient $\bar{g}$ is

$$\bar{g} = \left(\frac{\omega}{2c}\right)\frac{ecs_e[N_D - N^+(0)]}{2\varepsilon_0\varepsilon\Delta k}\, G(\theta)\, \frac{[\Gamma_{De}^2 + (\delta_p + \delta_n)^2]^{1/2}}{\mathrm{DEN}}\, e^{i(\sigma+\varPhi)} \tag{3.35a}$$

with

$$G(\theta_c) = (n_e^4 r_{33} - 2n_o^2 n_e^2)\sin\theta_c\cos^2\theta_c + n_o^4 r_{23}\sin^3\theta_c \ , \tag{3.35b}$$

$$\mathrm{DEN} = [(\delta^2 - \delta\delta_p - B)^2 + (\delta A - \delta_p\Gamma_{Ie} + \delta_n\Gamma_{Re})^2]^{1/2} \ , \tag{3.35c}$$

and

$$\sigma + \Phi = -\tan^{-1}\left(\frac{\delta_{p} + \delta_{n}}{\Gamma_{De}}\right)$$

$$+ \tan^{-1}\left(\frac{\delta A - \delta_{p}\Gamma_{Ie} + \delta_{n}\Gamma_{Re}}{d^{2} - \delta\delta_{n} - B}\right). \tag{3.35d}$$

The photovoltaic induced frequency shifts are $\delta_{p} = pN_{+}(0)\hat{c} \cdot \Delta k/e$ and $\delta_{n} = 2p\Gamma_{p}N_{D}\hat{c} P \Delta k/e\Gamma_{Re}$ where $\hat{c}$ is the unit vector oriented along the c-axis of the crystal.

The solutions show some very interesting features. First, the coupled wave equations (3.35) are similar to those found in stimulated Brillouin scattering, with the exception that the gain coefficient $\bar{g}$ is dependent on the sum of the intensities of the two counterpropagating waves. In the undepleted pump regime, the argument of *Zel'dovich* et al. [3.66] provides a quantitative explanation of the origin of phase conjugate waves. Secondly, due to the competition between linear losses and nonlinear gain, self-pumped phase conjugation has a threshold intensity behavior. Thirdly, Fig. 3.7 shows the dependence of the gain on the angle of incidence θ_{c}. For the case of barium titanate, the gain equals zero at approximately 86° and no backscatter is possible at this angle. Similar results in the absence of the photovoltaic field were obtained by *MacDonald* et al. [3.67]. Their results differ from Fig. 3.7 because

(a)

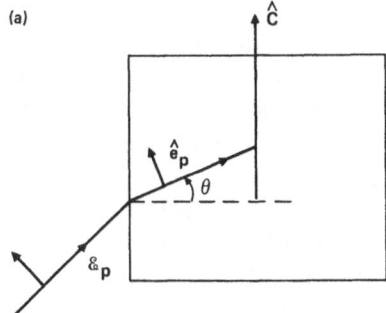

(b)

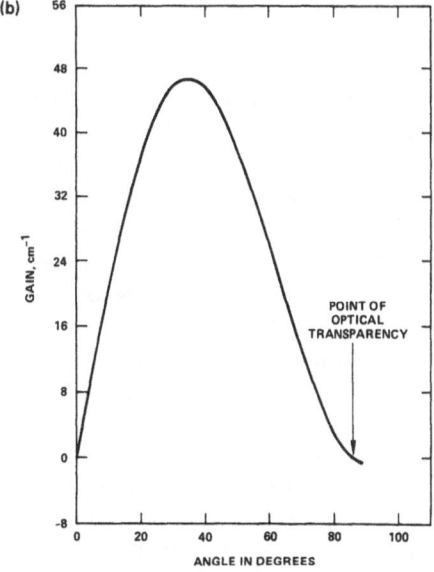

Fig. 3.7. (a) Interaction geometry of backward two-wave mixing relative to crystal c-axis. **(b)** Gain per unit length as a function of the angle of incidence θ_{c}

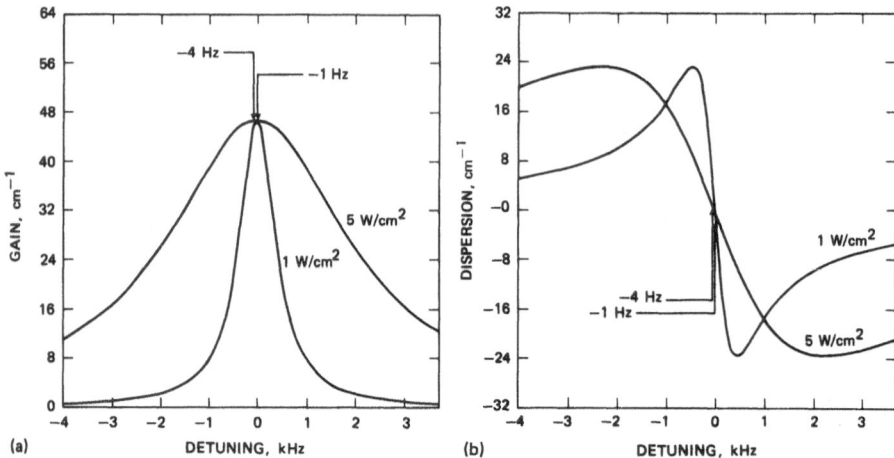

Fig. 3.8. (a) Gain per unit length as a function of detuning from the pump frequency. Maxima at $1\,\mathrm{W\,cm^{-2}}$ and $5\,\mathrm{W\,cm^{-2}}$ are shifted $-1\,\mathrm{Hz}$ and $-4\,\mathrm{Hz}$ from the pump frequency. **(b)** Dispersion per unit length as a function detuning. At zero detuning the dispersion is $0.27\,\mathrm{cm^{-1}}$

they used a static dielectric of the form $\varepsilon = \varepsilon_{par}\cos^2\theta_c + \varepsilon_{perp}\sin^2\theta_c$ where ε_{par} and ε_{perp} are the dielectric constants parallel and perpendicular to the c-axis, while $\varepsilon = 1000$ was used in Figs. 3.7 and 3.8.

Fourth, Figs. 3.8a, b show the spectral behavior of the nonlinear gain and dispersion. Note that as the pump intensity increases, the gain and dispersion bandwidths increase in a proportionate manner. This behavior is attributed to the intensity dependence of the relaxation rate of barium titanate, which is proportional to the dielectric response rate Γ_{die}. Furthermore, there is an intensity-dependent shift of the maximum of the gain and a corresponding shift in the dispersion. This effect is attributed to the photovoltaic induced frequency shift δ_n. It is important to note that the frequency shift cannot be cancelled with the application of a static electric field.

Lastly, the theory also predicts the existence of nonlinear dispersion, and this provides a mechanism for the observation of nonlinear diffraction of the pump beam [3.68]. This result has two important consequences. An incoming beam with asymmetrical transverse profile will experience beam bending, while phase perturbation at the incoming face of the crystal will lead to beam break up, in the same manner as observed in laser-induced filamentation. These two effects, collectively named beam fanning, can be explained by this theory if the fanning is accompanied by a counterpropagating beam that is shifted in frequency relative to the input beam.

In the preceding paragraphs we have discussed some results that bear on the physical processes in self-pumped phase conjugation. Much remains to be explained, however. This includes the detailed mechanism responsible for the frequency shifts observed in self-pumped conjugators and resonators [3.61–64] and the three-dimensional and time-dependent details of the start-up process that eventually leads to production of phase conjugate waves.

Acknowledgements. We are grateful to our colleagues at Hughes Research Laboratories for many discussions of photorefractive effects. Dr. M.B. Klein, Dr. R.A. Mullen and Dr. J. White provided helpful comments on the manuscript. We thank Prof. R.W. Hellwarth for helpful comments on the hopping model.

References

3.1   R.L. Townsend, J.T. LaMacchia: J. Appl. Phys. **41**, 5188 (1970)
3.2   F.S. Chen: J. Appl. Phys. **40**, 3389 (1969)
3.3   J.J. Amodei: Appl. Phys. Lett. **18**, 22 (1971)
3.4   W.D. Johnston, Jr.: J. Appl. Phys. **41**, 3279 (1970)
3.5   J. Feinberg, D. Heiman, A.R. Tanguay, Jr., R.W. Hellwarth: J. Appl. Phys. **51**, 1297 (1980)
3.6   N.V. Kukhtarev: Pis'ma Zh. Tekh. Fiz. **2**, 1114 (1976) [English transl.: Sov. Tech. Phys. Lett. **2**, 438 (1976)]
3.7   S. Ducharme, J. Feinberg: J. Opt. Soc. Am. B**3**, 283 (1986)
3.8   G.C. Valley: J. Appl. Phys. **59**, 3363 (1986)
3.9   F.P. Strohkendl, J.M.C. Jonathan, R.W. Hellwarth: Opt. Lett. **11**, 312 (1986)
3.10 F.S. Chen: J. Appl. Phys. **38**, 3418 (1967)
3.11 F.S. Chen, J.T. LaMacchia, D.B. Fraser: Appl. Phys. Lett. **13**, 223 (1968)
3.12 A.M. Glass, D. von der Linde, T.J. Negran: Appl. Phys. Lett. **25**, 233 (1974)
3.13 J.J. Amodei: RCA Rev. **32**, 185 (1971)
3.14 V.L. Vinetskii, N.V. Kukhtarev: Fiz. Tverd. Tela **16**, 3714 (1974) [English transl.: Sov. Phys.–Solid State **16**, 2414 (1975)]
3.15 D.L. Staebler, J.J. Amodei: J. Appl. Phys. **43**, 1042 (1972)
3.16 L. Young, W.K.Y. Wong, M.L.W. Thewalt, W.D. Cornish: Appl. Phys. Lett. **24**, 264 (1974)
3.17 S.F. Su, T.K. Gaylord: J. Appl. Phys. **46**, 5208 (1975)
3.18 G.A. Alphonse, R.C. Alig, D.L. Staebler, W. Phillips: RCA Rev. **36**, 213 (1975)
3.19 R. Magnusson, T.K. Gaylord: J. Appl. Phys. **47**, 190 (1976)
3.20 D.M. Kim, R.R. Shah, T.A. Rabson, F.K. Tittel: Appl. Phys. Lett. **28**, 338 (1976)
3.21 M.G. Moharam, L. Young: J. Appl. Phys. **47**, 4048 (1976)
3.22 V.L. Vinetskii, N.V. Kukhtarev: Kvantovaya Elektron. (Moscow) **5**, 405 (1978) [English transl.: Sov. J. Quantum Electron **8**, 231 (1978)]
3.23 N.V. Kukhtarev, V.B. Markov, S.G. Odulov, M.S. Soskin, V.L. Vinetskii: Ferroelectrics **22**, 949, 961 (1979)
3.24 N.V. Kukhtarev, V.B. Markov, S.G. Odulov: Opt. Commun. **23**, 338 (1977)
3.25 N.V. Kukhtarev, S.G. Odulov: Pis'ma Zh. Eksp. Teor. Fiz. **30**, 6 (1979) [English transl.: JETP Lett. **30**, 4 (1979)]
3.26 R.A. Mullen, R.W. Hellwarth: J. Appl. Phys. **58**, 40 (1985)
3.27 G.C. Valley: IEEE J. QE-**19**, 1637 (1983)
3.28 J.F. Lam: Appl. Phys. Lett. **42**, 155 (1983)
3.29 G.C. Valley: J. Opt. Soc. Am. B**1**, 868 (1984)
3.30 Ph. Refregier, L. Solymar. H. Rajbenbach, J.P. Huignard: J. Appl. Phys. **58**, 45 (1985)
3.31 J.F. Lam: Appl. Phys. Lett. **46**, 909 (1985)
3.32 M. Cronin-Golomb, B. Fischer, J.O. White, A. Yariv: IEEE J. QE-**20**, 12 (1984); Opt. Lett. **7**, 313 (1982)

3.33 Y.H.Ja: Opt. Quantum Electron. **14**, 547 (1982); ibid. **15**, 529, 539 (1983); Appl. Phys. B**33**, 51 (1984)
3.34 M.R.Belic: Phys. Rev. A**31**, 3169 (1985); Opt. Quantum Electron. **16**, 551 (1984)
3.35 P.Yeh: Opt. Commun. **45**, 323 (1983)
3.36 N.V.Kukhtarev, G.E.Dovgalenko, V.N.Starkov: Appl. Phys. A**33**, 227 (1984)
3.37 R.Orlowski, E.Krätzig: Solid State Commun. **27**, 1351 (1978)
3.38 M.B.Klein, G.C.Valley: J. Appl. Phys. **57**, 4901 (1985)
3.39 G.C.Valley: Appl. Opt. **22**, 3160 (1983)
3.40 R.Orlowski, E.Krätzig: Solid State Commun. **27**, 1351 (1978)
3.41 N.V.Kukhtarev, V.B.Markov, S.G.Odulov: Zh. Tekh. Fiz. **50**, 1905 (1980) [English transl.: Sov. Phys.–Tech. Phys. **25**, 1109 (1980)]
3.42 J.M.Spinhirne, D.Ang, C.S.Joiner, T.L.Estle: Appl. Phys. Lett. **30**, 89 (1977)
3.43 G.C.Valley, M.B.Klein: Opt. Eng. **22**, 704 (1983)
3.44 J.P.Huignard, A.Marrakchi: Opt. Commun. **38**, 249 (1982)
3.45 S.I.Stepanov, V.V.Kulikov, M.P.Petrov: Opt. Commun. **44**, 19 (1982)
3.46 H.Rajbenbach, J.P.Huignard, B.Loiseaux: Opt. Commun. **48**, 247 (1983)
3.47 R.A.Mullen, R.W.Hellwarth: paper ThH3, CLEO 1983 Digest, p. 174 (1983)
3.48 J.K.Tyminski, R.C.Powell: J. Opt. Soc. Am. B**2**, 440 (1985)
3.49 S.Ducharme, J.Feinberg: J. Appl. Phys. **56**, 839 (1984)
3.50 A.Rose: *Concepts in Photoconductivity and Allied Problems* (Robert E.Krieger, Huntington, NY 1978) pp. 38–43
3.51 C.-T.Chen, D.M.Kim, D. von der Linde: IEEE J. QE-**16**, 126 (1980)
3.52 C.-T.Chen, D.M.Kim, D. von der Linde: Appl. Phys. Lett. **34**, 321 (1979)
3.53 L.K.Lam, T.Y.Chang, J.Feinberg, R.W.Hellwarth: Opt. Lett. **6**, 475 (1981)
3.54 J.P.Hermann, J.P.Herriau, J.P.Huignard: Appl. Opt. **20**, 2173 (1981)
3.55 R.L.Abrams, J.F.Lam, R.C.Lind, D.G.Steel; In *Optical Phase Conjugation*, ed. by R.A.Fisher (Academic, New York 1983) Chap. 8
3.56 K.R.MacDonald, J.Feinberg: Phys. Rev. Lett. **55**, 821 (1985)
3.57 J.Feinberg, R.W.Hellwarth: Opt. Lett. **5**, 519 (1980)
3.58 J.O.White, M.Cronin-Golomb, B.Fischer, A.Yariv: Appl. Phys. Lett. **40**, 450 (1982)
3.59 J.Feinberg: Opt. Lett. **7**, 486 (1982)
3.60 K.R.MacDonald, J.Feinberg: J. Opt. Soc. Am. **73**, 548 (1983)
3.61 B.Fischer, S.Stenklar: Appl. Phys. Lett. **47**, 1 (1985)
3.62 W.B.Whitten, J.M.Ramsey: Opt. Lett. **9**, 44 (1984)
3.63 J.Feinberg, G.D.Bacher: Opt. Lett. **9**, 420 (1984)
3.64 M.Ewbank, P.Yeh: Opt. Lett. **10**, 496 (1985)
3.65 T.Y.Chang, R.W.Hellwarth: Opt. Lett. **10**, 408 (1985)
3.66 B.Ya. Zel'dovich, V.J.Popovichev, V.V.Ragulskii, F.S.Faisullov: Zh. Eksp. Teor. Fiz., Pis'ma Red. **15**, 160 (1972) [English transl.: JETP Lett. **15**, 109 (1972)]
3.67 K.R.MacDonald, J.Feinberg, Z.Z.Ming, P.Günter: Opt. Commun. **50**, 146 (1984)
3.68 J.Feinberg: J. Opt. Soc. Am. **72**, 46 (1982)

4. Dynamic Holographic Gratings and Optical Activity in Photorefractive Crystals

Nicolai V. Kukhtarev

With 10 Figures

Holographic recording in electro-optic photorefractive crystals (where refractive index changes are photoinduced) was first proposed by *Chen* et al. [4.1] and later by *Staebler* and *Amodei* [4.2]. Possible applications and the opportunity to explore an interesting new branch of modern optics subsequently attracted much attention [4.3–6].

This rapidly developing field, which may be called "dynamic holography" [4.4] or "real-time holgraphy" [4.5, 6], deals with nonlinear holography, where the read-write processes are performed simultaneously. Dynamic hologram writing in electro-optic materials may be described as follows: An interference pattern of the coherent light beams leads to the inhomogeneous generation of electrons or holes, which are subject to diffusion and drift in the electric field and finally are captured by traps. The resulting space-charge field modulates the refractive index by the electro-optic effect and thus leads to the formation of phase holograms. Diffraction of the writing wave by this dynamic hologram (self-diffraction) may change the wave amplitude, phase, polarization and frequency, providing a large variety of possible processing alternatives for coherent beams. Light-wave amplitude changes result in energy exchange and are now used for image amplification [4.7] and real-time interferometry [4.8]. Under some conditions the phase changes produce "phase conjugation" [4.6, 9, 10], which may be used for wave front correction and real-time processing [4.6, 10]. Polarization changes during self-diffraction can be used to improve the signal-to-noise ratio [4.8, 11, 12], and diffraction-induced frequency changes may be exploited for optical filtering [4.13].

A resurgence of interest in electro-optic materials was caused by the realization of holographic phase conjugation in a scheme such as Fig. 4.1. If a_1 is

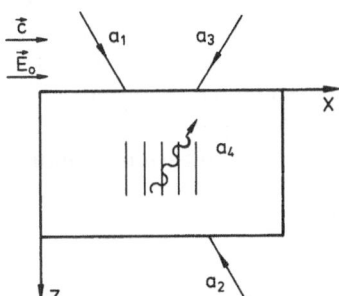

Fig. 4.1. Four-wave scheme of phase conjugation, where pump wave a_1 and signal wave a_3 write a transmission hologram, which is retrieved by pump wave a_2 with generation of the phase conjugate a_4

a plane wave coherent to the signal wave a_3, the holographic grating recorded by a_1 and a_3 leads to the appearance of a wave a_4 due to diffraction of the plane wave a_2, which is counterpropagating to a_1. The diffracted wave a_4 is phase conjugate to the signal wave: its phase φ_4 is connected with the phases of the other three waves by the simple relation [4.10]

$$\varphi_4 = \varphi_1 + \varphi_2 - \varphi_3 . \tag{4.1}$$

It may be seen from (4.1) that if a phase modulation identical to φ_3 is added to a_1 or a_2 then a_4 will appear as a plane wave. This property of phase conjugation in electro-optic crystals has been used for the construction of real-time correlators [4.14, 15].

The application of concepts from the field of critcial phenomena ("generation", optical bistability and hysteresis [4.6, 16, 17]) to photorefractive crystals has led to the realization of self-pumped resonators [4.18], operating with only one external wave.

After this brief historical account of the problem we shall now look more closely at the content of this contribution. The first section of this chapter is devoted to polar crystals in which isotropic diffraction has been demonstrated. After a brief formulation of the physical model of hologram recording in Sect. 4.1.1 the basic equations are presented in Sect. 4.1.2 where the photo-galvanic (photovoltaic) effect is also included. Two types of self-diffraction, the more-common isotropic case (Sect. 4.1.3) and the less-common anisotropic case (Sect. 4.1.4) are discussed. Phase transformation, multistability and diffraction-induced polarization rotation and ellipticity are described in Sects. 4.1.4–6. The special case of "vectorial" self-diffraction of orthogonally polarized waves interacting via a photogalvanic current is covered in Sect. 4.1.7.

The work discussed in Sect. 4.2 was stimulated by the pioneering results of *Huignard* et al. [4.8, 19] who investigated real-time holographic interferometry in the paraelectric crystal $B_{12}SiO_{20}$. In this crystal, which belongs to the symmetry group 23, the diffractive effects are more complicated due to a large natural optical activity [4.20]. Due to this optical activity the previously developed dynamical theory of isotropic self-diffraction [4.4, 7] is not directly applicable. The basic equations which include optical activity are introduced in Sects. 4.2.1, 2. The anisotropic nature of the self-diffraction leads to unusual and complex behavior of the wave interaction in these crystals: energy and polarization changes are discussed in Sects. 4.2.2–4. In Sect. 4.2.3 the change of the optical activity in an electric field (electrogyration) is dealt with. The influence of an external electric field on energy exchange is discussed in Sect. 4.2.5. Different configurations for real-time holographic interferometry are treated in Sect. 4.2.6 within the framework of the self-consistent dynamic theory.

4.1 Dynamic Gratings in Polar Crystals

4.1.1 Physical Model of Hologram Writing and Basic Equations

In our model we shall discuss a scheme of phototransitions of electrons and holes [4.7, 21] with corresponding subsystems of donor and acceptor impurities (Fig. 4.2).

In the donor subsystem we consider photoactive levels with concentration N_e^0, which are neutral in the ground state and after photoexcitation of electrons with concentration n_e become capture centers for electrons with recombination coefficient r_e and concentration N_e. We assume that some of the electrons from the donor levels are permanently captured by compensating centers with concentration C_e which are negative and take no part in the phototransitions.

In the acceptor subsystem we have the same processes but with holes replacing electrons: there are photoactive levels with concentration N_h^0, photoexcited holes with concentration n_h, capture centers for holes with recombination coefficient r_h and concentration N_h, and positive compensating centers with concentration C_h.

For the electrons and the impurities in the donor subsystem we have the following balance equations:

$$\frac{dn_e}{dt} = \frac{dN_e}{dt} - \frac{1}{e}(\operatorname{div} j_e) \; ,$$

$$\frac{dN_e}{dt} = (g_e + s_e I)(N_e^0 - N_e) - r_e n_e N_e \; , \tag{4.2}$$

where g_e is the thermal generation rate, s_e is the photon capture cross section, and $I = |\tilde{E}|^2$ is the light intensity, where $\tilde{E}$ is the electric field strength of the light wave. In the expression

$$j_e = e(\mu_e E - D_e \nabla)n_e + j^{\mathrm{ph}}$$

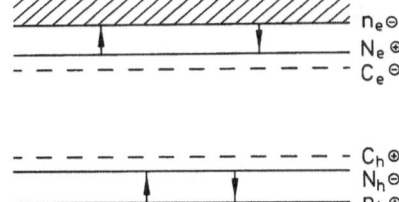

conduction band

$n_e \ominus$
$N_e \oplus$
$C_e \ominus$

$C_h \oplus$
$N_h \ominus$
$n_h \oplus$

valence band

Fig. 4.2. Phototransition scheme with photosensitive impurities of donor and acceptor type with concentrations N_e and N_h, and corresponding compensation centers with concentrations C_e and C_h

for the electric current in (4.2) we have a drift term (μ_e is the mobility of electrons) and a diffusion term (D_e is the diffusion constant for electrons) and also a photogalvanic term [4.22]

$$j_i^{\text{ph}} = \beta_{inl}^{\text{s}} \tilde{E}_n \tilde{E}_l^* + i\beta_{il}^{\text{a}} (\tilde{E} \times \tilde{E}^*)_i \ . \tag{4.3}$$

The tensor $\beta_{inl}^{\text{s}} = \beta_{iln}^{\text{s}}$ is similar to the electro-optic tensor, while the tensor β_{il}^{a} is similar to the tensor of the optical activity [4.22]. Similar equations may also be written for the acceptor subsystem where the corresponding equations have the index e replaced by h.

Using Poisson's equation

$$\text{div}(\varepsilon_0 \varepsilon E) = -e(n_e + N_h + C_e - n_h - N_e - C_h)$$

one gets the continuity equation ($\varepsilon \varepsilon_0$ is the static permittivity)

$$\frac{d}{dt}(\varepsilon_0 \varepsilon E) + j_e + j_h = J \ , \qquad \text{div} J = 0 \ , \tag{4.4}$$

where J is the total current including the displacement current. The time-averaged light intensity is

$$I = I_0 + \sum_{lm} [\Delta I_{lm} e^{i\eta_{lm}} + \text{c.c.}]$$

with

$$I_0 = 2 \sum_l |a_l|^2 \ ,$$

$$\Delta I_{lm} = 2a_l a_m^* \ ,$$

where a_l and η_l (and correspondingly a_m and η_m) are defined by

$$\tilde{E}_l = a_l e^{i\eta_l} + \text{c.c.} \ ,$$

$$\eta_l = \omega_l t - k_l r \ .$$

We shall assume that the quasistatic electric field E and the carrier concentrations n (n stands for n_e or n_h) are of the same form:

$$E = E_0 + \sum_{lm} [\Delta E_{lm} e^{i\eta_{lm}} + \text{c.c.}] \ ,$$

$$n = n_0 + \sum_{lm} [\Delta n_{lm} e^{i\eta_{lm}} + \text{c.c.}] \ . \tag{4.5}$$

From (4.2–5) we can find the electric field grating amplitude, which in turn modulates the light-frequency permittivity $\delta\varepsilon$. For small values of $\delta\varepsilon$ we obtain for the slowly varying amplitude of wave s

$$\mathbf{k}_s \times (\mathbf{V} \times \mathbf{a}_s) + \mathbf{V} \times (\mathbf{k}_s \times \mathbf{a}_s) = \mathrm{i}k_0^2 [\delta D \mathrm{e}^{-\mathrm{i}n_s}]_{t,r} \; . \tag{4.6}$$

Here δD is the electric displacement change at the optical frequency ω due to the electro-optic effect, to optical activity and to electrogyration [4.23, 24]

$$\delta D_i = \delta\varepsilon_{ij} \tilde{E}_j + \gamma_{ijl} V_l \tilde{E}_j + \tfrac{1}{2} \tilde{E}_j V_l \gamma_{ijl} \; . \tag{4.7}$$

In (4.6) $k_0 = \omega/c$ is the wave number and $[\ ]_{t,r}$ denotes a space and time average. The first term in (4.7) is due to the electro-optic effect

$$\delta\varepsilon_{ij} = -[\hat{\varepsilon}(\hat{r}E)\hat{\varepsilon}]_{ij} \; , \tag{4.8}$$

where $\hat{r}$ is the linear electro-optic tensor. The second and third terms describe optical activity in an inhomogeneous crystal. The tensor γ_{ijl} is antisymmetric with respect to the indices i and j and therefore can be written as [4.23]

$$k_0 \gamma_{ijl} = \delta_{ijs} g_{ls} \; ,$$

where δ_{ijs} is the completely antisymmetric tensor and g_{ls} is the pseudotensor of gyration. Furthermore, g_{ls} can be written in the form

$$g_{ls} = g_{ls}^0 + g_{lsn} E_n \; ,$$

where g_{ls}^0 describes the natural optical activity and g_{lsn} is the pseudotensor of electrogyration [4.24].

4.1.2 Calculation of the Quasi-Static Electric Field

In this section we investigate crystals with one type of impurity, say donors. We neglect the photogalvanic current and assume the quasi-static approximation $(dn_e/dt = 0)$ as well as linear recombination, which means that the recombination time is independent of the intensity. The last condition is fullfilled if [4.21]

$$C_e \gg n_0 = (g_e + s_e I) \frac{(N_e - C_e)}{C_e r_e} \; . \tag{4.9}$$

Omitting for simplicity the indices l and m, so that for instance $\Delta I_{lm} = \Delta I$, from (4.2–6) we obtain in the limit of small modulation

$$\frac{d(\Delta E)}{dt} = \frac{\Delta I}{(I_0 + I_d)} (P_\mathrm{n} + \mathrm{i}P_\mathrm{s}) - \Delta E \left[\frac{1}{\tau_\mathrm{r}} + \left(\mathrm{i}\Omega + \frac{1}{\tau_\mathrm{osc}} \right) \right] \; . \tag{4.10}$$

In this formula we have put

$$P_{\mathrm{n}} = -\frac{E_0 M}{\tau_{\mathrm{M}}} \,,$$

$$P_{\mathrm{s}} = \frac{M}{\tau_{\mathrm{M}}} \left[E_{\mathrm{D}} \left(1 + \frac{E_{\mathrm{D}}}{E_M} \right) + \frac{E_0^2}{E_M} \right] \,,$$

$$I_{\mathrm{d}} = g_{\mathrm{e}}/s_{\mathrm{e}}$$

(which is the "intensity" proportional to dark conductivity),

$$\frac{1}{\tau_r} = \frac{M}{\tau_{\mathrm{M}}} \left[\left(1 + \frac{E_{\mathrm{D}}}{E_q} \right) \left(1 + \frac{E_{\mathrm{D}}}{E_M} \right) + \frac{E_0^2}{E_q E_M} \right] \,,$$

$$\Omega = \omega_l - \omega_m \,,$$

$$\tau_{\mathrm{osc}} = \frac{M E_0}{\tau_{\mathrm{M}}} \left(\frac{1}{E_q} - \frac{1}{E_M} \right) \,.$$

Here E_0 is the component of the external electric field along the grating vector $\mathbf{q}$, while the diffusion field E_{D} and the fields E_M and E_q are defined by

$$E_{\mathrm{D}} = \frac{k_{\mathrm{B}} T}{eL} \,,$$

$$E_M = \frac{r_e N_e}{\mu_e q} \quad \text{with} \quad q = |k_m - k_l|$$

$$E_q = \frac{e C_e N_e}{\varepsilon \varepsilon_0 q (N_e - C_e)} \,.$$

Here $k_{\mathrm{B}} T$ is the temperature in energy units and L is the grating period. These fields are used to define M:

$$\frac{1}{M} = \left(1 + \frac{E_{\mathrm{D}}}{E_M} \right)^2 + \left(\frac{E_0}{E_M} \right)^2 \,.$$

Finally,

$$\tau_{\mathrm{M}} = \frac{\varepsilon \varepsilon_0}{\sigma}$$

is the Maxwell relaxation time where σ is the conductivity.

Equation (4.10) allows for a small frequency difference $\Omega = \omega_l - \omega_m$ of the interacting waves l and m which leads to 'running gratings" along the grating vector with velocity $v = \Omega/q$.

Let us discuss the physical meaning of the different terms in (4.10). The quantities P_{n} and P_{s} give the growth rate of the electric field grating amplitude at the initial stage of recording; P_n describes the nonshifted part and P_s describes the shifted part of the grating. The terms with the common factor ΔE describe the relaxation of the grating. For equal frequencies ω_1 and ω_m, τ_{osc} is the quasi-period of oscillation [4.4]. For running gratings with $\Omega = -\tau_{osc}^{-1}$, however, the oscillation is no longer present.

For a constant electric field E_0 and under the assumption that the intensity changes only very slowly during hologram writing, (4.10) can be integrated to obtain

$$\Delta E = \Delta I (F_n + i F_s)(I_0 + I_d)^{-1} \ . \tag{4.11}$$

Here the nonshifted part F_n and the shifted part F_s of the grating may be written as

$$
\begin{aligned}
F_n &= \left[\frac{P_n}{\tau_r} + \frac{P_s}{\tau} - \left(\frac{1}{\tau_r} \cos \frac{t}{\tau} - \frac{1}{\tau} \sin \frac{t}{\tau} \right) \exp\left(\frac{-t}{\tau_r} \right) \right] \frac{1}{\tau_r^2 + \tau^2} \ , \\
F_s &= \left[\frac{P_s}{\tau_r} - \frac{P_n}{\tau} + \left(\frac{1}{\tau} \cos \frac{t}{\tau} + \frac{1}{\tau_r} \sin \frac{t}{\tau} \right) \exp\left(\frac{-t}{\tau_r} \right) \right] \frac{1}{\tau_r^2 + \tau^2} \ ,
\end{aligned}
\tag{4.12}
$$

where

$$\frac{1}{\tau} = \Omega + \frac{1}{\tau_{Osc}} \ .$$

From (4.11 and 12) one sees that the amplitude of the electric field grows to the stationary value with a damped oscillation. The case of equal frequencies, $\Omega = 0$, has already been discussed in [4.21].

4.1.3 Isotropic Self-diffraction and Running Gratings for Crystals with Symmetry Group 3*m*

In this section we shall specialize the results of the preceding section to crystals with symmetry group 3*m* and to isotropic self-diffraction. Anisotropic self-diffraction will be dealt with in Sect. 4.1.4. We shall concentrate on the physics of running gratings.

For crystals with symmetry group 3*m* with the *c*-axis in the *z*-direction the change of the displacement field δD_s of wave *s* is

$$\delta D_s = \delta \varepsilon_{sj} d_j e^{i n_j} \ , \quad \text{with}$$

$$
\begin{aligned}
\delta \varepsilon_{xx} &= n_o^4 (r_{22} E_y - r_{13} E_z) \ , & \delta \varepsilon_{xy} &= n_o^4 r_{22} E_x \ , \\
\delta \varepsilon_{yy} &= -n_o^4 (r_{22} E_y + r_{13} E_z) \ , & \delta \varepsilon_{\alpha z} &= -(n_o n_e)^2 r_{51} E_\alpha \ , \\
\delta \varepsilon_{zz} &= -n_e^4 r_{33} E_z \ , & (\alpha &= x, y) \ .
\end{aligned}
\tag{4.13}
$$

Here n_o and n_e are the ordinary and the extraordinary refractive indices and the r_{lm} are the electro-optic coefficients.

From (4.13) it may be seen that the orientation of the electric field along the z-axis influences only the diagonal elements of $\delta\varepsilon$, leading to isotropic diffraction without polarization changes. Considering the interaction of two ordinarily polarized waves $a_{1,3} = I_{1,3}^{1/2} e(i\varphi_{1,3})$, in transmission geometry (see Fig. 4.1) and assuming $I_0 \gg I_d$ we obtain the intensity of wave 1 and the phase difference in the steady state as

$$I_1 = I_0(1 + me^{\Gamma z}) ,$$

$$\varphi_3 - \varphi_1 = \frac{rF_n}{\Gamma} \ln \left(\frac{(1+m)^2}{(1+me^{\Gamma z})^2} e^{\Gamma z} \right) ,$$

(4.14)

with

$$I_0 = I_{10} + I_{30} ,$$

$$m = I_{30}/I_{10} ,$$

$$\Gamma = -2rF_s ,$$

$$r = k_0 n_e^3 r_{33} (2\cos\theta)^{-1} ,$$

$$F_n = \left(\frac{P_n}{\tau_r} + \frac{P_s}{\tau} \right) \left(\frac{1}{\tau_r^2} + \frac{1}{\tau^2} \right)^{-1} ,$$

$$F_s = \left(\frac{P_s}{\tau_r} - \frac{P_n}{\tau} \right) \left(\frac{1}{\tau_r^2} + \frac{1}{\tau^2} \right)^{-1} .$$

Here k_0 is the wave number and 2θ is the angle between waves 1 and 3; I_{10} and I_{30} are the input intensities of waves 1 and 3.

As a function of Ω, the logarithmic gain Γ has two extrema

$$\Gamma_\pm = \frac{rP_n^2}{P_s \pm (P_s^2 + P_n^2)^{1/2}} \quad \text{at}$$

(4.15)

$$\Omega_\pm = -\frac{1}{\tau_{osc}} + \frac{P_s \pm (P_s^2 + P_n^2)^{1/2}}{\tau_r P_n} .$$

(4.16)

The direction of energy exchange differs for the two values of Ω. For details on the influence of running gratings on the energy exchange see [4.25, 26].

The possibility of selective suppression of the shifted or the nonshifted part of the grating by a running interference pattern is of special interest. The shifted part of the grating is suppressed for

$$\Omega\tau_{os} = -1 + \frac{P_s \tau_{osc}}{P_n \tau_r} .$$

(4.17a)

In this case there is no energy exchange. On the other hand, the nonshifted part of the grating is suppressed for [4.7]

$$\Omega \tau_{osc} = -1 - \frac{P_n \tau_{osc}}{P_s \tau_r} \ . \tag{4.17b}$$

4.1.4 Anisotropic Self-diffraction, Phase Conjugation, and Phase Doubling for Crystals with Symmetry Group $3m$

We shall now investigate anisotropic self-diffraction for crystals with symmetry group $3m$. In the process of anisotropic self-diffraction, the polarization of the diffracted beam is different from the polarization of the reading beam. This type of diffraction is less well known, but it possesses useful properties.

When an electric field is applied along the x-axis or the y-axis, equations (4.13) show that there is a nondiagonal element of the tensor of the dielectric permittivity proportional to r_{51}. Consequently, the effect has been observed in $BaTiO_3$ [4.27] and $LiNbO_3$ [4.28], where r_{51} is large. The geometry of Fig. 4.3 was used in both experiments: two extraordinarily polarized waves with an angle 2θ between them, with amplitudes a_1 and a_2 and wave vectors k_{1e} and k_{2e}, write a grating with grating vector $q = k_{1e} - k_{2e}$. When the phase-matching condition

$$q = k_{1o} - k_{1e} = k_{2e} - k_{2o}$$

is satisfied, two diffracted waves with ordinary polarization and with wave vectors k_{1o} and k_{2o} appear symmetrically about the x-axis. From the phase-matching condition one infers that the experiment only works if the angle between the extraordinarily polarized waves fulfils the condition

$$(8 \sin\theta)^2 = (n_o + n_e)(n_o - n_e) \ .$$

Then the angle 2ψ between the anisotropically diffracted, ordinarily polarized waves is given by

$$\sin\psi = 3\sin\theta \ .$$

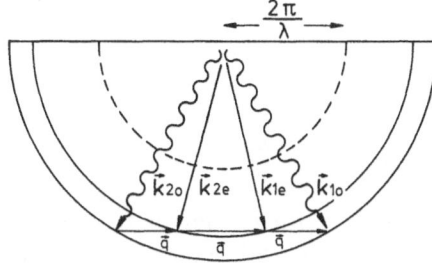

Fig. 4.3. Wave vector diagram for anisotropic self-diffraction with two extraordinarily polarized input waves with wave vectors k_{1e} and k_{2e}, and two ordinarily polarized generated waves with wave vectors k_{1o} and k_{2o} [27]

For a wavelength $\lambda = 496$ nm one finds $2\theta = 24°$ and $2\psi = 77.6°$ in BaTiO$_3$. In [4.27] the coupled wave equations have been solved for the geometry just described. The following conservation laws for the phases (the notation should be obvious) have been obtained:

$$2\varphi_{1e} - \varphi_{1o} - \varphi_{2e} = \text{const}_1 ,$$

$$\varphi_{2o} + \varphi_{1o} - 2\varphi_{2e} = \text{const}_2 . \tag{4.18}$$

For instance, when wave 1e is the plane reference wave and wave 2e is the signal wave, then wave 1o is phase conjugate to wave 2e. Alternatively, when wave 2e is the plane reference wave and wave 1e is the signal wave, then wave 1o is phase-doubled to wave 1e. Phase conjugation was demonstrated in [4.28] for LiNbO$_3$, and phase doubling, which manifests itself in a focus change, of lenses, was demonstrated in [4.27] for BaTiO$_3$.

4.1.5 Optical Multistability in Four-Wave Phase Conjugation

In the gometry of Fig. 4.1, when a_2 is incoherent to a_1 and a_3, the last two waves write a transmission grating. Diffraction of the pump wave a_2 creates the wave a_4 which is phase conjugate to the signal wave a_3.

When the photon cross sections for capture by donors and acceptors are equal ($s_e = s_h$), and when linear recombination dominates, the equations for the slowly varying amplitude may be written as [4.28, 29]

$$\frac{da_{1,2}}{dz} = -(\gamma_r \pm i\gamma_i)(I_{3,4}a_{1,2} + a_3 a_4 a_{2,1}^*)I_0^{-1} ,$$

$$\frac{da_{3,4}}{dz} = (\gamma_r \mp i\gamma_i)(I_{1,2}a_{3,4} + a_1 a_2 a_{3,4}^*)I_0^{-1} , \tag{4.19}$$

with

$$\gamma_r = ME_D[1 + E_D E_q^{-1} + E_0^2 (E_D E_q)^{-1}] ,$$

$$\gamma_i = \xi M E_0 ,$$

$$I_i = |a_i|^2 ,$$

$$I_0 = I_1 + I_2 + I_3 + I_4 .$$

Here

$$\xi = \frac{C_h - C_e}{C_h + C_e}$$

is the compensation parameter which equals -1 for crystals with electron photoconductivity ($C_e \gg C_h$) and $+1$ for crystals with hole conductivity

$(C_h \gg C_e)$. For waves polarized normal to the plane of incidence

$$M = \frac{r_{13} n_o^3 K_0}{2 \cos \theta} (E_0^2 + E_D^2) \left[(\xi E_0)^2 + E_D^2 \left(1 + \frac{E_D}{E_q} + \frac{E_0^2}{E_D E_q} \right)^2 \right]^{-1} , \qquad (4.20)$$

where E_0 is the external electric field along the z-axis, and $K_0 = \omega/c$ is the wave number.

For local response $\gamma_i \gg \gamma_r$ one can derive an analytical expression for $I_4(z)$ and the phase difference

$$\phi = \varphi_1 + \varphi_2 - \varphi_3 - \varphi_4 ,$$

where the phases φ_s are defined by $a_s = \sqrt{I_s} \exp(i\varphi_s)$ [4.28]. Using the boundary condition $I_4(z_0) = 0$, where z_0 is the thickness of the crystal, one obtains

$$I_4(z) = \frac{f_1}{f_2} \sin^2 \left(\frac{\gamma_i f_2^{1/2}}{I_0} (z - z_0) \right) , \qquad (4.21)$$

$$\cos \phi = \frac{I_4^{1/2} [A - B - C + 2(I_4 - I_{40})]}{[(A + I_4 - I_{40})(B - I_4)(C + I_{40} - I_4)]^{1/2}} .$$

Here $I_{40} = I_4(0)$, and A, B, and C are the constants of the motion defined by

$$A - I_{40} = I_1(z) - I_4(z) ,$$

$$B = I_2(z) + I_4(z) ,$$

$$A + C = I_1(z) + I_3(z) ,$$

the values of which can be determined at the crystal boundary if $I_4(0)$ is already known. Furthermore,

$$f_1 = B(C + I_{40})(A - I_{40}) ,$$

$$f_2 = \tfrac{1}{4} (B - A - C)^2 + B(A - I_{40}) .$$

To determine $I_4(0)$ one can solve (4.21) for $z = 0$. It is possible to find multiple solutions. Each of the solutions corresponds to a certain function $I_4(z)$ that oscillates in space with a certain period, which can be obtained from (4.21) as

$$z_p = \frac{\pi I_0}{\gamma_i \sqrt{f_2}} .$$

Changing from one stable solution to another means changing the period z_p of $I_4(z)$.

For $\gamma_r \neq 0$ the solutions of equations are given in [4.30, 31]. In Fig. 4.4 the reflectivity $R = I_{40}/C$ is shown as a function of E_0 for different values of ξ.

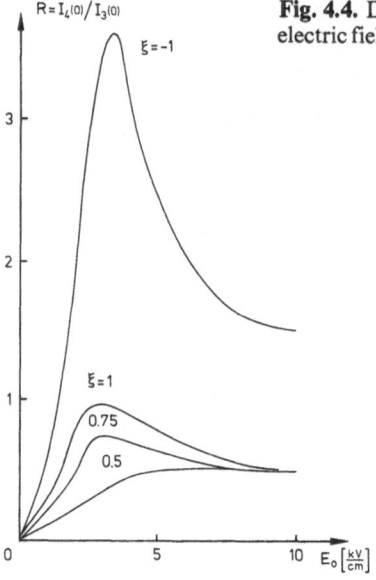

Fig. 4.4. Dependence of the phase-conjugate reflectivity on the electric field for different values of the compensation parameter ξ

The calculation is for $KNbO_3$: Fe which is often used for phase conjugation (PC) [4.32]. The parameters used are $\lambda = 592$ nm, $L = 2.6\ \mu m$, $z_0 = 0.33$ mm, $r_{33} = 64\ pm\ V^{-1}$, $n_e^3 = 10.65$, $E_q = 7\ kV\ cm^{-1}$, $E_D = 628\ V\ cm^{-1}$, $B/A = 0.6$, $A/C = 155$. The dependence of the PC efficiency upon the magnitude and the sign of the compensation parameter ξ provides an experimental method to determine the type of conductivity and the degree of compensation.

4.1.6 Self-diffraction Gyration

Now we shall discuss another scheme of PC which may be realized for two counterpropagating beams. Let the wave $\underline{a}\ (a_x \neq 0, a_y \neq 0, a_z = 0)$ propagate along the c axis, as shown in Fig. 4.5, and the wave b, polarized along the x-axis, be counterpropagating. In the crystal the components a_x and b_x write a holographic

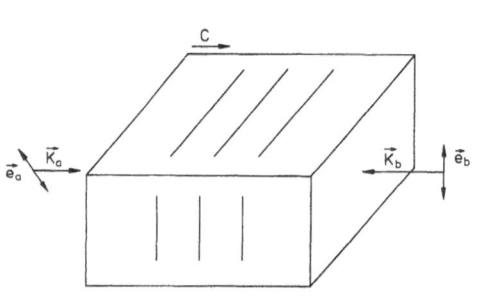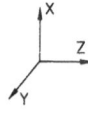

Fig. 4.5. Scheme of self-diffraction gyration with two counterpropagating beams with nonparallel polarization

reflection grating. The diffraction of a_y by this grating leads to the appearance of a component b_y, so that in effect we have a four-wave interaction between the four components of the waves $\underline{a}$ and b. We shall adapt our notation accordingly below. The process may be described by the vectorial equations [4.31, 33]

$$\frac{d\mathbf{a}}{dz} = -\frac{(\gamma_r + i\gamma_i)}{I_0}(\mathbf{a} \cdot \mathbf{b}^*)\mathbf{b} + \frac{ck_0 n_o^3}{2} r_{13}\mathbf{a} \,,$$

$$\frac{d\mathbf{b}}{dz} = \frac{(\gamma_r - i\gamma_i)}{I_0}(\mathbf{a}^* \cdot \mathbf{b})\mathbf{a} - \frac{ck_0 n_o^3}{2} r_{13}\mathbf{b} \,.$$

From these equations follows the conservation law $\mathbf{a} \times \mathbf{b} = \mathbf{L}$, where $\mathbf{L}$ is a constant vector. For sufficiently large nonlinear phase changes, $\phi = \Delta n k_0 z_0 > 10$ (Δn is the nonlinear refractive index change, k_0 is the wave number, and z_0 is the crystal thickness), there are multivalued solutions for the Cartesian components of the waves a and b [4.31].

In the absence of external and photogalvanic fields (i.e., $\gamma_i = 0$) one finds a conservation law for phases. To conform with the notation of Fig. 4.1 for pump waves 1 and 2, signal wave 3, and generated wave 4, we define

$$a_x = \sqrt{I_3} \exp(i\varphi_3) \,, \quad a_y = \sqrt{I_1} \exp(i\varphi_1) \,,$$

$$b_x = \sqrt{I_2} \exp(i\varphi_2) \,, \quad b_y = \sqrt{I_4} \exp(i\varphi_4) \,,$$

and find

$$\varphi_2 - \varphi_4 = \varphi_1 - \varphi_3 + \psi \,. \tag{4.22}$$

Here $\psi = 0$ or $\psi = \pi$, depending on the sign of γ_r, that is, depending on the type of photoconductivity. For linear polarization there are only two possibilities. One possibility is $\varphi_1 = \varphi_3$, that is, the polarization vector of wave a lies in the first or in the third quadrant. Then we find that wave b is rotated to the left for $\psi = 0$, and to the right for $\psi = \pi$. The other possibility is $\varphi_1 = \varphi_3 + \pi$, i.e., the polarization vector of wave a lies in the second or fourth quadrant. Then the rotation of wave b is the other way round.

The polarization changes just described may be called "self-diffraction gyration" [4.12]. Another scheme for obtaining self-diffraction gyration (Fig. 4.6) uses transmission gratings which are recorded by the ordinary wave b and the ordinary component of the wave a. The reading out by the extraordinary component of wave a leads to the generation of an extraordinary component of b. Estimates for LiNbO$_3$ crystals with a thickness of 0.1 cm show that the self-diffraction gyration angle may be of the order of about 1° [4.33].

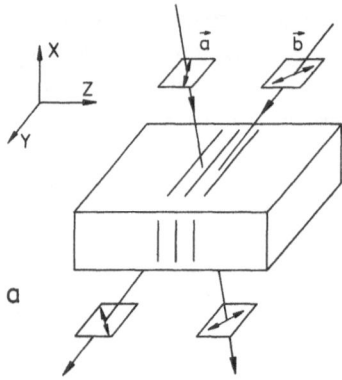

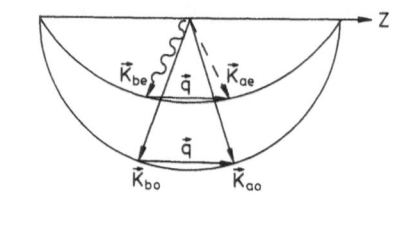

Fig. 4.6. (a) Scheme of two-wave self-diffraction gyration for transmission gratings. **(b)** Wave vector diagram

4.1.7 Vectorial Self-diffraction and Self-oscillation in Photogalvanic Crystals

a) Starting Equations

Another type of anisotropic self-diffraction has been realized in photogalvanic crystals, for instance in doped LiNbO$_3$ [4.34], and may be called "vectorial self-diffraction". It is mainly caused by the vectorial part

$$j_l^a = i\beta_{il}^a (E \times E^*)_l$$

of the photogalvanic response [4.22] which allows a hologram to be written even with orthogonally polarized waves. In crystals with $3m$ symmetry only the imaginary part of the element

$$\beta = \beta_{15} = \beta_s + i\beta_a$$

of the photovoltaic tensor contributes to j^a. If the c-axis is perpendicular to the x-axis one obtains

$$j_x^a = -i\beta_a (\tilde{E}_x \tilde{E}_z^* - \text{c.c.}) \ ,$$

$$j_y^a = i\beta_a (\tilde{E}_z \tilde{E}_y^* - \text{c.c.}) \ ,$$

$$j_z^a = 0 \ .$$

We shall investigate the writing of a hologram with orthogonally polarized waves for the interaction geometry of Fig. 4.7, where $n_o > n_e$ is assumed. The extraordinarily polarized wave a_2 propagates in the y-direction and a_1 is an ordinarily polarized wave. Anisotropic self-diffraction is obtained in this

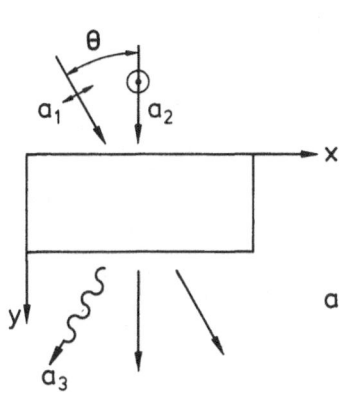

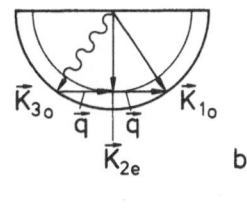

Fig. 4.7. (a) Scheme of vectorial self-diffraction. a_1: ordinarily polarized input wave; a_2: extraordinarily polarized input wave; θ: angle between the wave vectors. The c-axis is in the z-direction. (b) Wave-vector diagram

geometry if the grating vector $q = k_{1o} - k_{2e}$ is parallel to the x-axis. In this case an ordinarily polarized third wave a_3 is generated by the wave a_2, because the phase matching condition

$$q = k_{1o} - k_{2e} = k_{2e} - k_{3o}$$

for a_3 is fulfilled if a_3 emerges symmetrically to a_1. The angle θ is obtained as

$$\sin^2 \theta = n_o^2 - n_e^2 \ .$$

We start from

$$\tilde{E}_x = a_1 e^{-i\eta_1} + a_3 e^{-i\eta_3} + \text{c.c.} \ ,$$

$$\tilde{E}_y = 0 \ ,$$

$$\tilde{E}_z = a_2 e^{-i\eta_2} + \text{c.c.} \ ,$$

where we have assumed $\theta \ll 1$. The electric current is entirely due to the photogalvanic effect. Only its x-component is different from zero:

$$j_x = (\beta a_2 a_1^* + \beta^* a_3 a_2^*) e^{iqx} + \text{c.c.} \tag{4.23}$$

The exponential factor in this formula indicates that j is spatially oscillating. Furthermore, the symmetric part of the photogalvanic tensor also contributes to the current since it is not proportional to the unit tensor.

We write the space-charge field in the form

$$E_x = \Delta E e^{iqx} + \text{c.c.}$$

Using the relation $r_{51} \gg r_{22}$, which is correct for LiNbO$_3$, we obtain the following starting equations

$$\varepsilon\varepsilon_0 \frac{d\Delta E}{dt} + e\mu n_0 \Delta E = -(\beta a_2 a_1^* + \beta^* a_3 a_2^*) \ ,$$

$$\frac{da_1}{dy} = il\Delta E^* a_2 \ ,$$

$$\frac{da_2}{dy} = il(\Delta E a_1 + \Delta E^* a_3) \ ,$$ (4.24)

$$\frac{da_3}{dy} = il\Delta E a_2 \ .$$

Here n_0 is the concentration of electrons (we assume that our crystal is of the donor type) and

$$l = \frac{\omega n_0 n_e r_{51}}{c \cos\theta} \quad \text{with}$$

$$\cos\theta = \frac{n_e}{n_0} \ .$$

The system of equations (4.24) describes the kinetics of the intensity and phase changes during vectorial self-diffraction due to the photogalvanic current in crystals with $3m$ symmetry. Orthogonal polarization of the incident beams has been assumed.

b) Kinetics of the Amplitude Changes

We linearize (4.24) in the parameter $l = r_{51}$

$$a_i = a_{i0} + a_i^{(1)} \ , \quad i = 1,2,3 \ .$$

Using the boundary conditions

$$a_1(y=0) = a_{10} \ ,$$
$$a_2(y=0) = a_{20} \ ,$$
$$a_3(y=0) = 0 \ ,$$

we obtain the result

$$a_1^{(1)} = -ilf\beta^* I_{20} a_{10} \ ,$$

$$a_2^{(1)} = -ilf\beta I_{10} a_{20} \ ,$$ (4.25)

$$a_3^{(1)} = -\beta ilf\beta a_{20}^2 a_{10}^* \ ,$$

where

$$f=\frac{1}{\sigma}\left[1-\exp\left(-\frac{t}{\tau_M}\right)\right]y \quad \text{with}$$

$$\sigma=e\mu n_0 , \quad \tau_M=\frac{\varepsilon\varepsilon_0}{\sigma} .$$

The intensity and phase changes may be written as

$$\delta I=I_2-I_{20}=I_{10}-I_1=2l\beta_a fI_{10}I_{20} ,$$

$$I_3=(lfI_{20})^2(\beta_a^2+\beta_s^2)I_{10} , \tag{4.26}$$

$$\delta\varphi_{12}=\varphi_1-\varphi_2=lf\beta_s(I_{10}-I_{20}) .$$

From (4.25) one can see that if a_2 is a plane reference wave, wave a_3 is phase conjugate to the signal wave a_1. If a_1 is a plane reference wave, then phase doubling of the phase of the signal wave a_2 is obtained in the wave a_3. From the energy relations (4.26) we conclude that for the initial part of the kinetics, when $f=yt/\varepsilon\varepsilon_0$, the change of the amplitudes a_1 and a_2 is caused by the antisymmetric part of the photogalvanic tensor β_a and the direction of the energy exchange depends on the sign of the product $\beta_a r_{51}$. On the other hand, I_3 is proportional to $r_{51}^2(\beta_s^2+\beta_a^2)$. Taking into account that the phase difference $\delta\varphi_{12}$ in (4.26) is proportional to $\beta_a r_{51}$, one can see that it is possible to use vectorial self-diffraction to determine the symmetric and antisymmetric parts of the photogalvanic tensor element β_{15}.

c) Steady-State Solutions

In the steady-state limit $t\gg\tau_M$ we use the undepleted-pump approximation $I_1,I_2\gg I_3$, with constant intensities I_1 and I_2, to obtain from (4.24) the solution

$$I_3(y)=\frac{4I_1I_2^2}{2I_1-I_2}\left[e^{[-2\gamma_a(2I_1-I_2)y]}-2e^{[-\gamma_a(2I_1-I_2)y]}\cos\left(\frac{\gamma_s}{2}(2I_1-I_2)y\right)+1\right],$$
$$\tag{4.27}$$

where

$$\gamma_{a,s}=\frac{\beta_{a,s}}{\sigma} .$$

We see that the symmetric part β_s of β_{15} causes the oscillation of the intensity of the diffracted beam. The amplitude of the intensity oscillation depends on the antisymmetric part β_a of β_{15} and also upon the thickness of the crystal.

We can get away from the undepleted-pump approximation if the symmetric part β_s of β_{15} may be neglected compared with the antisymmetric part β_a of β_{15}

and if $\beta_a < 0$. The solution of (4.24) is in this case

$$4I_{1,3} = I_{10}\left[\left(\frac{1+2m}{1+2me^{-\gamma y}}\right)^{1/2} \pm 1\right]^2 ,$$

$$(4.28)$$

$$I_2 = I_{20}\frac{2m+1}{2m+e^{\gamma y}} ,$$

where

$$m = I_{20}/I_{10} , \quad \gamma = -2\gamma_a(I_{10} + 2I_{20}) .$$

For large energy transfer the solution (4.28) gives

$$I_{1,3}(d) = \tfrac{1}{4}\left(\sqrt{2I_{20}+I_{10}} \pm \sqrt{I_{10}}\right)^2 ,$$

where d is the thickness of the crystal. One has output intensities

$$I_{1,3}(d) = \tfrac{1}{2} I_{20}$$

for vanishing input intensity I_{10}, which means self-oscillation.

Feedback due to reflection from the crystal boundaries may be described by

$$I_{10} = RI_1(d) , \quad \text{where}$$

$$R = M_1 M_2$$

and M_1 and M_2 are reflection coefficients. In this case we have

$$e^{-\gamma d} > (2\sqrt{R}-1)$$

as the condition for self-oscillation. This may also be regarded as a condition for the generation of scattered noise which tends to spoil holographic recording, for instance in $LiNbO_3$ [4.25].

4.2 Holographic Gratings in Paraelectric Crystals with 23 Symmetry

As discussed in Sect. 4.1, holographic recording in polar crystals shows many interesting features, but there are some disadvantages:

1) large light scattering in doped materials with high sensitivity,
2) slow response time.

Experimental results for hologram recording in paraelectric crystals like $Bi_{12}SiO_{20}$(BSO) [4.8, 11] shows that these crystals exhibit high sensitivity and fast response.

Anisotropic self-diffraction in paraelectric crystals, however, is not without problems because the transmitted and diffracted beams in most practical cases overlap to a high degree, in contrast to anisotropic diffraction in polar crystals where the beams are angularly separated. This is due to the fact that for polar crystals the angular separation between the transmitted wave a_1 and the part of the wave a_2 diffracted in the same direction is proportional to the linear birefringence $(n_e - n_o) = 10^{-2}$. But the same angular separation in optically active crystals is determined by the circular birefringence $\Delta n = (n_+ - n_-) = 10^{-4}$, which is very small, and therefore the beams overlap if Δn is less then the angular selectivity of the hologram. On the other hand, this very fact allows anisotropic self-diffraction to be studied for a wide range of grating periods, in contrast to polar crystals where phase matching conditions must be fulfilled.

Unfortunately, the dynamic theory of hologram recording [4.4, 7] developed mainly for polar crystals is not directly applicable to optically active paralelectric crystals. For instance in these crystals there may be steady-state energy exchange controlled by the electric field even for small drift lengths, the direction of the energy exchange depending on the light-wave polarization and the direction of the external electric field.

4.2.1 Basic Equations

For a description of hologram writing in cubic crystals we choose the general model of bipolar photoconductive materials described in Sect. 4.1.1. We can use the calculation of the quasi-static electric field (Sect. 4.1.2) also for cubic crystals.

For the slowly varying amplitudes we also obtain (4.7), but for the explicit form of the induction change δD we must include not only the electro-optic effect but also optical activity.

In some applications the crystals are cut in the orientation $(1\bar{1}0, 110, 001)$, which may be obtained from the main crystallographic orientation by a rotation of $-\pi/4$ around the axis (001). In this case the change δD of the induction is

$$\delta D = \left[n^4 r_{14} \begin{pmatrix} E_z & 0 & E_x \\ 0 & -E_z & -E_y \\ E_x & -E_y & 0 \end{pmatrix} + \frac{\lambda g}{2\pi} \begin{pmatrix} 0 & \dfrac{\partial}{\partial z} & -\dfrac{\partial}{\partial y} \\ -\dfrac{\partial}{\partial z} & 0 & \dfrac{\partial}{\partial x} \\ \dfrac{\partial}{\partial y} & -\dfrac{\partial}{\partial x} & 0 \end{pmatrix} \right.$$

$$\left. + \frac{\lambda g_{14}}{4\pi} \begin{pmatrix} 0 & \eta_{12} & \eta_{13} \\ -\eta_{12} & 0 & \eta_{23} \\ -\eta_{13} & -\eta_{23} & 0 \end{pmatrix} \right] \tilde{E} , \qquad (4.29)$$

where the first matrix describes the electro-optic effect, the second the optical activity (g is the parameter of optical activity), and the third electrogyration (g_{14} is the electrogyration coefficient [4.24]). The relevant elements of the tensor η are given by

$$\eta_{12} = \frac{\partial E_y}{\partial y} + 2 E_y \frac{\partial}{\partial y} - \frac{\partial E_x}{\partial x} - 2 E_x \frac{\partial}{\partial x} \ ,$$

$$\eta_{23} = -\frac{\partial E_y}{\partial x} - 2 E_z \frac{\partial}{\partial x} - \frac{\partial E_x}{\partial z} - 2 E_x \frac{\partial}{\partial z} \ ,$$

$$\eta_{13} = -\frac{\partial E_z}{\partial y} - 2 E_z \frac{\partial}{\partial y} - 2 E_y \frac{\partial}{\partial z} - \frac{\partial E_y}{\partial z} \ .$$

4.2.2 Calculation of the Light-Wave Amplitudes for a Geometry Without Electrogyration

For the calculation of the light-wave amplitudes we shall assume the crystal orientation $(1\bar{1}0, 110, 001)$ and the writing geometry in which the holographic grating vector q is along the x-axis. Then anisotropic self-diffraction will occur. In the interaction geometry shown in Fig. 4.8 we get for the amplitudes of the interaction waves a and b, assuming a small angle between a and b,

$$\frac{da_{x,z}}{dy} = -\mathrm{i}\,[rEb_{z,x} + (rE_0 \pm \mathrm{i}G)a_{z,x}] \ , \tag{4.30}$$

$$\frac{db_{x,z}}{dy} = -\mathrm{i}\,[rE^*a_{z,x} + (rE_0 \pm \mathrm{i}G)b_{z,x}] \ .$$

Here the amplitude E of the grating running with velocity $v = \Omega q$ (Ω is the light-wave frequency difference) is determined from the equation

$$\frac{dE}{dt} = \frac{(a \cdot b^*)}{I_0 + I_d}\,(P_\mathrm{n} + \mathrm{i}P_\mathrm{s}) - E[\tau_\mathrm{r}^{-1} + \mathrm{i}(\Omega + \tau_\mathrm{osc}^{-1})] \ . \tag{4.31}$$

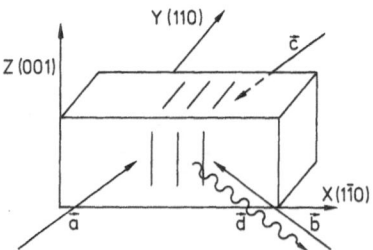

Fig. 4.8. Scheme of anisotropic self-diffraction in crystals with 23 symmetry

Here

$$r=\frac{k_0 n^3 r_{14}}{2\cos\theta} , \qquad G=\frac{1}{2} g k_0$$

is the specific optical activity, and the remaining notation is explained in Sect. 4.1.1. From (4.30) we conclude that the polarization vectors of waves a and b will rotate in this geometry. There is no electrogyration, however.

First we shall solve (4.30 and 31) in the simple case when no optical activity is present, $G=0$, and when there is no external field, $E_0=0$. We assume that initially waves a and b are linearly polarized along the z-direction. Using the notation

$$a_{x,z}=A_{x,z}e^{i\alpha_{x,z}} , \qquad \Delta\alpha=\alpha_z-\alpha_x ,$$
$$b_{x,z}=B_{x,z}e^{i\beta_{x,z}} , \qquad \Delta\beta=\beta_z-\beta_x ,$$

we obtain the conservation laws

$$\sin\Delta\alpha=0 , \qquad F_D\cos\Delta\alpha>0 ,$$
$$\sin\Delta\beta=0 , \qquad F_D\cos\Delta\beta>0 , \tag{4.32}$$

with

$$F_D=\{-E_D[1-e^{-t/\tau_M}\cos\Omega t]+\delta E_D e^{-t/\tau_M}\sin\Omega t\}[(1+\delta^2)(I_0+I_d)]^{-1} .$$

Here δ is defined by $\delta=\Omega\tau_M$. From (4.32) we conclude that for $F_D>0$, $\Delta\alpha=0$ and $\Delta\beta=\pi$ so that wave a is polarized in the first or in the third quadrant while wave b is polarized in the second or in the fourth quadrant. For $F_D<0$ we obtain $\Delta\alpha=\pi$ and $\Delta\beta=0$ and the roles of $\Delta\alpha$ and $\Delta\beta$, and of waves a and b are interchanged (Table 4.1).

Let us discuss the direction of the energy exchange. From (4.30, 31) it may be shown that wave a is amplified for $F_D\cos\Delta\alpha>0$. For $F_D\cos\Delta\alpha<0$, wave b is

Table 4.1. Dependence of the polarization rotation of two waves during anisotropic self-diffraction on the sign of F_D

Polarization of the input waves	F_D	Amplified wave
$\alpha = \beta = 0$	$+$	a
	$-$	b
$\alpha = \beta = \pi$	$+$	b
	$-$	a

Table 4.2. Dependence of the energy exchange in anisotropic self-diffraction on the sign of F_D

amplified (Table 4.2). Therefore, the dependence of polarization rotation and energy exchange upon the sign of F_D may be used to determine the type of photocarriers.

Next we discuss the amplitudes given by

$$A_x^2(y) = B_{z0}^2 \frac{\sinh^2(\phi_D/2)}{\cosh \phi_D} \ , \qquad B_x^2 = A_{z0}^2 - A_z^2 \ ,$$

$$A_z^2(y) = A_{z0}^2 \frac{\cosh^2(\phi_D/2)}{\cosh \phi_D} \ , \qquad B_z^2 = B_{z0}^2 - A_x^2 \ ,$$

(4.33)

where

$$\phi_D = k_0 y \Delta n_D A_{z0} B_{z0} (I_0 + I_d)^{-1} \ ,$$

and

$$\Delta n_D = -\frac{r_{14} n^3 E_D}{2\cos\theta} \ .$$

For the total intensity A^2 of wave a and the rotation angles θ_a, θ_b of the polarization vectors of waves a, b we obtain

$$A_2 = A_x^2 + A_x^2$$

$$= A_{z0}^2 + (B_{z0}^2 - A_{z0}^2) \frac{\sinh^2(\phi_D/2)}{\cosh \phi_D} \ ,$$

$$\tan\theta_a = \frac{A_x}{A_z} = \frac{B_{z0}}{A_{z0}} \tanh\left(\frac{\phi_D}{2}\right) \ ,$$

(4.34)

$$\tan\theta_b = \frac{B_x}{B_z} = \frac{A_{z0}}{B_{z0}} \tanh\left(\frac{\phi_D}{2}\right) \ .$$

The index zero designates input values at the crystal boundary $y = 0$.

We see that the low-intensity wave is amplified, that the direction of the polarization rotation for the two waves is different, and that the angle of rotation is larger for the lower-intensity wave. For BSO with $r_{14} = 5 \times 10^{-10}$ cm V^{-1}, $\lambda = 0.63$ μm, $E_D = 10^3$ V cm^{-1}, $n = 2$, $\Delta n_D = 3 \times 10^{-6}$ we get $\phi_D = 0.1 y$ [cm] and the polarization rotation per unit length due to diffraction will be about 3 degrees/cm.

Next we give some results for crystals with optical activity ($G \neq 0$) in an external field ($E_0 \neq 0$). The reuslts are obtained from (4.30, 31) after linearization with respect to P_n and P_s [4.35].

1) For two waves initially linearly polarized along the z-axis we find

$$|a_z|^2 = A_{z0}^2 (\cos^2 \gamma y - ry\beta P_n \gamma^{-1} B_{z0}^2 \sin 2\gamma y) ,$$

$$|a|^2 = |a_x|^2 + |a_z|^2 \tag{4.35}$$

$$= A_{z0}^2 (1 + 2r B_{z0}^2 GP_s \gamma^{-2} \sin^2 \gamma y) ,$$

where

$$\gamma = \sqrt{G^2 + \beta^2} \quad \text{with}$$

$$\beta = rE_0 .$$

2) For two waves initially linearly polarized along the x-axis we find

$$|a_z|^2 = A_{z0}^2 (\cos^2 \gamma y - ry\beta P_n \gamma^{-1} B_{z0}^2 \sin 2\gamma y) ,$$

$$|a|^2 = A_{x0}^2 (1 - 2r B_{x0}^2 GP_s \gamma^{-2} \sin^2 \gamma y) . \tag{4.36}$$

One can see that the sign of the energy exchange is different for these different input polarizations, and that optical activity leads to an oscillating dependence of the energy exchange upon the crystal thickness. This explains the experimental observation that the transmitted energy may be larger for thinner crystals [4.36].

3) For two waves initially circularly polarized we obtain [4.35]

$$|a|^2 = 2A^2 \{ 1 - 2r B^2 \beta GP_s \gamma^{-3} [2\gamma y (1 + \sin^2 \gamma y) - \sin 2\gamma y] \} . \tag{4.37}$$

Here A and B are defined by

$$A = A_{x0} = A_{z0} , \quad B = B_{x0} = B_{z0} ,$$

and we have assumed

$$\alpha_{z0} - \alpha_{x0} = \frac{\pi}{2} , \quad \beta_{z0} - \beta_{x0} = \frac{\pi}{2} .$$

The direction of the energy exchange depends upon the sign of the external electric field. It has been shown in [4.35] that the direction of the energy exchange is also altered by changing the circularity to $\alpha_{z0} - \alpha_{x0} = \beta_{z0} - \beta_{x0} = -\pi/2$.

We should like to add some comments on self-diffraction gyration. If only input wave a is present, initially linearly polarized at an angle φ_a to the x-axis, then behind the crystal and behind an analyzer rotated about an angle ψ to the x-axis, we obtain for the intensity

$$I(\psi) = A^2 \cos^2\left(\frac{Gy}{\cos\theta} - \psi + \varphi_a\right) , \qquad (4.38)$$

where θ is the angle of incidence. We get zero intensity if the angle ψ of the analyzer is equal to

$$\psi_d = \varphi_a + \frac{Gy}{\cos\theta} \pm \frac{\pi}{2} .$$

If input wave b is also present, a holographic grating will build up and the diffracted part of wave b will add to wave a. From (4.30, 31) we obtain for $E_0 = 0$ (which means $P_n = 0$) that in this case

$$I(\psi_d) = [A \, BrF_s G^{-1} \sin Gy \cos(\varphi_a - \varphi_b) \sin(\varphi_a + \varphi_b - Gy)]^2 . \qquad (4.39)$$

We may speak of self-diffraction bleaching (SDB). The effect can be used for improving the signal-to-noise ratio in image amplification and dynamical interferometry [4.37, 38]. A comprehensive review of the polarization properties of photorefractive diffraction in BSO crystals is given in [4.39]. Self-diffraction effects, however, are not considered there.

Oscillating behavior of the energy exchange is found for reflection gratings, which are described in [4.12] for two waves propagating along the (001) axis.

4.2.3 Natural and Photoinduced Electrogyration

Let us consider another configuration which has been used for hologram writing, for which the grating vector is along the (001) axis as shown in Fig. 9. In this case the coupled wave equations are

$$\frac{da_x}{dy} = Ga_z - i\beta a_x + (gb_z - irb_x)\Delta E ,$$

$$\qquad\qquad\qquad\qquad\qquad\qquad\qquad\qquad (4.40)$$

$$\frac{da_z}{dy} = -Ga_z - g\Delta E b_x ,$$

where

$$G = G_0 + gE_0 , \qquad \beta = rE_0 .$$

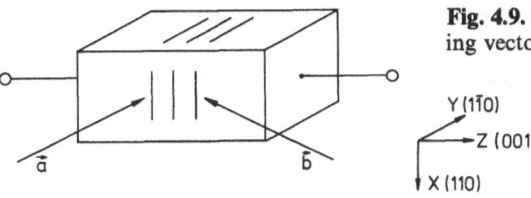

Fig. 4.9. Scheme of electrogyration. The grating vector q lies in the (001)-direction

Y (1$\bar{1}$0)

Z (001)

X (110)

Here

$$G_0 = \frac{k_0^2 \lambda g_0}{4\pi}$$

is the optical rotatory power and

$$g = \frac{k_0 g_{14}}{2n}$$

is the electrogyration factor. The equations for wave b can be obtained from (4.40) by the substitutions $a_i \to b_i$, $\Delta E \to \Delta E^*$. We can see from these equations that the electrogyration, which is proportional to g, influences self-diffraction and optical activity and that therefore there are two gratings: an electro-optic one and a grating due to modulation of the optical activity by the photoinduced electric field, which may be called the electrogyration grating.

To estimate the influence of electrogyration on the polarization properties, let us analyze the propagation of one single wave in the geometry of Fig. 4.9. Solving (4.40) for one wave initially polarized along the z-axis, we obtain for the intensity

$$I = I_M \left[\cos^2(\gamma y - \psi) + \left(1 - \frac{G}{\gamma}\right) \sin \gamma y \left(\sin(\gamma y - 2\psi) + \frac{G}{\gamma} \sin \gamma y \cos 2\psi \right) \right],$$
$$(4.41)$$

where ψ is the angle of the analyzer behind the crystal with respect to the z-axis and I_M is the maximum value of the intensity. From (4.41) one can see that electrogyration leads to a nonsymmetric dependence of the intensity on the electric field.

We have tried to fit the experimental results for a $Bi_{12}TiO_{20}$ crystal with the theoretical expression (4.41) and obtained a natural optical rotatory power of $G_0 = 66$ degrees/cm at $\lambda = 0.633\,\mu m$ and an electrogyration factor of $g_{14} = (1.3 \pm 0.26)\,10^{-9}\,cm\,V^{-1}$ which is 28% of the natural rotatory power in an electric field of the order of $10\,kV\,cm^{-1}$ [4.35].

4.2.4 Anisotropic Phase Conjugation by Four-Wave Mixing

We shall now discuss phase conjugation in the usual four-wave mixing geometry [4.9] of Fig. 4.8. If the angle between the waves a and b is small, and therefore also the angle between the waves c and d, and if wave a is coherent with wave b, and wave c is coherent with wave d, we have transmission geometry, and the starting equations are

$$\frac{da_{x,z}}{dy} = -ir(E_{13} + E_{12})b_{z,x} - (irE_0 \mp G)a_{z,x} ,$$

$$\frac{db_{x,z}}{dy} = -ir(E_{31} + E_{24})a_{z,x} - (irE_0 \mp G)b_{z,x} ,$$

$$\frac{dc_{z,x}}{dy} = ir(E_{31} + E_{24})d_{z,x} + (irE_0 \pm G)c_{z,x} ,$$ (4.42)

$$\frac{d_{x,z}}{dy} = ir(E_{13} + E_{12})c_{z,x} + (irE_0 \pm G)d_{z,x} ,$$

where

$$E_{13} = E_{31}^* , \qquad E_{24} = E_{42}^* .$$

The time dependence is obtained from

$$\frac{dE_{13,42}}{dt} = \frac{\Delta I_{13,42}}{I_0 + I_d} (iP_s - P_n) + E_{13,42} \left[\frac{1}{\tau_r} + i\left(\frac{1}{\tau_{osc}} + \Omega \right) \right]$$

with

$$\Omega = \omega_1 - \omega_3 = \omega_2 - \omega_4 ,$$

$$I_{13} = ab^* , \qquad I_{24} = cd^* .$$

From (4.42) it can be deduced that the interaction of the plane pump waves a and c with the signal wave b results in the appearance of the phase-conjugate wave d. The polarization dependence of the direction of energy exchange allows the realization of high-contrast gratings with equal intensity waves, subsequent retrieval in the phase conjugation (PC) scheme, and amplification by self-diffraction. This may be achieved for instance by writing with two waves a and b of equal intensity, polarized along the z-axis. For $E_0 = 0$ and $G = 0$ there is no energy exchange between waves a and b and the grating contrast is large. Retrieval by the wave c polarized in the second or fourth quadrant (for $P_s > 0$) leads to amplification by self-diffraction of the wave d. In the absence of an external electric field E_0 and for the frequency-degenerate case $\Omega = 0$ we obtain $\tau_{osc}^{-1} = P_n = 0$, and the time development of the holographic grating amplitude may be calculated in the quasi-stationary approximation

$$E_{13,42} = iM_s \Delta I_{13,42} , \quad \text{where}$$

$$M_s = \frac{P_s \tau_r}{I_0 + I_d} (1 - e^{-t/\tau_r}) .$$

In an approximation quadratic in the electro-optic coefficient r, we obtain for the intensity of the phase-conjugate wave

$$|d_{x,z}|^2 = |rM_s(\mathbf{a} \cdot \mathbf{b}^*)c_{z,x}|^2 = \left| rE_D \frac{(1 + E_D/E_M)}{(1 + E_D/E_q)} \frac{(\mathbf{a} \cdot \mathbf{b}^*)c_{z,x}}{(I_0 + I_d)} \right|^2 . \quad (4.43)$$

From (4.43) one can see that the polarization of the phase-conjugate wave d is orthogonal to that of the read-out wave c.

4.2.5 Modulation of Beam Coupling by an Electric Field

We shall now discuss the influence of an external electric field on beam coupling by anisotropic self-diffraction. We start from a temporal modulation of the electric field E_0 of the form

$$E_0 = E_s \sin \Omega_0 t + E_c \cos \Omega_0 t ,$$

where the frequency Ω_0 is large compared with τ_r^{-1}, $\Omega_0 \gg \tau_r^{-1}$. We assume that waves a and b are polarized in the first quadrant, $\alpha_{x,z} = \beta_{x,z} = 0$, that the angle between the polarization vectors and the x-axis is φ, and that the input amplitudes of waves a and b are A and B. In an approximation linear in r and G we obtain

$$|a|^2 = A_x^2 + A_z^2$$
$$+ \frac{2ry(\delta J_{00} + 2J_{21}) [J_0(m) AB]^2 \sin 2\varphi}{(I_0 + I_d)(\delta^2 + \gamma^2)} [E_D J_{00} - (J_{10} E_s + J_{01} E_c)] , (4.44)$$

where

$$\delta = 1 + \frac{E_D}{E_q} ,$$

$$\gamma = \Omega \tau_M ,$$

$$J_{00} = J_0(P_c) J_0(P_s) ,$$

$$J_{01} = J_0(P_c) J_1(P_s) ,$$

$$J_{10} = J_1(P_c) J_0(P_s) , \quad \text{and}$$

$$J_{21} = J_2(-P_c) J_1(-P_s) .$$

Here J_m is the mth order Bessel function and

$$P_{c,s} = \frac{E_{c,s}}{\Omega_0 \tau_M E_q}$$

are dimensionless electric field amplitudes. They grow with decreasing frequency Ω_0. For small diffusion field $E_D \ll E_q$ and taking into account that $J_0 > J_1 > J_2$, assuming $P_{c,s} \lesssim 1$ we obtain in the frequency-degenerate case $\gamma = 0$ an expression for the intensity change $\Delta I_{osc} = a^2 - A_x^2 - A_z^2$:

$$\Delta I_{osc} = -\frac{ry[J_0(m)AB]^2 \sin 2\varphi}{(I_0 + I_d)} (P_c E_s + E_c P_s) \ . \tag{4.45}$$

Comparing this expression with the intensity change in a constant electric field E_0 we obtain

$$f = \frac{\Delta I_{osc}}{\Delta I} = (P_c + P_s) \frac{2}{\Omega_0 \tau_M} \left[1 + \left(\frac{E_0}{E_q} \right)^2 \right] \ .$$

It follows that modulation of the electric field leads to an increase of the energy exchange. For instance, for $\Omega_0 \tau_M = 2$ and $E_0 = E_q$ we get $f = 2$. The ratio f increases with E_0/E_q. Therefore it decreases with the grating period L because E_q is proportional to L.

To compare the given theoretical results with experiments on $Bi_{12}TiO_{20}$ [4.40], where the development of a hologram is studied in the presence of an alternating electric field, one needs more information about the dynamics of the formation of space charge in the vicinity of the electrodes.

4.2.6 Real-Time Holographic Interferometry

In this section we briefly report on an analytic approach to real-time interferometry [4.19]. Let us consider the case that the object vibrates, that is, that the phase of the object wave is modulated in time in the form

$$\varphi = m \cos \nu t \ , \quad \text{where}$$

$$m = \frac{2\pi}{\lambda} d(\cos \theta_i + \cos \theta_r) \ .$$

Here d characterizes the amplitude of the vibration, λ is the wave length of the light wave, ν is the frequency of the modulation, θ_i is the angle of incidence and θ_r is the angle of reflection from the surface of the vibrating object. The calculations of [4.38] show that in the approximation of a slow nonlinear response $\Omega^{-1}, \tau_M \gg \nu^{-1}$, the gain coefficient Γ for two-wave mixing of Sect. 4.1.3 is to be multiplied by $J_0^2(m)$ where J_0 is the zero-order Bessel function. For real-time interferometry in four-wave mixing the intensity of the phase conjugated wave is to be multiplied by the same factor. Since the Bessel function $J_0(m)$ has a maximum for $m=0$ [$J_0(0)=1$] the brightest lines of the interferogram correspond to the nonvibrating parts of the object, that is, to the nodes of the vibration, while the dark regions in the interferogram correspond to the zeros of $J_0(m)$. The first zero at $m=2.4$ [4.41] corresponds to an amplitude $d=0.38\lambda$ if $\theta_i=\theta_r=0$.

4.3 Conclusion

In conclusion we shall briefly discuss some applications of dynamic gratings in photorefractive crystals. As we can see from the previous discussion, a variety of different possibilities are offered by dynamic gratings. Some applications, such as holographic storage [4.2], light amplification [4.4], and novel types of self-pumped resonators [4.17] are discussed in the literature. Here we concentrate on the description of novel and less well-known applications: (1) material studies from the kinetics, (2) phase doubling, (3) optical hysteresis and bistability, and (4) dynamic interferometry.

4.3.1 Materials Studies from the Kinetics

From practical hologram recording in ferroelectric crystals [4.21, 42, 43] it was learned that the true steady state predicted by the theory of two or four-wave mixing is very difficult to achieve for most doped crystals. The is mainly due to the large effect of light scattering which leads to amplification of the scattered "light" or "noise". In practice only the initial stage of recording can be described adequately.

The validity of the kinetic method for obtaining parameters of materials was demonstrated in [4.42], where a photovoltaic field $E_{ph}=7\,kV\,cm^{-1}$ was found from isotropic self-diffraction. From experiments during the initial stage of vectorial self-diffraction in $LiNbO_3$ the author, in collaboration with M. Lobanov and A. Knyazkov, has obtained estimates for the antisymmetric part of the photogalvanic tensor. For $LiNbO_3:Fe$ (0.06%) and light of wavelength $\lambda=0.44\,\mu m$ we found $\beta_a=10^{-11}$ A/W, and for $\lambda=0.6328\,\mu m$ we obtained $\beta_a=10^{-13}$ A/W. For both wavelengths the extraordinary wave was amplified, which means that $\beta_a r_{15}>0$.

4.3.2 Phase Doubling

This effect was described in Sects. 4.1.4 and 7. One of the consequences of phase doubling is the dynamic transformation of the lens focal distance for the diffracted wave; this was demonstrated in [4.27] for anisotropic self-diffraction in $BaTiO_3$.

Another interesting application of phase doubling may be to improve the sensitivity of measurement of refractive index changes in real-time interferometry.

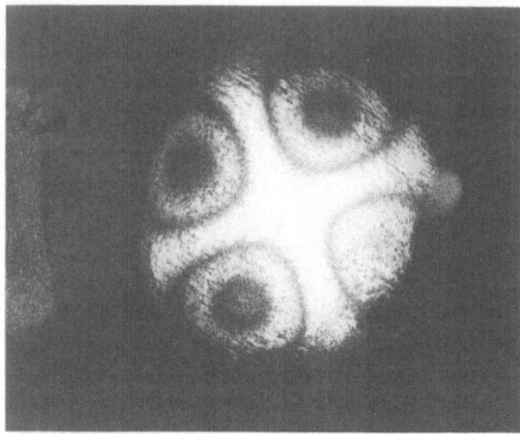

Fig. 4.10. Time averaged vibrating modes of a metal membrane with a vibration frequency of 6.5 kHz obtained in a SBN crystal at $\lambda = 0.44\ \mu m$ [4.35]

4.3.3 Optical Hysteresis and Bistability

It is interesting to note that of the predicted "critical phenomena" which may be observed for large diffraction efficiency, namely oscillation [4.6], and optical hysteresis and bistability [4.16], oscillation has already been realized experimentally [4.18] and therefore is a candidate for the construction of novel types of resonators. Some preliminary results on bistability in a self-pumped phase-conjugate mirror with feedback were reported in [4.44, 45].

4.3.4 Dynamic Interferometry

A short review of the results of Huignard et al. on making vibrating structures visible is presented in [4.46]. The first use of $Bi_{12}TiO_{20}$ crystals for real-time interferometry was reported in [4.38].

That hydrodynamical flows around a moving fish can be made visible was demonstrated in [4.47] for $\lambda = 0.633\ \mu m$ in $Bi_{12}TiO_{20}$. In Fig. 4.10 the time-averaged vibrating modes are shown which were obtained for $\lambda = 0.44\ \mu m$ in the polar crystal strontium-barium niobate (SBN) [4.35].

Acknowledgements. I gratefully thank Professor K.H. Ringhofer for valuable help with the preparation of this chapter, and colleagues from the Leningrad Ioffe Physical Technical Institute and the Institute of General Physics (Moscow) for providing photorefractive crystals.

References

4.1 F.S.Chen, J.T. La Macchia, D.B.Fraser: Appl. Phys. Lett. **13**, 223 (1968)
4.2 D.L.Staebler, J.J.Amodei: Ferroelectrics **3**, 107 (1972)
4.3 P.Günter, U.Flückiger, J.P.Huignard, F.Micheron: Ferroelectrics **13**, 297 (1976)

4.4   N. Kukhtarev, V. Markov, S. Odulov: Opt. Commun. **23**, 230 (1977)
4.5   A. M. Glass: Opt. Eng. **17**, 470 (1978)
4.6   A. Yariv: IEEE J. QE-**14**, 650 (1978)
4.7   N. Kukhtarev, V. Markov, S. Odulov, M. Soskin, V. Vinetskii: Ferroelectrics **22**, 949, 961 (1979)
4.8   A. Marrakchi, J. P. Huignard, J. P. Herriau: Opt. Commun. **34**, 15 (1980)
4.9   R. Hellwarth: J. Opt. Soc. Am. **67**, 1 (1977)
4.10 N. Kukhtarev, S. Odulov: JETP Lett. **30**, 4 (1979)
4.11 S. V. Miridonov, M. P. Petrov, S. I. Stepanov: Sov. Tech. Phys. Lett. **4**, 393 (1978)
4.12 N. Kukhtarev, G. Dovgalenko, V. Starkov: Appl. Phys. A **33**, 227 (1984)
4.13 D. Pepper, R. Abrams: Opt. Lett. **3**, 212 (1978)
4.14 J. White, A. Yariv: Appl. Phys. Lett. **37**, 5 (1980)
4.15 C. R. Petts, M. W. McCall, L. C. Laycock: Electron. Lett. **20**, 32 (1984)
4.16 N. Kukhtarev, A. Borshsch, M. Brodin, V. Volkov, T. Semetets: J. de Phys., Suppl. **44**, C2-C2-5-14 (1983)
4.17 J. O. White, M. Cronin-Golomb, B. Fischer, A. Yariv: Appl. Phys. Lett. **40**, 450 (1982)
4.18 M. Cronin-Golomb, B. Fischer, J. O. White, A. Yariv: IEEE J. QE-**20**, 12 (1984)
4.19 J. P. Huignard, A. Marrakchi: Opt. Lett. **6**, 622 (1981)
4.20 A. Yariv, J. Lotspeich: J. Opt. Soc. Am. **72**, 273 (1977)
4.21 N. Kukhtarev: Pis'ma Zh. Tekh. Fiz. **2**, 1114 (1976)
4.22 B. Belinicher, B. Sturman: Usp. Fiz. Nauk **130**, 415 (1980)
4.23 V. Agranovich, V. Ginzburg: *Kristallooptica s uchetom prostranstvennoi dispersii* (Nauka, Moscow 1979)
4.24 O. V. Vlokh, J. Zheludev: Izv. Akad. Nauk SSSR, Ser. Fiz. **41**, 470 (1974)
4.25 G. Valley: J. Opt. Soc. Am. **1**, 868 (1984)
4.26 P. Refregier, L. Solymar, H. Rajbenbach, J.-P. Huignard: J. Appl. Phys. **58**, 45 (1985)
4.27 N. Kukhtarev, E. Krätzig, H. C. Külich, R. A. Rupp, J. Albers: Appl. Phys. B**35**, 17 (1984)
4.28 N. Kukhtarev, S. Odulov: Pis'ma Zh. Tekh. Fiz. **6**, 1176 (1980)
4.30 N. Kukhtarev, T. Semenets: Ukr. Fiz. Zh. (Russ. Ed.) **28**, 1874 (1983)
4.29 N. Kukhtarev, S. Odulov: Opt. Commun. **32**, 183 (1980)
4.31 N. Kukhtarev, B. Pavlik, S. Sorokina, T. Semenets: Kvantovaya Electron. (Moscow) **13** (12), 326 (1986)
4.32 P. Günter: Opt. Lett. **7**, 10 (1982)
4.33 N. Kukhtarev: Ukr. Fiz. Zh. (Russ. Ed.) **288**, 612 (1983)
4.34 S. Odulov: JETP Lett. **35**, 11 (1982)
4.35 N. Kukhtarev, V. Muraviev, T. Semenets: Preprint No. 19, 22 p. (in Russian) (Institute of Physics. Academy of the Ukrainian SSR, Kiev 1986)
4.36 P. Refregier, L. Solymar, H. Rajbenbach, J. P. Huignard: Electron. Lett. **20**, 656 (1984)
4.37 J. P. Herriau, J. P. Huignard, A. G. Apostolidis, S. Mallick: Opt. Comm. **56**, 141 (1985)
4.38 A. Kamslulin, E. Mokrushina: SPIE **433**. Symposium OPTIKA'84 (1984 Budapest, Hungary), p. 83–86
4.39 A. Marrakchi, R. V. Johnson, A. R. Tangway: J. Opt. Soc. Am. B**3**, 321 (1986)
4.40 S. Trofimov, S. Stepanov: Pis'ma Zh. Tekh. Fiz. **11**, 615 (1985)
4.41 G. Korn, T. Korn: *Mathematical Handbook for Scientists and Engineers* (McGraw-Hill, New York 1961)
4.42 K. Belabaev, V. Markov, B. Kondilenko, N. Kukhtarev, S. Odulov, M. Soskin: Zh. Tekh. Fiz. **50**, 2560 (1980)
4.43 M. G. Moharam, T. G. Gaylord, R. Magnusson, L. Young: J. Appl. Phys. **50**, 5642 (1979)
4.44 S.-K. Kwong, M. Cronin-Golomb, A. Yariv: Appl. Phys. Lett. **45**, 1016 (1984)
4.45 S.-K. Kwong, A. Yariv: Opt. Lett. **11**, 377 (1986)
4.46 P. Günter: Phys. Rep. **93**, 200 (1982)
4.47 N. Kukhtarev, G. Dovgalenko: Preprint No. 5, 19 p. (in Russian) (Institute of Physics, Kiev 1985)

5. Photorefractive Centers in Electro-optic Crystals

Eckhard Krätzig and Ortwin F. Schirmer

With 22 Figures

Light-induced refractive index changes in electro-optic crystals – so-called photorefractive effects – were discovered in 1966 by *Ashkin* et al. [5.1] with the study of electro-optic and nonlinear properties of $LiNbO_3$ and $LiTaO_3$ and were successfully utilized by *Chen* et al. [5.2, 3] for the storage of volume phase holograms. Since then many novel basic research lines and new applications have been developed, as can be seen from the scope of this book.

Photorefractive effects are based on the transposition of a light pattern into a refractive index pattern. Interfering light beams generate bright and dark regions in an electro-optic crystal. When light of suitable wavelength is chosen charge carriers–electrons or holes – are excited in the bright regions and become mobile. The charge carriers migrate in the lattice and are subsequently trapped at new sites. By these means electrical space-charge fields are set up which give rise to a modulation of the refractive index via the electro-optic effect. Light-induced index changes up to about 10^{-3} are obtained. The trapped charge can be released and the field pattern erased by uniform illumination or by heating. Various materials have been utilized for photorefractive measurements. The most intensively studied examples are the trigonal crystals $LiNbO_3$ and $LiTaO_3$, but other crystals have also been the subjects of numerous investigations.

Optimization of the material properties for photorefractive studies requires detailed knowledge of the microscopic processes involved. Of particular interest is information on the centers supplying charge carriers, on the trapping centers and on optical excitation and transport processes. These questions are discussed in the present chapter. Among the many defects studied in electro-optic materials we shall confine our attention to those centers proven to be involved in photorefractive processes.

After this introduction, in Sect. 5.1 a short survey of experimental methods with which photorefractive defects have been investigated will be given.

Section 5.2 deals with extrinsic impurity centers in $LiNbO_3$ and $LiTaO_3$ and their influence on the light-induced charge transport that determines photo-refractive effects. For these crystals it is well known that the light-induced charge transport can be greatly affected by transition metal dopants. Photovoltaic effects, photoconductivity and thermal fixing are discussed. The properties of intrinsic centers, comprising only ions belonging to the normal crystal composition, are described in Sect. 5.3. There are indications that they can also take part in photorefractive processes. In Sect. 5.4 centers in various electro-optic crystals are discussed. Among these crystals are the ferroelectric perovskites $KNbO_3$,

$KTa_{1-x}Nb_xO_5$ and $BaTiO_3$, the tungsten-bronze-type crystals $Ba_2NaNb_5O_{12}$ and $Sr_{1-x}Ba_xNb_2O_6$, the nonferroelectric sillenites $Bi_{12}SiO_{20}$ and $Bi_{12}GeO_{20}$ and the semiconductors GaAs and InP.

5.1 Methods of Investigation

5.1.1 Electron Spin Resonance

In favorable cases, electron spin resonance (ESR) can lead to the identification of paramagnetic defects in solids by yielding the following information:

- symmetry of defect
- identity of nuclei within the defect
- charge state of paramagnetic ions involved
- delocalisation of defect wave function
- concentration of defects
- change of charge state under external influences (illumination, change of temperature).

In photorefractive materials, ESR has been used to identify paramagnetic transition metal (mainly $3d$) ions, trapped hole centers and electrons close to conduction states. Numerous reports on the technique and merits of ESR on defects in solids have been published (e.g. [5.4–8]), so we shall give here only a short overview of how the above information can be obtained.

In an ESR experiment high frequency (usually microwave) transitions are induced between levels close to the ground state of a paramagnetic system, which are split by one or more of the following influences: external magnetic field, electron-nuclear (hyperfine) interaction, crystal field effects (generally caused by spin-orbit coupling in higher order). These interactions are usually collected in a (spin) Hamiltonian, describing the dynamics of only the lowest-lying states:

$$\mathscr{H} = \beta B \underline{g} S + I \underline{A} S + S \underline{D} S + \dots ,$$

where only three representative terms are given for illustration. The first one describes the Zeeman interaction, the second, the magnetic dipole hyperfine interaction, and the last one, a crystal field interaction. Here β is the Bohr magneton, B the external magnetic field, $\underline{g}$ the Zeeman interaction tensor, S the spin operator ($2S+1$ is the multiplicity of the lowest levels considered), I the nuclear spin operator, $\underline{A}$ the tensor of the magnetic dipole hyperfine interaction, and $\underline{D}$ the crystal field tensor. Other possible terms include the interaction with nuclear quadrupole moments or crystal field and Zeeman interactions which are of higher order in the spin operator S. In order to describe lower than cubic symmetry the various couplings generally have to be described by tensors rather than scalars.

The analysis of a defect proceeds by guessing (using the results of experience) the Hamiltonian and the electron spin/nuclear spin manifold which are assumed to describe the envisaged model of the center. Solving the resulting eigenvalue problem for the energies of allowed transitions, one can either reproduce the observed spectra by adjusting the tensor components or one has to look for a more appropriate Hamiltonian. For simple cases analytic solutions of the eigenvalue problem are available.

The *geometry* of a defect is related to the symmetry of the coupling tensors, which in many cases can be derived by observing the change of the spectra on varying the angle between crystal axes and magnetic field. The *identity of nuclei* in the center can be derived from their hyperfine splittings, generally into $2I+1$ lines for interaction with one kind of nuclei having spin I. The hyperfine patterns resulting from all isotopes of a certain element generally display a characteristic "fingerprint". The structure of a spin Hamiltonian and the values of the coupling parameters are highly specific for a given *charge state* of an ion. Interpretation of these coupling parameters can yield information on the *delocalisation* of the defect wave function. Employing necessary corrections, the *concentration* of defects is obtained by comparing the intensity of its ESR spectrum with that of a standard of a known number of spins. The decay or increase of a spectrum under illumination is an indication of a *change of charge state* of the corresponding paramagnetic ion. By looking for the dependence of such effects on the wavelengths of the exciting light, information on the optical absorption of a center can be obtained.

To conclude this section the ESR of Fe^{3+} ($3d^5$) in $LiNbO_3$ is taken as an example. The spin Hamiltonian of this $^6S_{5/2}$ ion in axial symmetry is

$$\mathcal{H} = \beta \boldsymbol{B}\underline{g}\boldsymbol{S} + D\left(S_z^2 - \frac{S(S+1)}{3}\right) + \cdots ,$$

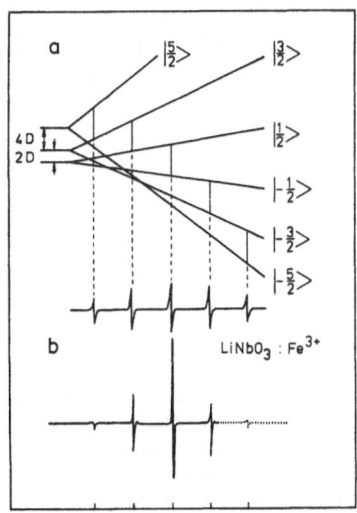

Fig. 5.1. (a) Level scheme of the lowest states of Fe^{3+} ($^6S_{5/2}$) in $LiNbO_3$ (trigonal axial field) for $\boldsymbol{B} \parallel c$. Allowed transitions are indicated and compared to those observed (b) [5.9]

where D expresses the action of the trigonal crystal field on the $S = 5/2$ ground state. Eigenvalues, transitions and the corresponding ESR spectrum are shown in Fig. 5.1 for B parallel to the trigonal axis.

5.1.2 Optical Absorption

Optical absorption of photorefractive defects in the near uv, visible and near ir spectral ranges has been used to gather the following information:

- identity of defect ion and its charge state
- energy levels
- nature of transition.

Figure 5.2 schematically indicates various possible processes. One can discriminate between internal transitions at a defect, which leave its charge state unchanged, and charge transfer transitions. The former generally take place between levels originating via the crystal-field interaction from free ion terms. They tend to be weaker and narrower than charge transfer transitions, which are usually electric dipole allowed. For an overview see, for example, [5.10]. Examples of such absorptions are found in Figs. 5.4-6.

Optical transitions involving free or bound small polarons can be considered as special types of charge transfer processes. A small polaron is formed when a charge carrier is trapped at one of at least two equivalent ions by distortion of the surrounding lattice. If the trapping ions are close to an attracting defect and arranged in such a way that they are still equivalent, the polaron is bound. Optical transfer of free and bound small polarons occurs by excitation of the trapped charge from its self-induced potential well to one of the neighboring equivalent ones. Since the process takes place as a Franck-Condon transition in a fixed nuclear framework a finite photon energy is needed for the transition in spite of the equivalence of the ions involved. For literature, see, e. g., [5.11, 12]. An example of bound small polaron absorption is contained in Fig. 5.22. The band at 1.6 eV in Fig. 5.20 is ascribed to absorption of free small polarons.

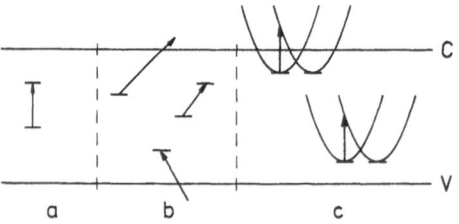

Fig. 5.2a–c. Various defect absorption processes. (**a**) Vertical arrow indicating a crystal field transition. (**b**) Oblique arrows representing charge transfer or intervalence transitions. (**c**) Small polaron transitions. In such transitions the lattice distortion is essential to obtain finite transition energies, so parabola are shown which mark the dependence of the defect energy on lattice distortion. The levels deep in the gap indicate a bound polaron

Near infrared optical absorption can yield information on the vibrational properties of light defect ions. In this way it was shown [5.13] that protons can be incorporated into $LiNbO_3$ in the form of OH^- ions. It was established that they are responsible for thermal fixing of phase holograms in this material [5.14].

5.1.3 Mössbauer Effect

In the investigation of photorefeactive centers the Mössbauer effect (ME) was instrumental in elucidating the properties of defects containing Fe. The effect has furnished information on the charge state of Fe, defect symmetry, etc. Similarly to ESR the method can determine splittings of the electronic ground state arising from crystal-field, Zeeman and hyperfine interactions. In addition, the isomer shift can be monitored. This is caused by the difference in Coulomb interaction (spherical part) between nucleus and surrounding electrons in the excited and ground nuclear states. A useful feature of the ME is its ability to monitor charge states not accessible to ESR, e.g., Fe^{2+} in $LiNbO_3$. For reviews of the method see, for example, [5.15, 16]. The latter is particularly relevant to the study of $LiNbO_3$: Fe.

The method is a high resolution spectroscopy exploiting, for the Fe centers concerned, the narrow natural linewidth of the 14.4 keV γ-fluorescence transition between the lowest, $I = 3/2 \rightarrow I = 1/2$, nuclear states of ^{57}Fe. If this isotope is prepared in the excited, $I = 3/2$, state while incorporated in a solid, a fraction of all emitted γ-quanta do not suffer energy losses. These can thus be resonantly absorbed by an ^{57}Fe nucleus in the crystal to be studied. The highly monochromatic radiation can be used for spectroscopy by varying the speed between source and absorber. Figure 5.3 gives examples of the ME of Fe in $LiNbO_3$. Spectra originating from $^{57}Fe^{3+}$ in electronic environments with $m_s = \pm 5/2$ and $m_s = \pm 3/2$ and from $^{57}Fe^{2+}$ can be distinguished.

5.2 Extrinsic Photorefractive Centers in $LiNbO_3$ and $LiTaO_3$ and Their Influence on Light-Induced Charge Transport

In some cases the relations between extrinsic impurity centers and the light-induced charge transport properties are well established. This is especially true for $LiNbO_3$ and $LiTaO_3$ crystals, which will be treated in some detail in this section. We start with the results of ESR, Mössbauer effect and optical absorption measurements identifying various transition metal ions in these crystals. Then we discuss photovoltaic effects and photoconductivity, which are largely determined by impurities in certain valence states. We compare the results of conventional photoconductivity measurements with those of holographic beam coupling experiments. A discussion of protons and thermal fixing concludes this section.

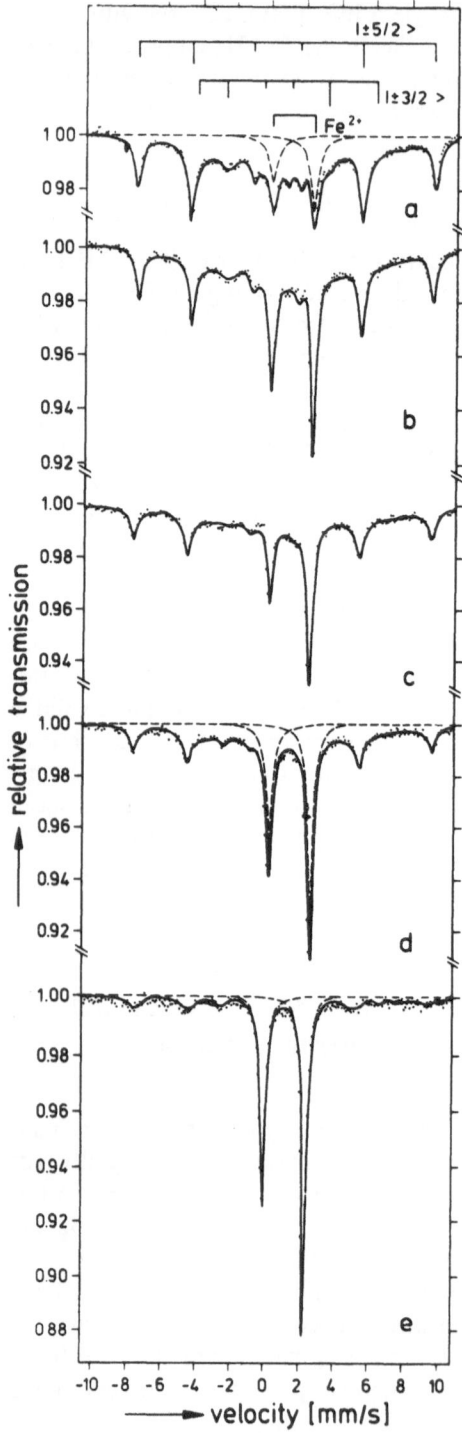

Fig. 5.3. Mössbauer spectra of LiNbO$_3$: ^{57}Fe at 80 K, (**a**) in the as-grown state, (**b**) after annealing at 1000 °C in 760 Torr Ar atmosphere for 50 h, (**c**) after additional annealing at 1000 °C in 760 Torr Ar atmosphere for 50 h and rapid cooling, (**d**) after additional annealing at 1000 °C and 50 Torr Ar atmosphere for 10 h, (**e**) after additional annealing at 1000 °C and 0.2 Torr Ar atmosphere for 20 h. Source 50 mCi ^{57}Co in Rh [5.17]

5.2.1 Transition Metals in LiNbO₃ and LiTaO₃

During recent years most photorefractive investigations have been performed with the isomorphous crystals $LiNbO_3$ and $LiTaO_3$. Large single crystalline samples can be grown from the melt with typically a diameter of about 30 mm and a length of 60 mm. $LiNbO_3$ and $LiTaO_3$ are pyroelectric with trigonal symmetry ($3m$) at room temperature. The trigonal c-axis is identical with the direction of spontaneous polarization.

However, the formulas $LiNbO_3$ and $LiTaO_3$ do not completely describe these crystals. The congruently melting composition occurs at 48.5 mol % LiO_2 in the case of $LiNbO_3$, and for $LiTaO_3$ the situation is similar. For this reason melt-grown crystals may deviate significantly from the stoichiometric composition.

The crucial influence of transition metal impurities on the photorefractive properties of $LiNbO_3$ and $LiTaO_3$ was discovered very early [5.18, 19]. Transition metal dopants may occur in different valence states, making possible the generation of space-charge fields. Usually the dopants are added to the melt, for example as oxides. Distribution coefficients between 0.1 and 1.9 and concentrations of dopants up to about 1 mol% have been reported. In addition, thin layers (10–100 μm) may be doped by diffusion [5.20]: Metal films are vacuum deposited on crystal wafers which are then heated to temperatures above 1000 °C for several days.

The valence states of transition metal dopants in $LiNbO_3$ and $LiTaO_3$ can be greatly influenced by suitable thermal annealing treatments [5.18, 20]. Heating in an oxygen atmosphere to temperatures of about 1000 °C for several hours tends to oxidize the impurities, for example to Fe^{3+} or Cu^{2+}. This process is reversible, and heating in an argon atmosphere or vacuum (low oxygen partial pressure) tends to reduce the impurities, for example to Fe^{2+} and Cu^+. In the case of $LiTaO_3$, however, the crystals have to be poled again after these annealing treatments because the Curie temperature T_c of about 660 °C is exceeded. In the case of $LiNbO_3$ ($T_c \approx 1200$ °C) this is usually not necessary.

Certain valence states of impurities have been identified by various methods described in Sect. 5.1. Electron spin resonance (ESR) experiments have been used to determine the spectra of the transition metal ions Fe^{3+} [5.18, 21], Mn^{2+} [5.21, 22], Cu^{2+} [5.23], and Cr^{3+} [5.24]. An especially powerful method for the determination of valence states of Fe is Mössbauer spectroscopy. Both Fe^{3+} and Fe^{2+} spectra are observed [5.15, 17, 25], thus yielding the Fe^{2+}/Fe^{3+} ratio. By these means the changes of the valence states in $LiNbO_3$:Fe and $LiTaO_3$:Fe have been carefully studied, showing that only Fe^{2+} and Fe^{3+} centers are present in the crystals, and the decrease of Fe^{3+} on heating the crystals in a reducing atmosphere corresponds to the increase of Fe^{2+}. Experimental results are illustrated in Fig. 5.3.

For photorefractive studies small impurity concentrations below 0.1 mol% are preferred. In this case all results available point to isolated centers. No systematic deviations from local symmetry have been detected and for this reason the details of the charge compensation mechanism are still unclear. For larger impurity concentrations there is some evidence for the formation of pair centers [5.26].

A further problem unsolved as yet is which lattice site is occupied by the impurities. The crystal structure can be described as oxygen layers forming distorted octahedra with their centers between the layers. The octahedra are arranged in a string-like fashion and their centers are occupied in the sequence Nb^{5+}/Ta^{5+}, Li^+, structural vacancy, Nb^{5+}/Ta^{5+}, etc. In the case of Eu^{3+}, optical absorption spectra point to two different trigonal sites (substitution for Li^+ and Nb^{5+}/Ta^{5+}) of this ion [5.27]. For Cr^{3+}, the Nb^{5+}/Ta^{5+} site has been proposed from estimations of the ligand field parameters. Field annealing experiments favor the Li^+ site for Fe^{2+} and Fe^{3+} [5.28]. A further possibility has also been suggested [5.29] which is that the impurity ions form a new site and shift the whole string Nb^{5+}/Ta^{5+}, Li^+, structural vacancy, etc. In this case the impurity cannot be regarded as being substituted in a specific site.

In any case, two statements should be emphasized: The small linewidths of the Mössbauer spectra point to a single lattice site of the Fe centers. The change of Fe^{2+} to Fe^{3+} and vice versa by optical excitations requires the same lattice site for Fe^{2+} and Fe^{3+}.

Of special importance for photorefractive effects are optical absorption processes. Pure $LiNbO_3$ and $LiTaO_3$ crystals are transparent in the near ir, the visible and the near uv region up to about 3.8 eV ($LiNbO_3$) and 4.6 eV ($LiTaO_3$), where the fundamental absorption begins. Impurities cause characteristic bands.

Figure 5.4a shows the spectrum of a $LiNbO_3$:Fe crystal. The band at 1.1 eV is assigned to crystal field transitions $^5A-^5E$ of Fe^{2+} ions, the broad band centered at about 2.6 eV to intervalence transfers $Fe^{2+}-Nb^{5+}$ (leading to the creation of electrons in the conduction band formed by Nb^{5+} ions), the band beginning at about 3.1 eV and extending to higher photon energies to charge transfers from oxygen π-orbitals to Fe^{3+} ions (leading to the creation of holes in the valence band formed by O^{2-} ions), and, finally, the two small lines at 2.55 and 2.95 eV to spin-forbidden $d-d$ transitions of Fe^{3+} ions [5.28]. The absorption spectrum is characteristically changed by reducing treatments (Fig. 5.4b): Bands associated with Fe^{2+} increase while bands associated with Fe^{3+} decrease.

The absorption spectrum of $LiTaO_3$:Fe is very similar to that of $LiNbO_3$:Fe, however, the charge transfer band $O^{2-}-Fe^{3+}$ is quite well separated from the fundamental absorption. For this reason the determination of the Fe^{2+}/Fe^{3+} ratio from absorption measurements is possible [5.30]. In strongly oxidized samples about 99% of the Fe ions are present as Fe^{3+}. The decrease of the Fe^{3+} charge transfer band by reducing treatments reflects the fraction of Fe^{3+} ions transferred into the divalent state. From these measurements the oscillator strengths of the absorption bands have been derived. Results are summarized in Table 5.1.

The absorption spectra of Cu-doped $LiNbO_3$ crystals (Fig. 5.5) have also been utilized for the determination of the Cu^+/Cu^{2+} ratio and the oscillator strengths of the corresponding bands [5.31]. The band at about 1.2 eV results

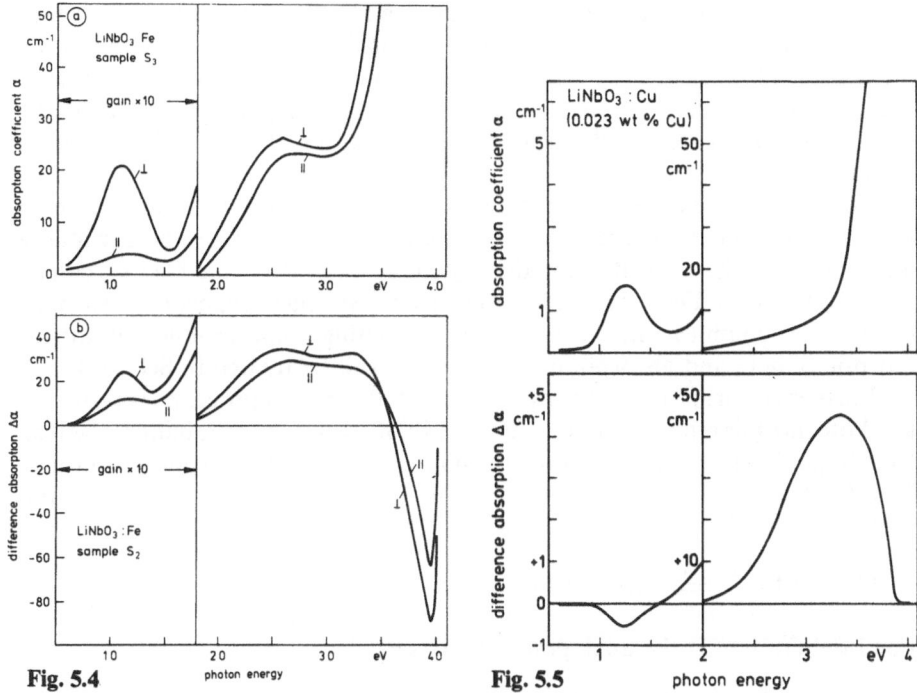

Fig. 5.4   photon energy   **Fig. 5.5**   photon energy

Fig. 5.4. Optical absorption spectra of $LiNbO_3$: Fe, (a) sample S_3 (342 wt. ppm Fe) reduced by annealing in Ar atmosphere, 760 Torr, (b) difference absorption $\Delta\alpha$ between reduced and oxidized sample S_2 (142 wt. ppm Fe) [5.17]

Fig. 5.5. Optical absorption spectra of $LiNbO_3$: Cu for light polarized perpendicular to the c-axis [5.31]. Top: Absorption coefficient a of an oxidized sample ($c_{Cu^+}/c_{Cu^{2+}} < 0.05$). Bottom: Difference absorption $\Delta\alpha$ between partially reduced ($c_{Cu^+}/c_{Cu^{2+}} = 0.6$) and oxidized ($c_{Cu^+}/c_{Cu^{2+}} < 0.05$) samples

Table 5.1. Oscillator strengths of Fe and Cu bands in $LiNbO_3$ and $LiTaO_3$ [5.17, 30, 31]

Crystal	Peak [eV]	Transition	Oscillator strength
$LiNbO_3$: Fe	1.1	$^5A-^5E(Fe^{2+})$	4×10^{-4}
	2.6	$Fe^{2+}-Nb^{5+}$	1×10^{-2}
	~ 4	$O^{2-}-Fe^{3+}$	5×10^{-2}
	2.55 and 2.95	$d-d(Fe^{3+})$	$< 10^{-5}$
$LiTaO_3$: Fe	1.1	$^5A-^5E(Fe^{2+})$	5×10^{-4}
	3	$Fe^{2+}-Ta^{5+}$	1×10^{-2}
	4	$O^{2-}-Fe^{3+}$	5×10^{-2}
	2.6 and 2.9	$d-d(Fe^{3+})$	$< 10^{-5}$
$LiNbO_3$: Cu	1.2	$^2E-^2T_2(Cu^{2+})$	2×10^{-4}
	3.3	Cu^+-Nb^{5+}	4×10^{-2}

from ^2E–$^2T_{2g}$ transitions of Cu^{2+} ions. The band at about 3.3 eV on the other hand is caused by Cu^+–Nb^{5+} intervalence transfers. As can be seen from the lower part of Fig. 5.5 the Cu^{2+} band decreases and the Cu^+ band increases with reducing treatments, the decrease of Cu^{2+} again being proportional to the increase of Cu^+. These results clearly indicate that only Cu^+ and Cu^{2+} centers are involved.

The experiments discussed up to now relate to relatively moderate reductions that influence only the extrinsic impurities and leave the host crystals essentially unchanged. By these means more than about 80% of the Fe and Cu ions can be transformed into Fe^{2+} or Cu^+, respectively. Stronger reductions, such as heating in vacuum to temperatures near the melting point, produce additional intrinsic centers and absorption bands, which will be treated in Sect. 5.3.

Further confirmation of the assignment of absorption processes is provided by photochromic effects in double-doped $LiNbO_3$ crystals containing Fe/Mn or Cu/Mn [5.32, 33]. Figure 5.6 shows the photocurrent and absorption spectra of $LiNbO_3$: Fe/Mn after illumination with uv or visible light. The change of the spectra is described by the processes

$$Mn^{2+} + Fe^{3+} \underset{\text{vis.}}{\overset{\text{uv}}{\rightleftharpoons}} Mn^{3+} + Fe^{2+} \ .$$

For $LiNbO_3$: Fe/Cu the analogous relations read

$$Cu^+ + Fe^{3+} \underset{\text{vis.}}{\overset{\text{uv}}{\rightleftharpoons}} Cu^{2+} + Fe^{2+} \ .$$

The increase and decrease of the corresponding absorption bands can be clearly seen in the spectra.

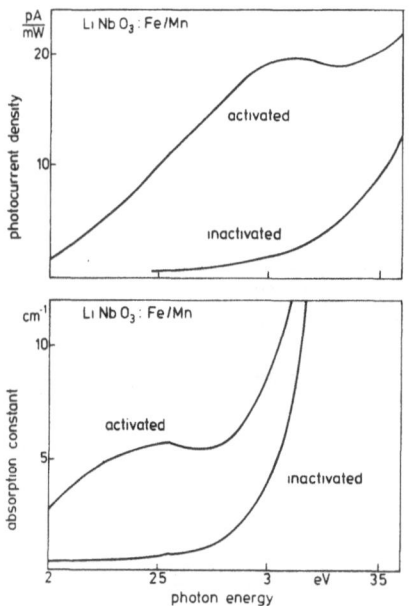

Fig. 5.6. Photocurrent and absorption spectra of a $LiNbO_3$: Fe/Mn crystal after illumination with uv light (activated) and visible light (inactivated) [5.33]

5.2.2 Photovoltaic Effects

The discovery of new bulk photovoltaic effects characteristic of pyroelectrics additionally stimulated the investigation of the light-induced charge transport in these crystals. Already rather early photoinduced currents and voltages have been observed in the absence of external fields in noncentrosymmetric crystals [5.2, 34]. *Glass* et al. [5.35] pointed out in 1974 that new bulk photovoltaic effects are indeed involved which cannot be explained by conventional photovoltaic processes in crystals containing macroscopic inhomogeneities or by Dember voltages.

Homogeneous illumination of single-domain $LiNbO_3$ crystals in thermal equilibrium yields steady-state currents in a short-circuit configuration and steady-state voltages and fields up to 10^7 Vm^{-1} in an open circuit [5.33]. Photovoltaic effects are induced not only by visible and uv light but also by x rays [5.36]. The effects do not depend on electrode material or on illumination of the electrodes.

In first investigations a photovoltaic current has been observed along the polar c-axis only with the $+c$ end negative. The magnitude of the current density j_{ph} is proportional to the absorbed power density:

$$j_{ph} = \kappa \alpha I \ ,$$

where I is the light intensity, α the absorption coefficient, and κ a constant describing the anisotropic charge transport.

Since the photovoltaic current density j_{ph} is linear in the light intensity it must be bilinear in the polarization (unit) vector e of the light, as pointed out by *Belinicher* et al. [5.37]:

$$j_{phi} = \beta_{ijk} I e_j e_k \ .$$

The quantities β_{ijk} represent the photovoltaic tensor. This tensor has four nonvanishing independent components in the case of $LiNbO_3$ and $LiTaO_3$ ($3m$):

$$j_{ph} = \beta_{31} I(e_x^2 + e_y^2)z + \beta_{33} I e_z^2 z + \beta_{22} I[(e_y^2 - e_x^2)y - 2e_x e_y x]$$
$$+ 2\beta_{15} I(e_x e_z x + e_y e_z y) \ .$$

The (unit) vectors x, y and z describe the corresponding axes which are chosen in accordance with the *Standards of Piezoelectric Crystals* [5.38]. The photovoltaic constants β_{31}, β_{33}, β_{22} and β_{15} depend on crystal properties only.

The tensor description of photovoltaic effects has been completely confirmed experimentally [5.39, 40]. Though the largest current densities are obtained along the c-axis, the currents perpendicular to the c-axis, which are about an order of magnitude smaller, are also unambiguously identified by the dependence on polarization angle. Only the tensor element β_{15} could not be obtained by these means because the corresponding current is modulated on account of birefringence effects [5.40].

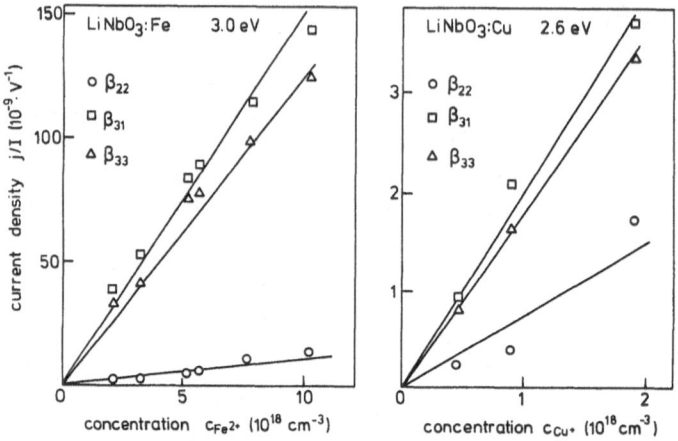

Fig. 5.7. Dependence of the tensor elements β_{22} (O), β_{31} (□), and β_{33} (△) on concentration of Fe^{2+} ions for 3.0 eV and Cu^+ ions for 2.6 eV [5.40]

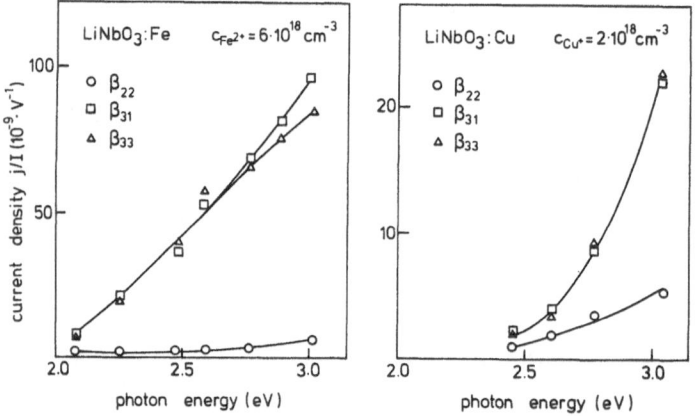

Fig. 5.8. Spectral dependence of the tensor elements β_{22} (O), β_{31} (□), and β_{33} (△) in $LiNbO_3$: Fe for $c_{Fe^{2+}} = 6 \times 10^{18}$ cm^{-3} and $LiNbO_3$: Cu for $c_{Cu^+} = 2 \times 10^{18}$ cm^{-3} [5.40]

Photovoltaic effects are completely determined by the number of filled impurity traps, e. g., Fe^{2+} or Cu^+ ions. Experimental results are summarized in Fig. 5.7 for $LiNbO_3$: Fe and $LiNbO_3$: Cu. Crystals with concentrations of filled traps up to about 10^{19} cm^{-3} have been investigated [5.40]. In all cases a linear dependence of the current density on $c_{Fe^{2+}}$ or c_{Cu^+} is observed. The elements β_{31} and β_{33} are of the same order of magnitude and both elements are considerably larger than β_{22}.

The spectral dependence of the tensor elements is shown in Fig. 5.8. For photon energies larger than 3.1 eV the measurements are seriously impeded by the influence of charge transfer transitions from oxygen orbitals of the valence bands to empty impurity traps, e.g., Fe^{3+} or Cu^{2+} ions. The tensor elements increase with increasing photon energy. Measurable differences between β_{31} and β_{33} can be observed only for concentrations of filled traps larger than 10^{18} cm^{-3} and for photon energies larger than 2.6 eV.

The linear dependence of the current density on concentration of filled traps supports the phenomenological model proposed by *Glass* et al. [5.35, 41]. The main point is the asymmetric charge transfer from the absorbing center to the neighboring ions during optical excitation: The charge carriers are excited with preferred momentum. When they are scattered they lose their directional properties and contribute afterwards to drift and diffusion currents but not to the photovoltaic current. Recombination may also be asymmetric but in general the recombination current will not cancel the excitation current since different states are involved.

The current density may be written [5.35, 41]

$$\frac{j_{ph}}{I} = \kappa\alpha = \frac{\Phi r_{ph}\alpha e}{\hbar\omega} = \mu\tau' E_{ph}\frac{\alpha e}{\hbar\omega} \ ,$$

where Φ is the quantum efficiency, r_{ph} the mean free path, e the charge, μ the mobility, τ' the time for which the excited carriers contribute to the anisotropic charge transport, and E_{ph} a phenomenologically introduced local field acting on the charge carriers.

The photovoltaic constants of Fe-doped crystals have been determined at 3 eV [5.38]:

$$\kappa_{22} = 0.3 \times 10^{-9} \ , \quad \kappa_{31} = 3.3 \times 10^{-9} \ , \quad \kappa_{33} = 2.7 \times 10^{-9} \ cm\,V^{-1} \ .$$

These values yield for the products of quantum efficiency and mean free path

$$(\Phi r_{ph})_{22} = 0.08 \ , \quad (\Phi r_{ph})_{31} = 1.0 \ , \quad (\Phi r_{ph})_{33} = 0.8 \ \text{Å} \ .$$

The corresponding values of Cu-doped crystals at 3 eV read [5.9].

$$\kappa_{22} = 0.07 \times 10^{-9} \ , \quad \kappa_{31} = 1.0 \times 10^{-9} \ , \quad \kappa_{33} = 0.6 \times 10^{-9} \ cm\,V^{-1}$$

and

$$(\Phi r_{ph})_{22} = 0.02 \ , \quad (\Phi r_{ph})_{31} = 0.3 \ , \quad (\Phi r_{ph})_{33} = 0.2 \ \text{Å} \ .$$

According to this interpretation it seems reasonable to assume that the field E_{ph} depends only on the local asymmetriy of filled traps and on the polarization of incident light. The mobility μ is determined by the band structure. Then the experimental results indicate that the time τ' is also independent of the concentration of filled and empty traps. This behavior of photovoltaic effects is completely different from that of usual photoconductivity, which will be described in the next section.

The maxima of the photovoltaic constant and of the mean free path differ from the absorption maximum, which is observed at 2.6 eV for $LiNbO_3$: Fe. The unidirectional charge transport properties increase with increasing photon energy even in the spectral region in which the transition probability decreases.

Photovoltaic effects have been observed only for intervalence transfers from the filled traps to the conduction band. No photovoltaic currents have been detected which result from charge transfers and subsequent hole migration. An estimate of the measuring accuracy indicates that photovoltaic effects due to hole migration – if they exist at all – are at least three orders of magnitude smaller than that of electron migration [5.42].

Theoretical descriptions of these photovoltaic effects and of the underlying physical charge transport mechanism have been given by *Heyszenau* [5.43], *von Baltz* et al. [5.44–48] and *Belinicher* and *Sturman* [5.49].

5.2.3 Photoconductivity

A further mechanism responsible for the light-induced charge transport in electro-optic crystals is the drift of optically excited charge carriers in electric fields, which is characterized by the photoconductivity σ_p. This quantity is given by

$$\sigma_p = ne\mu = g\tau e\mu \ ,$$

where n is the density, e the charge, μ the mobility, g the generation rate and τ the lifetime of excited carriers.

We consider the case that extrinsic impurity centers act as filled and empty traps and that the number of these traps is not much altered by excitation and trapping processes. Then the generation rate is proportional to the light intensity and to the concentration c_f of filled traps supplying the charge carriers. The lifetime on the other hand is inversely proportional to the concentration c_e of empty traps capturing the charge carriers. The mobility is mainly determined by the band structure and is independent of the impurity centers in this approximation. We thus obtain

$$\sigma_p \sim \frac{Ic_f}{c_e} \ .$$

This simple model is completely confirmed by measurements of Fe- and Cu-doped $LiNbO_3$ and $LiTaO_3$ crystals [5.31, 50, 51]. A relation $\sigma_p \sim I$ is found, in contrast to the relation $\sigma_p \sim I^{1/2}$ which is expected for intrinsic excitation and recombination processes.

In the visible spectral region Fe^{2+} and Cu^+ centers act as filled traps; empty traps are Fe^{3+} and Cu^{2+} centers. Figures 5.9–11 show experimental results indicating a relation $\sigma_p \sim c_f/c_e$. Furthermore, these figures illustrate that a similar behavior is observed in the near uv region for $c_{Fe^{2+}}/c_{Fe^{3+}}$ and $c_{Cu^+}/c_{Cu^{2+}}$

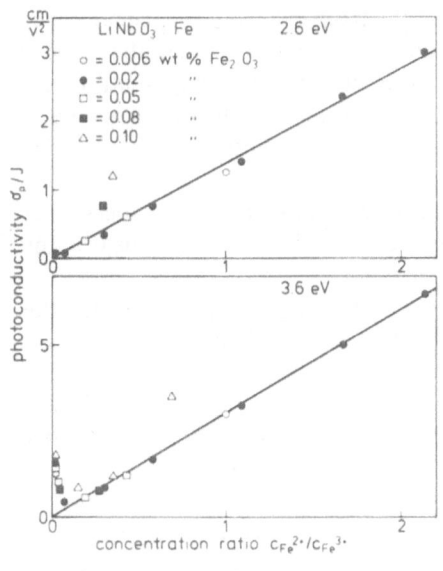

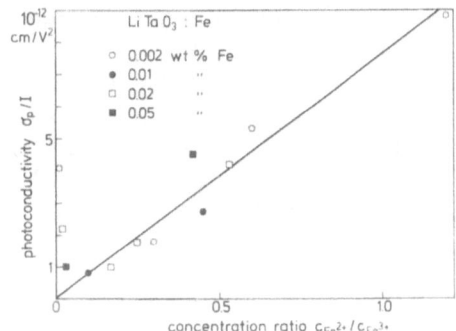

◀ **Fig. 5.9.** Concentration dependence of photoconductivity in $LiNbO_3$: Fe at 2.6 and 3.6 eV [5.50]

Fig. 5.10. Concentration dependence of photoconductivity in $LiTaO_3$: Fe at 3.5 eV [5.51]

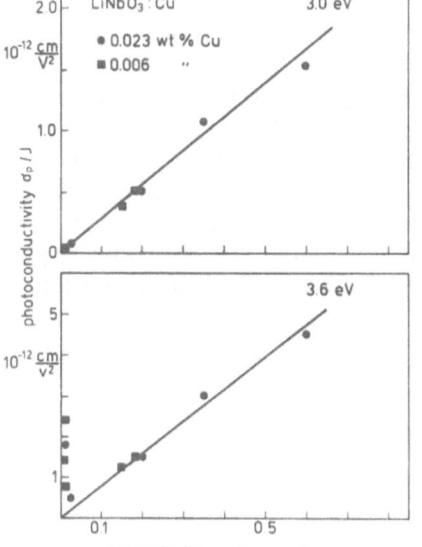

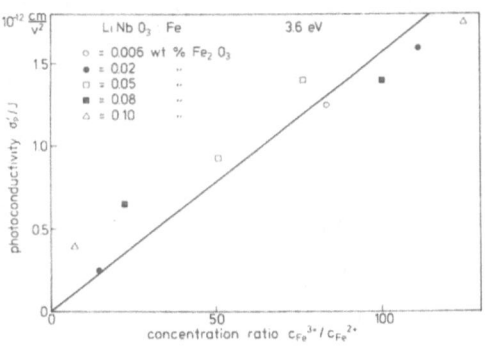

◀ **Fig. 5.11.** Concentration dependence of photoconductivity in $LiNbO_3$: Cu at 3.0 and 3.6 eV [5.31]

Fig. 5.12. Concentration dependence of additional photoconductivity contribution in $LiNbO_3$: Fe [5.51]

ratios larger than about 0.1. For smaller ratios a considerable increase of σ_p is measured, indicating additional contributions σ_p'. This quantity σ_p' is obtained by subtracing the contribution proportional to $c_{Fe^{2+}}/c_{Fe^{3+}}$ or $c_{Cu^+}/c_{Cu^{2+}}$ from the entire photoconductivity. In Fig. 5.12 the quantity σ_p' for $LiNbO_3$: Fe crystals is plotted versus the ratio $c_{Fe^{3+}}/c_{Fe^{2+}}$. Within the measuring accuracy a linear dependence $\sigma_p' \sim c_{Fe^{3+}}/c_{Fe^{2+}}$ is observed. This experimental result confirms the suggestion [5.52] that the contribution σ_p' results from electronic charge transfer processes from oxygen π-orbitals to Fe^{3+} ions. The holes migrate in the valence band until they are trapped by Fe^{2+} ions. In this case the following relations are valid: $g \sim c_{Fe^{3+}}$, $\tau \sim 1/c_{Fe^{2+}}$ and hence $\sigma_p' \sim c_{Fe^{3+}}/c_{Fe^{2+}}$. With increasing Fe^{2+} content σ_p' decreases and for $c_{Fe^{2+}}/c_{Fe^{3+}} > 1$ excitations on Fe^{2+} ions predominate.

The experimental results yield information on the transport properties of electrons and holes. For an equal number of Fe^{2+} and Fe^{3+} centers one obtains $\sigma_p(\text{holes})/\sigma_p(\text{electrons}) = g_h\mu_h\tau_h/g_e\mu_e\tau_e = 0.005$.

The photoconductivity increases with increasing temperature in the range investigated between 25 and 250 °C [5.31]. Experimentally a relation $\sigma_p \sim \exp(-W/k_BT)$ is observed. The activation energy W is nearly independent of the dopants as verified for Fe, Cu, Mn and Co. The values 0.20 eV for $LiNbO_3$ and 0.16 for $LiTaO_3$ have been measured. The activation energy is mainly influenced by the temperature dependence of the mobility μ.

The measurements described above have been performed with unpolarized light to obtain as much intensity as possible. The photoconductivity values of doped $LiNbO_3$ and $LiTaO_3$ are only weakly influenced by light polarization and by crystal geometry. The exact dependences, however, are not yet known in detail.

Furthermore it should be noted that the crystals used for these investigations had been grown from the congruent melt. A few samples have been analyzed and the results indicate that the crystals exhibit the congruent composition with a Li deficit as well.

The measurements lead to the following model of the light-induced charge transport for $LiNbO_3$:Fe crystals (Fig. 5.13): One observes on the one hand optical excitations of electrons from Fe^{2+} centers to the conduction band giving rise to a bulk photovoltaic effect. After a time τ' the electrons lose their directional properties, diffuse according to concentration gradients, and can be

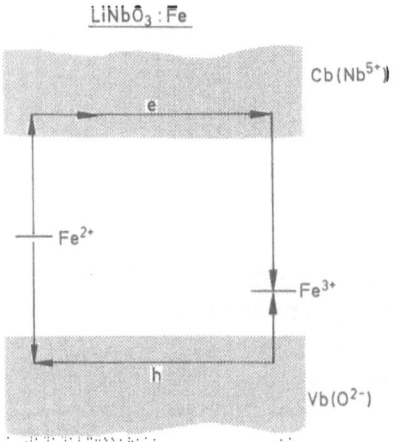

Fig. 5.13. Optically induced charge transport in $LiNbO_3$:Fe [5.31]

strongly influenced by external electric fields. For this migration the lifetime τ is inversely proportional to the Fe^{3+} concentration. On the other hand, in the near uv region one also observes excitations from the oxygen states of the valence band to Fe^{3+} centers and migration of the generated holes. The hole lifetime is inversely proportional to the concentration of Fe^{2+} centers annihilating the holes.

These conclusions have been derived from conventional photovoltaic and photoconductivity measurements using incoherent light. The results are in perfect agreement with those of holographic investigations of the light-induced charge transport which are described in the next section.

The model of the light-induced charge transport can also be transferred to two-photon experiments. These experiments were suggested and first performed by *von der Linde* et al. [5.53, 54] to avoid unwanted erasure of stored holograms during read-out. The holograms are recorded by the use of two photons, can be read nondestructively by one-photon processes and can be erased utilizing two photons again.

Investigations of the light-induced charge transport confirm the occurrence of processes analogous to those described above. A detailed study of $LiTaO_3 : Fe$ crystals containing Fe^{2+} and Fe^{3+} ions has been carried out [5.55]. Under the influence of the crystal field the ^5D level of the free Fe^{2+} ion splits into a $^5T_2$ ground state and a ^5E excited state, giving rise to a broad absorption band centered at 1.1 eV. The 1.06 μm (1.17 eV) emission line of a Nd:YAG laser fits this absorption band well. Simultaneous irradiation with the doubled laser frequency (0.53 μm) causes electron transitions from the ^5E state to the conduction band. The excited electrons migrate until they are trapped by Fe^{3+} centers.

This model is confirmed by measurements of the concentration dependence of the two-photon sensitivity

$$S^{TP} = \frac{\partial(\Delta n_3)}{\partial(I_{1.06} I_{0.53} t)}\bigg|_{t=0} ,$$

where $I_{1.06}$ and $I_{0.53}$ are the light intensities at 1.06 and 0.53 μm. The quantity S^{TP} has been derived to be proportional to the absorption coefficient $\alpha_{1.06}$ at 1.06 μm and therefore to $c_{Fe^{2+}}$ [5.55]. Figure 5.14 shows experimental results supporting this relation.

In addition, $LiNbO_3 : Cr$ crystals have been investigated [5.56], yielding the following picture for the light-induced charge transport. The crystals contain essentially Cr^{3+} ions and presumably a small number of Cr^{4+} ions. By two-step processes electrons of Cr^{3+} ions are excited from the $^4A_2$ ground state to the $^4T_2$ excited state and then to the conduction band. The electrons are captured by Cr^{4+} ions forming Cr^{3+} ions again or by other unknown traps forming color centers.

Recently a hundredfold increase in the photoconductivity of $LiNbO_3$ has been reported on doping the crystals with 4.5 at.% MgO or more [5.57]. By these

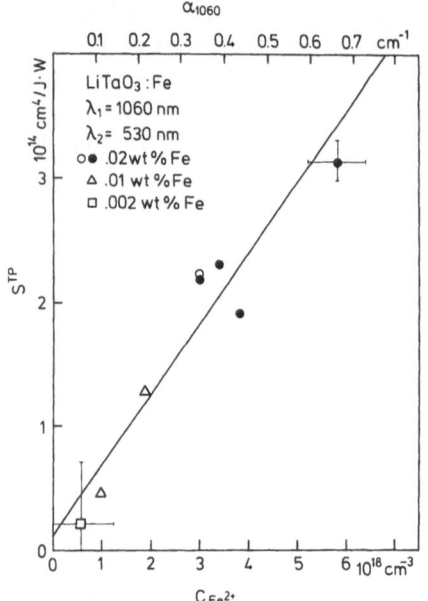

Fig. 5.14. Two-photon sensitivity S^{TP} versus Fe^{2+} concentration $c_{Fe^{2+}}$. The absorption constant at $\lambda = 1.06 \, \mu m$ is shown on the upper scale. S^{TP} values represented by full symbols are determined by multiple measurements at various intensity levels, the open symbols denote S^{TP} values determined by one measurement only [5.55]

means the resistance of the crystals to unwanted photorefractive damage has been greatly improved. The details of the mechanisms involved are not yet understood.

5.2.4 Holographic Investigation of Light-Induced Charge Transport

Detailed information on light-induced charge transport properties of electro-optic crystals can be obtained from holographic experiments. In particular, electron and hole contributions to photoconductivity can be determined. In this section we present an approximate description [5.42]; more complete treatments that also take dynamical effects into account have been given, too [5.58–60].

The method [5.42, 61] proposed to determine photoinduced carriers is based on the generation of space-charge fields by diffusion currents alone. If oppositely charged charge carriers (electrons and holes) contribute to diffusion, the current density is reduced compared to single-species diffusion because electrons and holes migrate in the same direction along the concentration gradient. Therefore the value of the photoinduced space-charge field, and hence the refractive index change, is reduced, too, which allows the ratio of electrons and holes involved in the diffusion process to be determined. The dominant polarity can be obtained from the phase shift between refractive index and light patterns and the sign of the electro-optic coefficient.

For a quantitative evaluation we consider the current density $j(z)$ along the c-axis of an electro-optic crystal induced by illumination with a sinusoidal light intensity distribution $I(z) = I_0 (1 + m \cos Kz)$ resulting from the superposition of

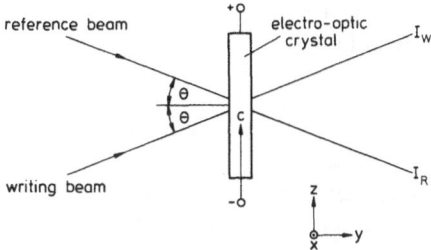

reference beam

electro-optic crystal

θ
θ

I_w

c

writing beam

I_R

+o

-o

z
x
y

Fig. 5.15. Experimental setup for the holographic measurements [5.42]

two plane waves (writing and reference beam, Fig. 5.15). The quantity m is the modulation index and $K = 4\pi \sin\theta/\lambda$ is the spatial frequency, determined by the light wavelength λ and the half interference angle θ.

In the absence of external drift fields there remain two current contributions:

1) The photovoltaic current density $j_{ph}(z) = \kappa\alpha I(z)$, where κ is the photovoltaic constant and α the absorption coefficient.
2) The diffusion current density $j(z) = eD\partial n(z)/\partial z$, where n is the photoinduced free carrier concentration and D the diffusion constant determined by the mobility $\mu = eD/k_B T$ according to the Einstein relation.

The redistribution of charge carriers due to the generation of internal space-charge fields $E_{sc}(z)$ can be described by the drift current density $j(z) = [\sigma_p(z) + \sigma_d]E_{sc}(z)$, where $\sigma_p(z) = \sigma_{p0}(1 + m\cos Kz)$ is the photoconductivity and σ_d the dark conductivity.

The photovoltaic current can be completely compensated by applying an external electric field E_{sc} opposite but equal in mangnitude to the photovoltaic saturation field $E_{ph} = \kappa\alpha I_0/\sigma_{p0}$ [5.52]. In this case only diffusion currents have to be taken into account for the generation of space-charge fields.

Generally, electrons and holes may contribute to diffusion. For small diffusion lengths ($r_d \ll K^{-1}$) the photoinduced free carrier concentrations n_e and n_h for electrons and holes, respectively, can be written

$$n_{e,h} = n_{0e,h}(1 + m\cos Kz) \ .$$

Then the current density along the z-direction reads

$$j(z,t) = -\frac{k_B T}{e}(\sigma_{p0e} - \sigma_{p0h})mK\sin Kz$$

$$+ [\sigma_{pe}(z) + \sigma_{ph}(z) + \sigma_d]E_{sc}(z,t) \ . \tag{5.1}$$

The time dependence of $E_{sc}(z,t)$ is derived using the continuity and Poisson equations [5.62]:

$$E_{sc}(z,t) = \frac{k_B T\hat{\sigma}_p m'K\sin Kz}{e(1 + m'\cos Kz)}\left[1 - \exp\left(-\frac{[\sigma_{pe}(z) + \sigma_{ph}(z) + \sigma_d]t}{\varepsilon\varepsilon_0}\right)\right] \tag{5.2}$$

with $\hat{\sigma}_p = (\sigma_{p0e} - \sigma_{p0h})/(\sigma_{p0e} + \sigma_{p0h})$ and $m' = m/[1 + \sigma_d/(\sigma_{p0e} + \sigma_{p0h})]$. The space-charge field $E_{sc}(z, t)$ yields a refractive index change

$$\Delta n_3(z, t) = -\tfrac{1}{2} n_3^3 r E_{sc}(z, t) , \tag{5.3}$$

where r is the corresponding electro-optic coefficient.

In the following only the saturation value Δn_{3s} is considered. The fundamental component Δn_{3fs} is obtained by a Fourier expansion of (5.2):

$$\Delta n_{3fs}(z) = -\frac{1}{2} n_3^3 r \frac{k_B T}{e} \hat{\sigma}_p m' f(m') K \sin Kz \tag{5.4}$$

with $f(m') = 1 + m'^2/4 + m'^4/8 + \dots$.

The corresponding first-order diffraction efficiency η of a volume phase grating reads

$$\eta = \frac{1}{2} \left(1 - \frac{1}{\cosh\left(\dfrac{2\Delta n_{3fs} \pi d}{\lambda \cos\theta}\right)} \right) . \tag{5.5}$$

Two points are essential: the refractive index change is reduced if both types of charge carriers are present and the refractive index pattern is phase shifted by $\pm\pi/2$ with respect to the intensity distribution of the incident light if either electrons or holes are the dominant charge carriers.

This phase shift causes interactions between the incident beams, as was discovered by *Staebler* and *Amodei* [5.63]. For the geometry shown in Fig. 5.15 a shift by $+\pi/2$ leads to an increase of the transmitted intensity I_W and a decrease of I_R, a shift by $-\pi/2$ to a decrease of I_W and an increase of I_R.

Using the results of a nonlinear coupled wave theory of *Vahey* [5.64], the saturation value of the change ΔI_R for a phase shift of $\pm\pi/2$ is given by

$$\frac{\Delta I_R}{I_R} = \mp \tanh\left(\frac{2\Delta n_{3fs} \pi d}{\lambda \cos\theta}\right) . \tag{5.6}$$

Thus the measurement of the sign of $\Delta I_R/I_R$ yields the sign of $\hat{\sigma}$. The value of $\hat{\sigma}$ can be obtained from the measurement of Δn_{fs} according to (5.4 and 5) or from the measurement of $\Delta I_R/I_R$ according to (5.4 and 6).

For an experimental test of the holographic method LiNbO$_3$:Fe crystals have been used [5.42] because the impurity centers and the optical transitions are well known. The measurements were performed at a wavelength $\lambda = 350.7$ nm using a krypton laser and a holographic setup that was similar to that schematically outlined in Fig. 5.15. The intensities of the reference and writing beams were equal, the modulation index was reduced to $m = 0.35$ by a noninterfering third beam, yielding $f(m) \approx 1$ in (5.4). The crystals were illuminated homogeneously, and the impedance of the voltage source was much smaller than the crystal impedance.

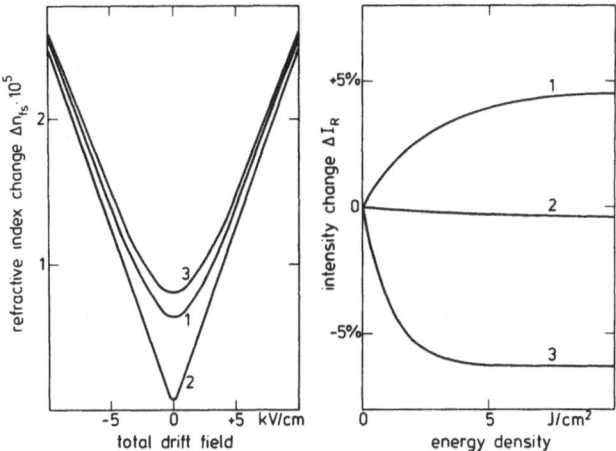

Fig. 5.16. Saturation value of refractive index change versus total drift field (left) and change of reference beam intensity versus incident energy density (right) for three $LiNbO_3$: Fe crystals of different Fe^{2+}/Fe^{3+} ratios (thickness $d = 0.37$ mm, light polarization perpendicular to the c-axis). Curve 1: $Fe^{2+}/Fe^{3+} = 0.012$; curve 2: $Fe^{2+}/Fe^{3+} = 0.08$; curve 3: $Fe^{2+}/Fe^{3+} = 2.1$ [5.42]

Experimental results are shown in Fig. 5.16 for three $LiNbO_3$: Fe (0.005 wt% Fe) crystals with various Fe^{2+}/Fe^{3+} ratios between 0.012 and 2.1. The saturation value of refractive index change has a minimum for a total drift field $E = E_0 - E_{ph} = 0$. The remaining saturation index change results from diffusion currents only. The corresponding intensity change $\Delta I_R/I_R$ at the compensation point is also shown in Fig. 5.16. The change of the sign $\Delta I_R/I_R$ clearly indicates a change of the majority carriers. For Fe^{2+}/Fe^{3+} ratios larger than 0.08, electron processes predominate. The relative contribution of electrons and holes may be evaluated using the values of Δn_{3fs} or $\Delta I_R/I_R$ at the compensation point.

A quantitative comparison of the results with those of direct photoconductivity measurements is presented in Fig. 5.17. From these measurements and an analysis outlined in Sect. 5.3.3 electron and hole contributions in $LiNbO_3$: Fe have been derived [5.50]:

$$\sigma_{p0} = \sigma_{p0e} + \sigma_{p0h} = I_0 \left(f_1 \frac{Fe^{2+}}{Fe^{3+}} + f_2 \frac{Fe^{3+}}{Fe^{2+}} \right)$$

with $f_1 = 3 \times 10^{-14}$ mV^{-2} and $f_2 = 1.7 \times 10^{-16}$ mV^{-2} at 350 nm.

These relations yield the solid and dashed curves in Fig. 5.17. The full and open circles represent the results of the holographic method, and good agreement is found. The picture of the light-induced charge transport developed in Sect. 5.3.3 is completely confirmed.

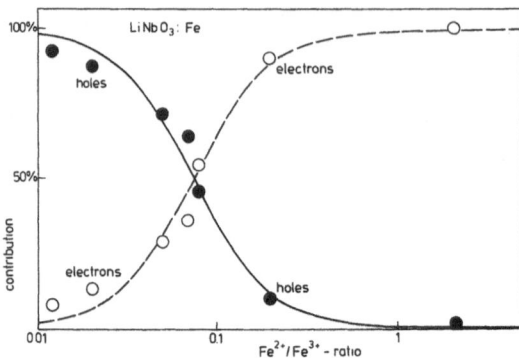

Fig. 5.17. Electron and hole contributions to photoconductivity versus Fe^{2+}/Fe^{3+} ratio in $LiNbO_3$: Fe crystals. Solid and dashed lines represent results of direct photoconductivity measurements, full and open circles represent results of the holographic method [5.42]

The determination of the different contributions to the charge transport from direct photoconductivity measurements requires precise knowledge of the electronic centers and transitions involved. Furthermore, many measurements on crystals with various Fe concentrations and various Fe^{2+}/Fe^{3+} ratios have to be performed [5.50]. The holographic method does not suffer from these disadvantages and can be applied to crystals where details of centers and transitions are not known.

Both $LiNbO_3$: Cu and $LiTaO_3$: Fe crystals have also been investigated by the holographic method [5.42, 61]. In the visible spectral region photoconductivity contributions due to mobile electrons generated by transitions from the impurity centers to the conduction band have been found.

In the near uv region at 350 nm additional hole contributions have been obtained, indicating transitions from the oxygen valence band to the impurity centers. The photoconductivity due to holes is small in strongly reduced crystals and increases considerably with decreasing number of Fe^{2+} and Cu^+ centers. The largest hole transport is measured in strongly oxidized crystals which have been annealed in a pure oxygen atmosphere at temperatures about 150 °C below the melting point for 20 h. These crystals contain less than 1% of the impurity centers as Cu^+ or Fe^{2+}, respectively. Hole contributions up to 80% for $LiNbO_3$: Cu and 90% for $LiTaO_3$: Fe have been measured.

5.2.5 Protons and Thermal Fixing

The storage of volume phase holograms in electro-optic crystals like $LiNbO_3$ and $LiTaO_3$ suffers from the disadvantage of destructive read-out: The crystal is illuminated homogeneously and the space charge fields carrying the information are partly compensated.

To avoid unwanted erasure during read-out two-photon processes may be utilized with hologram recording as described in Sect. 5.2.3. A further possibility for the stabilization of stored information is thermal fixing, which was discovered by *Amodei* and *Staebler* [5.65]. A LiNbO$_3$: Fe crystal is heated during hologram recording or afterwards to temperatures between 100 and 200 °C. By these means thermally activated ions neutralize the electronic space-charge fields. Upon cooling to room temperature and redistributing the electronic pattern by homogeneous illumination, a hologram due to the ionic charge pattern remains which is stable against illumination and can be erased only by further heating.

The nature of the mobile ions which are responsible for the thermal fixing process has been the subject of several investigations. *Williams* et al. [5.66] have ruled out Fe and Nb ions, and have found a change of the Si concentration in the illuminated region of the crystal. However, no influence of the Si ions on the fixing rate could be ascertained. *Bollmann* and *Stöhr* [5.67] have suggested OH$^-$ ions on account of activation energy measurements, but no experimental evidence for the participation in thermal fixing could be given. *Meyer* et al. [5.68] have concluded that Si and OH$^-$ ions are in disagreement with their results for the mobility and the concentration of the ions of interest. These authors favor Li, Nb, or O vacancies. Finally, *Vormann* et al. [5.14] have shown that protons neutralize the electronic charge pattern during thermal fixing in LiNbO$_3$: Fe. The protons form OH$^-$ centers with oxygen ions and can be detected via vibrational absorption bands.

Protons and OH$^-$ ions are incorporated into LiNbO$_3$ crystals presumably due to the humidity of the growth atmosphere. A stretching vibrational band in the infrared region is observed near 2.87 μm [5.13, 69], polarized perpendicular to the polar *c*-axis.

Various crystals, nominally pure and containing Fe, Cu and Mn impurities, have been investigated [5.14]. In the as-grown state, absorption constants between 1 and 2 cm^{-1} have been measured at 2.87 μm.

The determination of the OH$^-$ concentration from absorption measurements requires knowledge of the oscillator strength. Using the value derived by *Johnson* et al. [5.70] a OH$^-$ content between 1.5 and 3×10^{18} cm^{-3} is deduced. The results of *Bollmann* et al. [5.71] yield $(3-6) \times 10^{19}$ cm^{-3}.

By annealing treatments in a water-free atmosphere the OH$^-$ content is considerably reduced [5.14]. A treatment of 20 h at 1000 °C in an oxygen or argon atmosphere diminishes the absorption constant by a factor of about 10. A considerable increase (factor of 30) of the OH$^-$ band could be obtained by heating in a water atmosphere.

The fixing ability of LiNbO$_3$: Fe decreases considerably if the OH$^-$ content is decreased by annealing the crystals in a water-free atmosphere. To demonstrate this, several facts have to be taken into account [5.14]. The experimental procedure was begun by recording holograms at room temperature using two beam interference of plane waves. Subsequently, the crystal was heated to 160 °C, cooled down to room temperature again and then illuminated homogeneously. By this method rather low read-out efficiencies were obtained which were insufficient for quantitative interpretation. A considerable increase of efficiency is possible by heating during hologram writing. The saturation value

Table 5.2. Decrease of saturation index changes Δn_{3s} and Δn_{3fs} at 488 nm with decreasing OH^- absorption constant α_{OH^-} at 2.87 μm for a $LiNbO_3$: Fe crystal containing 0.014 wt.% Fe [5.14]

α_{OH^-} [cm^{-1}]	Fe^{2+}/Fe^{3+} ratio	Δn_{3s} [$\times 10^{-5}$]	Δn_{3fs} [$\times 10^{-5}$]
1.60 ± 0.03	0.11 ± 0.02	15 ± 1.5	7.2 ± 0.5
0.30 ± 0.03	0.16 ± 0.02	14 ± 1.5	3.0 ± 0.7
0.17 ± 0.03	0.19 ± 0.02	15 ± 1.5	< 0.6

Δn_{3fs} of the refractive index change of the fixed hologram was measured in this way.

A further important point for quantitative experiments is the dependence of the space-charge field on the Fe^{2+} and Fe^{3+} concentration [5.50]. For this reason these quantities were kept constant during the experimental series. A crystal with an Fe content of 0.014 wt.% and an Fe^{2+}/Fe^{3+} ratio of about 0.1 in the as-grown state was used. The OH^- absorption constant, the saturation value of the index change Δn_{3s} at room temperature and the value Δn_{3fs} were determined. Then the crystal was annealed in oxygen for several hours at 1000 °C and afterwards reduced in argon until the Fe^{2+}/Fe^{3+} ratio of 0.1 was approximately attained again. Subsequently α_{OH^-}, Δn_{3s} and Δn_{3fs} were measured and a further annealing treatment begun. The results are summarized in Table 5.2. They demonstrate the decrease of Δn_{3fs} with decreasing OH^- content.

To substantiate the role of hydrogen in thermal fixing the spatial distribution of OH^- ions in fixed patterns has also been investigated [5.14]. A Fourier IR spectrometer was used and measurements of OH^- absorption as a function of position were performed with a $1.0 \times 4.0 \times 4.6$ mm^3 crystal plate. A 0.25 mm slit of 2 mm height was placed in front of the large crystal surface oriented perpendicular to the c-axis. The crystal was moved relatively to the slit with a micrometer positioning device in the direction of the c-axis. The reproducibility was better than 0.1 mm.

The spatial distribution of the OH^- absorption before illumination was carefully measured and showed small variations at different positions. Then the crystal was illuminated for 100 min with a 3 W, 514 nm laser beam. A cylindrical lens imaged the beam in the form of a strip with a width of about 1 mm and 10 mm length onto the crystal perpendicular to the c-axis. During illumination the crystal was heated to 160 °C. Then the OH^- absorption was measured again.

A typical result of the absorption change as a function of crystal position is shown in Fig. 5.18. The illuminated region is indicated by dashed lines. Near the $+c$ border of this region the OH^- content has increased, whereas it has decreased at the $-c$ border.

The OH^- distribution has also been calculated under the assumption that the mobile ions neutralizing the electronic charge are protons [5.14]. The results

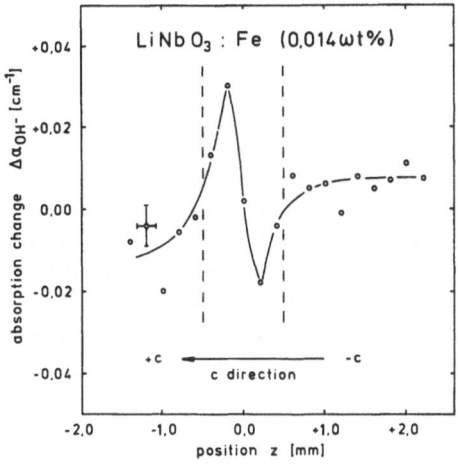

Fig. 5.18. Spatial variation of OH⁻ absorption change at 2.87 μm after illumination of a strip 1 mm wide (indicated by dashed lines) [5.14]

of this calculation are in qualitative agreement with the experimental results given in Fig. 5.18. In particular, the positions of maximum and minimum concentration are well described. This is an indication that the mobile ions are protons and not OH⁻ ions. The migration of the protons may occur in various ways, for instance, either just by hopping of the protons or by an exchange of the positions of OH⁻ and O^{2-} ions. Migration of the OH⁻ ions alone would result in a reversed pattern with a maximum OH⁻ absorption at the $-c$ border of the illuminated region and a minimum concentration at the $+c$ border.

The results described above allow the following interpretation of thermal fixing processes in $LiNbO_3$:Fe. During hologram writing electronic space-charge fields are generated by inhomogeneous illumination. At elevated temperatures (100°–200 °C) protons migrate and neutralize these fields. The protons form OH⁻ centers which can be observed via vibrational absorption.

Protons are not only important for thermal fixing processes in $LiNbO_3$ and $LiTaO_3$. The influence of protons on the photorefractive properties of electro-optic crystals was discovered fairly early [5.69]. In proton-exchanged $LiNbO_3$ waveguides optical damage effects are largely suppressed [5.72, 73]. The stabilization of Fe^{2+} ions by Ti^{4+} ions, which is observed in $LiNbO_3$:Ti waveguides [5.74], is also considerably reduced by an additional annealing of the waveguides in a H_2O atmosphere. Obviously the indiffusion of protons partially cancels the influence of Ti ions [5.74].

5.3 Intrinsic Centers in LiNbO₃ and LiTaO₃

In this section intrinsic defects will be treated, i.e. defects that consist only of constituents present in the normal chemical composition of the crystals.

5.3.1 Overview

In $LiNbO_3$ two such defects are known. We shall first give a rough description of the ESR properties of these centers, then describe ways of producing them, analyze their structural properties, discuss their optical features and finally give information about their influence on the photorefractive behavior of $LiNbO_3$.

These centers were first noticed by ESR following two-photon absorption of $LiNbO_3$ [5.75]. Table 5.3 lists their ESR parameters and Fig. 5.19 gives examples of their ESR spectra with B parallel to the crystal c-axis. Generally, ESR lines in $LiNbO_3$ are rather wide due to unresolved hyperfine interaction with Nb and Li nuclei, which all have magnetic moments. This prevents resolution of closely spaced ESR lines.

Table 5.3. ESR parameters of intrinsic centers in $LiNbO_3$

Center I:	$g = 2.030$	$B \parallel c$	[5.75]
	$g = 2.0294$		[5.76]
	$g = 2.025$	$B \perp c$	[5.75]

Remark: These values are taken from ESR measurements at ~ 9 GHz microwave frequency. Preliminary results obtained at 35 GHz (Schirmer, unpublished), allowing higher resolution, have identified several component lines with g values spreading between 2.003 and 2.047. The symmetry of the g tensors has not yet been ascertained.

Center II:	$g_\parallel = 1.90$,	$A_\parallel (^{93}Nb) = 110 \times 10^{-4}$ cm^{-1}	
	$g_\perp = 1,72$,	$A_\perp (^{93}Nb) = 230 \times 10^{-4}$ cm^{-1}	[5.75]

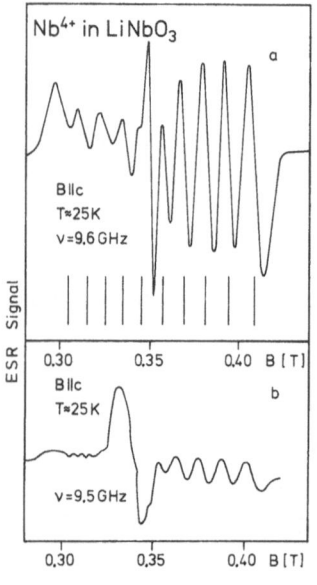

Fig. 5.19a, b. ESR of intrinsic centers in $LiNbO_3$. (a) Reduced undoped $LiNbO_3$ after low temperature illumination with unfiltered Xe arc light. The ten line signal is attributed to Nb^{4+}. The strong line in the middle part of the spectrum does not belong to this center. (b) Undoped $LiNbO_3$ after low temperature, in situ x-irradiation. The strong structure accompanying the Nb^{4+} signal is due to O^--trapped holes

The g values of center I are characteristic of holes trapped at O^{2-} sites in $LiNbO_3$ forming the paramagnetic entity O^-, as seen for example by comparison with the multitude of such centers in other oxides in the literature [5.77]. These examples and the underlying theory show that the hole is subject to a strong, low symmetry crystal field originating from a nearby crystal perturbation charged negatively with respect to the lattice. We shall discuss below that in the present case the field is likely to be caused by a Li vacancy.

Center II is characterized by its ten hyperfine lines identifying interaction of the electron magnetic moment with one 100% abundant $I=9/2$ nucleus, conditions met by ^{93}Nb. The ESR parameters point to a single electron outside closed shells as the origin of the paramagnetism: Nb^{4+} situated in nearly axial environment. We shall first deal with this center and then treat center I.

5.3.2 Nb^{4+} in $LiNbO_3$

As mentioned before, this charge state was first produced by two-photon absorption of undoped $LiNbO_3$ [5.75]. It was also shown [5.75] that x irradiation at low temperatures yields Nb^{4+} – and also the trapped hole center (I). Both excitation processes result in the formation of electron-hole pairs. The hole is subsequently trapped at the defect site to be discussed, whereas the electron can be thought of as being self-trapped as a small polaron at a Nb site. These defects are stable up to temperatures in the region of 100–200 K, where they vanish simultaneously [5.78].

A third production mechanism has been discovered by *Ketchum* et al. [5.79]. They showed that reduction of undoped $LiNbO_3$ in vacuum at elevated temperatures and subsequent illumination with light of $\lambda < 600$ mm [5.80] causes very strong Nb^{4+} ESR, identical to the signals induced the other ways. It is remarkable that no ESR is observed after reduction prior to illumination. Apparently the crystal ground state after reduction is diamagnetic, all spins being paired off.

The symmetry of the hyperfine and Zeeman (g) tensors and the values of their components give some information on the structure of the Nb^{4+} center. The spectra are nearly axially symmetric, consistent with an electron trapped at a Nb site of the lattice. The origin of the slight deviation from axiality is not yet known. The large width of the ESR lines has prevented a full analysis so far in magnetic fields corresponding to ~ 9 GHz microwave frequency. Further information can be expected from measurements at higher frequencies and from ENDOR investigations. The center can still be classified as intrinsic because it results from undoped $LiNbO_3$ in concentrations (up to 4.5×10^{18} cm$^{-3} = 240$ ppm [5.80]) above the background doping.

The g values of the center, $g_\parallel \sim g_e$ (free electron), $g_\parallel > g_\perp$, point to a $(3z^2 - r^2)$-type groundstate with z along the c-axis [5.75]. This wave function is occupied by a single electron. Otherwise different ESR spectra, corresponding to $S=1$ or $S=3/2$ in an axial crystal field, would have been observed. Also the

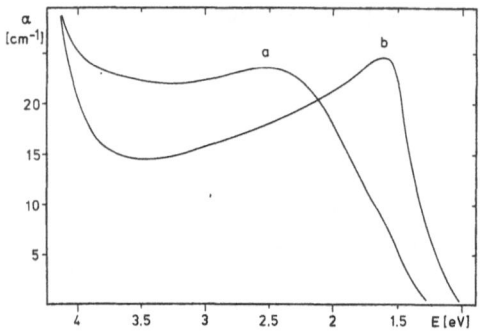

Fig. 5.20. Optical absorption of reduced undoped LiNbO$_3$ (1000 °C in vacuum, 1 h) at 10 K. *(a)* Before illumination; *(b)* after illumination with unfiltered 150 W Xe arc light

hyperfine interaction is in accord with this assignment; it is rather large, pointing to rather tight localization of the unpaired electron on one Nb site.

Sweeney and *Halliburton* [5.80] have shown that the rise of the Nb^{4+} ESR under short wavelength illumination is accompanied by the simultaneous appearance of a wide (Fig. 5.20) absorption band peaked at 1.6 eV. This band has also been observed by *Arizmendi* et al. [5.78] after x irradiation at low temperatures. It has features expected for the optical behavior of small polarons. The assignment of the band to small polaron absorption is supported by the fact that the peak energy is about four times the activation energy for electron transport in reduced LiNbO$_3$ [5.81], in accordance with small polaron theory [5.82]. The band at 2.5 eV, observed in reduced LiNbO$_3$ [5.78 (and refs. theirein), 5.79] was previously ascribed by *Clark* et al. [5.28] to a Nb^{4+} crystal field transition. The recent findings described here exclude its identification with isolated Nb^{4+}, since no ESR is correlated with this optical band. Sweeney and Halliburton propose the assignment to *F* centers, i.e. to oxygen vacancies filled with two electrons paired diamagnetically. The changeover during illumination into the 760 nm band is then interpreted [5.80] as the transfer of one of these electrons onto a Nb site leading to the ESR of the Nb^{4+} small polaron. This raises the difficulty of explaining why no ESR of an unpaired electron left back in the oxygen vacancy is observed. Furthermore, the presence of oxygen vacancies in LiNbO$_3$ is questioned by *Smyth* [5.83] for chemical reasons. It should be added that ESR of transition metal ions associated with oxygen vacancies has, to the best of our knowledge, not been observed so far in LiNbO$_3$. The presence of oxygen vacancies in oxygen perovskites has been proved in this way [5.8].

A way out of this impasse would be to assign the 500 nm band to bipolarons (diamagnetic pairs of polarons on neighboring Nb sites). This avoids the difficulties of the F-center model since now both electrons, that transferred by optical absorption as well as that left behind, will lead to the same Nb^{4+} spectrum. It is remarkable that the situation in reduced LiNbO$_3$ is rather similar to that found in reduced WO$_3$ [5.84]: At low temperatures, its ground state is diamagnetic and accompanied by a wide band at 1.1 eV. After illumination a new band at 0.7 eV arises, correlated with the ESR of W^{5+}. *Schirmer* and *Salje* [5.84] explain this situation by proposing a bipolaron crystal ground state.

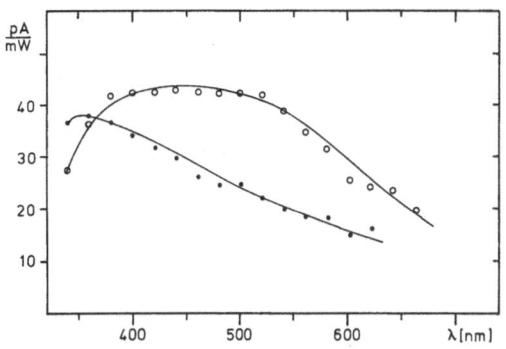

Fig. 5.21. Dependence of the photo-voltaic current in reduced undoped LiNbO$_3$ (1000 °C in vacuum, 1 h) on wavelength. (●:) Measurement at 295 K (Nb^{4+} not stable); (○:) at 80 K after illumination with unfiltered 150 W Xe arc light (Nb^{4+} stable)

The Nb^{4+} state has been found to be involved in the photorefractive response of LiNbO$_3$. *Von der Linde* et al. [5.85] showed that two-photon photorefractive recording in undoped LiNbO$_3$ is accompanied by the appearance of Nb^{4+} ESR and the adjoining optical absorption. Since Nb^{4+} is not stable at room temperature, it is only the transient component of the photorefractive effect that is correlated to Nb^{4+} at 300 K.

Recent findings by *Juppe* et al. [5.86] indicate that optical absorption in the diamagnetic ground state of reduced LiNbO$_3$ as well as in the illumination-induced paramagnetic state causes bulk photovoltaic currents, which have distinctly different spectral dependence in the two cases (see Fig. 5.21).

5.3.3 The Trapped Hole Center in Undoped LiNbO$_3$

The trapped hole center in undoped LiNbO$_3$ was found with ESR after two-photon and x irradiation at low temperatures [5.86]. In addition, *Halliburton* et al. [5.76] have identified it in undoped LiNbO$_3$ irradiated with 1.7 MeV electrons at 77 K. Because of the large inhomogeneous linewidths only a single ESR line is observed at 9 GHz microwave frequency, see Fig. 5.20. Its shape, however, indicates that it is composed of several components. Indeed, preliminary measurements at 35 GHz [5.87] have resolved several lines with g values lying between 2.003 and 2.047. The information available so far however is not yet sufficient to fix the geometry of the relevant g tensors and of the centers correlated with them.

In spite of this, the sign and sizes of the deviations Δg of the measured g values from that of the free electron, $\Delta g = g - g_e$, are clear indications of a hole trapped at O^{2-} close to a defect charged negatively with respect to the lattice [5.75, 77]. Lithium vacancies, V_{Li}, in LiNbO$_3$ are of such a type. The following observations make it likely that O$^-$ is associated with such vacancies. Firstly, all models proposed so far to explain the Li deficit of LiNbO$_3$ [5.88] postulate, among other features, the existence of Li vacancies. Secondly, the formation of these centers under high energy electron irradiation, as observed by ESR [5.76], can be understood as the consequence of Li ions being ejected from their initial sites. The much heavier Nb ions do not get sufficient energy to move from their

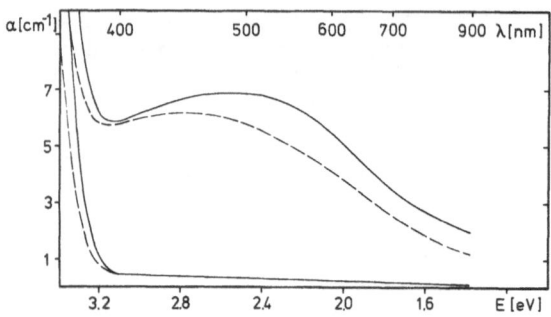

Fig. 5.22. Optical absorption of O^--trapped holes in low temperature (10 K) x-irradiated undoped $LiNbO_3$. The 1.6 eV Nb^{4+} band was not present in this crystal. Solid lines: σ-polarization; dashed lines: π-polarization

sites and therefore no increase of Nb^{4+} under this treatment is observed. Thirdly, maximal g values of the order of the observed ones (2.047) are expected for O^- near a Li vacancy in $LiNbO_3$. For O^- next to a Mg vacancy in MgO, V_{Mg} (twofold negative with respect to the lattice), a maximal g value of 2.0385 is observed; for O^- next to Li^+ on a Mg site, Li_{Mg}, 2.0542 [5.77]. The size of g decreases with increasing electric field at the O^- site [5.77]. It is expected to be higher for V_{Li} in $LiNbO_3$ (Li–O distance 2.01 Å) than for Li_{Mg} in MgO (distance 2.10 Å). A value of 2.047, as observed, thus appears to be consistent with the model.

The ESR of the present trapped hole center is accompanied by a wide optical absorption band with a peak near 500 nm (Fig. 5.22) (fortuitously close to the positions of the absorption maxima in both undoped reduced and Fe-doped reduced $LiNbO_3$). This band has been clearly identified by *Halliburton* et al. [5.76]. Previous absorption measurements [5.75] of $LiNbO_3$ after two-photon and x-ray absorption at low temperatures yielded spectra which can be understood as superpositions of the Nb^{4+} and trapped hole absorptions. A detailed investigation of such spectra has been reported by *Arizmendi* et al. [5.78].

Such absorptions are typical for trapped holes in oxides and have been investigated for several such centers by *Schirmer* [5.89]. The hole in each case is self-trapped as a small polaron at one of several equivalent oxygen sites near a defect. Light can induce transitions from one site to neighboring equivalent sites. The positions and widths of such bands are determined by the stabilization energy accompanying hole trapping. They are generally found in the region between 2 and 3 eV (MgO: V_{Mg}, for example, has an absorption peak at 2.3 eV). The creation of the trapped hole center in undoped $LiNbO_3$ accompanies two-photon photorefractive recording in this material [5.84]. Kurz and von der Linde have claimed (mentioned in [5.85]) that these defects can act as photorefractive centers (see also [5.90]).

5.3.4 Intrinsic Defects in $LiTaO_3$

With respect to $LiTaO_3$ the situation regarding intrinsic defects is rather similar to that encountered in $LiNbO_3$.

Recently *Kappers* et al. [5.91] have investigated reduced $LiTaO_3$ and observed that a broad optical absorption band at 460 nm increases with increasing degree of reduction; its features are similar to that of the 500 nm band in reduced $LiNbO_3$: again no paramagnetism accompanies the 460 nm band. It can be bleached at 77 K with light of a Xe arc in favor of absorption at 570 nm, which is joined by Ta^{4+} ESR. Again *Kappers* et al. [5.91] attribute the 460 nm absorption to oxygen vacancies containing two electrons, from which one electron is transferred to a host lattice Ta^{5+} ion, forming Ta^{4+}. The authors recognize the difficulty in this model that ESR of the F^+ center, one electron in the oxygen vacancy, is not observed. Optical absorption of reduced $LiTaO_3$ has been investigated in less detail by *Antonov* et al. [5.92]. The photoelectric properties of reduced $LiTaO_3$, e.g., photovoltaic current, photoconductivity and photorefraction, have been studied by *Avakyan* et al. [5.93].

Trapped hole centers have also been observed in $LiTaO_3$ [5.94]. Irradiation with 1.7 MeV electrons at 80 K leads to ESR at $g = 2.0224$ ($B \parallel c$), which, however, shows hyperfine splitting due to two equivalent $I = 1/2$ nuclei, possibly hydrogen. This spin resonance is accompanied by an absorption band at 470 nm, typical for holes trapped at O^{2-}.

5.4 Centers in Various Photorefractive Crystals

Beside the standard examples $LiNbO_3$ and $LiTaO_3$, a large number of photorefractive crystals have been studied. Of particular interest are the ferroelectric perovskites $KNbO_3$, $KTa_{1-x}Nb_xO_3$ and $BaTiO_3$, the tungsten-bronze-type crystals $Sr_{1-x}Ba_xNb_2O_6$ and $Ba_2NaNb_5O_{12}$, the nonferroelectric sillenites $Bi_{12}SiO_{20}$ and $Bi_{12}GeO_{20}$, and finally the semiconductors GaAs and InP. Several centers have been identifed, but the influence on the light-induced charge transport and the involvement in photorefractive processes remain largely unclear in many cases.

The photorefractive properties of doped and undoped $KNbO_3$ crystals have been investigated in detail by *Günter* et al. [5.95–98]. For the light-induced charge transport in $KNbO_3$:Fe a model analogous to that in $LiNbO_3$:Fe has been proposed. Electrons of Fe^{2+} ions are optically excited and migrate in the conduction band until they are finally trapped by Fe^{3+} ions. With reducing treatments a characteristic absorption band at 2.55 eV grows, which has been attributed to intervalence transfers from $d\varepsilon$ orbitals of Fe^{2+} ions to $d\varepsilon$ niobium orbitals. The photocurrent is also increased by reducing treatments [5.95]. A quantitative correlation is seriously impeded by the rather complicated reduction and poling procedures necessary in the case of $KNbO_3$. With the help of ESR measurements two centers have been identified in Fe-doped $KNbO_3$ [5.99, 100]: Fe^{3+} ions at Nb^{5+} sites without local charge compensation and possibly $Fe^{3+} - V_0$ centers with the Fe^{3+} ions also at Nb^{5+} sites. However, the above model does not completely describe the light-induced charge transport in

$KNbO_3$. Intrinsic centers may also be involved. An indication of this is the measured dependence of the space-charge field on light intensity [5.97]. The photoconductivity does not depend linearly on light intensity in contrast to the results obtained with doped $LiNbO_3$.

Excellent photorefractive properties have been reported for $KTa_{1-x}Nb_xO_3$ (KTN) crystals [5.101–104]. Furthermore, linear as well as quadratic electro-optic effects can be utilized with these crystals [5.105]. In Fe-doped samples Fe^{3+} centers have also been found substituting Ta^{5+} or Nb^{5+} ions [5.106]. However, no improvement of the photorefractive properties by transition metal doping has been obtained [5.104]. The results indicate that intrinsic centers are of special importance. In undoped samples the charge carriers for excitation with 458 nm light are electrons, as deduced from beam coupling experiments [5.105]. It has also been suggested [5.107] that the photorefractive properties of KTN crystals are influenced by effects of impurities on the phase-transition temperature.

The situation for $BaTiO_3$ is similar to that of KTN and is discussed in more detail in Chap. 7 of this volume. The crystals have been utilized for holographic storage [5.108, 109] and for four-wave mixing [5.110]. The bulk photovoltaic effect has been investigated [5.111]. By ESR measurements Fe^{3+} ions at Ti^{4+} sites have been discovered [5.112]. It has been shown that these Fe^{3+} ions remain at the center of the octahedra in the ferroelectric phases [5.113], however, the influence of transition metal ions on photorefractive properties could not be elucidated. Electrons and holes have been found to contribute to the charge transport [5.114, 115]. Acceptor concentrations of $(1.9 \pm 0.2) \times 10^{16}$ cm^{-3} [5.116] and $(2.9 \pm 0.6) \times 10^{16}$ cm^{-3} [5.117] have been derived.

Little information is available on the photorefractive centers in tungsten-bronze-type crystals. It is very difficult to obtain samples of sufficiently good quality exhibiting no inhomogeneities. Crystals of $Ba_2NaNb_5O_{15}$ doped with Fe and Mo have been studied with respect to holographic storage and thermal fixing [5.118, 119]. Crystals of $Sr_xBa_{1-x}Nb_2O_6$ (SBN) are of special interest for the electric field control of photorefractive effects [5.120–123]. Considerable improvements of the storage properties have been obtained by Ce doping [5.124, 125].

The nonferroelectric sillenites $Bi_{12}SiO_{20}$ and $Bi_{12}GeO_{20}$ are attractive candidates for real-time optical processing because of their high sensitivity and their fast response [5.124–126]. The photoinduced carriers have been found to be electrons for excitation with green light; holes have also been induced by uv light in heavily Al-doped samples [5.127–129]. Models for charge transport have been proposed [5.127, 130]. The concentration of the trapping centers has been derived to be much smaller than that of the absorbing centers, but the nature of the centers is not known.

Recently photorefractive effects and four-wave mixing have been investigated in the semiconductors InP:Fe, GaAs:Cr [5.131] and GaAs:$EL2$ [5.132, 133]. [In undoped GaAs the dominant deep levels are donor states designated as $EL2$ and presently assigned to arise from As_{Ga} antisite (As on a Ga site [5.134]) defects.] The results of these measurements will be discussed in Chap. 8 of this volume. It has been shown that the dominant charge carriers are electrons. For the light-induced charge transport in InP:Fe a model analogous to that in $LiNbO_3$:Fe has been suggested: Fe^{2+} ions act as filled and Fe^{3+} ions as empty traps [5.131].

References

5.1   A. Ashkin, G.D. Boyd, J.M. Dziedzic, R.G. Smith, A.A. Ballmann, K. Nassau: Appl. Phys. Lett. **9**, 72 (1966)
5.2   F.S. Chen, J.T. LaMacchia, D.B. Fraser: Appl. Phys. Lett. **13**, 223 (1968)
5.3   F.S. Chen: J. Appl. Phys. **40**, 3389 (1969)
5.4   G.W. Ludwig, H.H. Woodbury: In *Solid State Physics*, Vol. 13, ed. by F. Seitz, D. Turnbull (Academic, New York 1962) pp. 223–304
5.5   G.D. Watkins: "Electron Paramagnetic Resonance of Point Defects in Solids, with Emphasis on Semiconductors" in *Point Defects in Solids*, Vol. 2, ed. by J.A. Crawford, L.M. Slifkin (Plenum, New York 1975) pp. 333–392
5.6   J.E. Wertz, J.R. Bolton: *Electron Spin Resonance* (McGraw-Hill, New York 1972)
5.7   B. Henderson, J.E. Wertz: *Point Defects in Alkaline Earth Oxides* (Taylor & Francis, London 1977)
5.8   K.A. Müller: J. de Phys. **42**, 551 (1981)
5.9   J.B. Herrington, B. Dischler, J. Schneider: Solid State Commun. **10**, 509 (1972)
5.10 B. di Bartolo: *Optical Interactions in Solids* (Wiley, New York 1968)
5.11 I.G. Austin, N.F. Mott: Adv. Phys. **18**, 41 (1969)
5.12 O.F. Schirmer: J. de Phys. **41**, C6-479 (1980)
5.13 J.R. Herrington, B. Dischler, A. Räuber, J. Schneider: Solid State Commun. **12**, 351 (1973)
5.14 H. Vormann, G. Weber, S. Kapphan, E. Krätzig: Solid State Commun. **40**, 543 (1981)
5.15 U. Gonser (ed.): *Mössbauer Spectroscopy*, Topics Appl. Phys., Vol. 5 (Springer, Berlin, Heidelberg 1975)
5.16 W. Keune, S.K. Date, I. Deszi, U. Gonser: J. Appl. Phys. **46**, 3914 (1975)
5.17 H. Kurz, E. Krätzig, W. Keune, H. Engelmann, U. Gonser, B. Dischler, A. Räuber: Appl. Phys. **12**, 355 (1977)
5.18 G.E. Peterson, A.M. Glass, T.J. Negran: Appl. Phys. Lett. **19**, 130 (1971)
5.19 J.J: Amodei, W. Phillips, D.L. Staebler: IEEE J. QE-7, 321 (1971)
5.20 W. Phillips, J.J. Amodei, D.L. Staebler: RCA Rev. **33**, 94 (1972)
5.21 J.B. Herrington, B. Dischler, J. Schneider: Solid State Commun. **10**, 509 (1972)
5.22 T. Takeda, A. Watanabe, K. Sugihara: Phys. Lett. **27**A, 114 (1968)
5.23 B. Dischler, A. Räuber: Solid State Commun. **17**, 953 (1975)
5.24 N.F. Evlanova, L.S. Kornienko, L.N. Rashkovich, A.O. Rybaltovskii: Sov. Phys. – JETP **26**, 1090 (1968)
5.25 W. Keune, S.K. Date, U. Gonser, M. Bunzel: Ferroelectrics **13**, 443 (1976)
5.26 V.N. Belogurov, P.E. Senkov: Appl. Phys. **21**, 195 (1980)
5.27 L. Arizmendi, F. Abella, J.M. Cabrera: Ferroelectrics **56**, 75 (1984)
5.28 M.G. Clark, F.J. DiSalvo, A.M. Glass, G.E. Peterson: J. Chem. Phys. **59**, 6209 (1973)
5.29 H. Bunzel, U. Gonser, A. Trautwein: Ferroelectrics **38**, 789 (1981)
5.30 E. Krätzig, R. Orlowski: Appl. Phys. **15**, 133 (1978)
5.31 E. Krätzig, R. Orlowski: Ferroelectrics **27**, 241 (1980)
5.32 D.L. Staebler, W. Phillips: Appl. Phys. Lett. **24**, 268 (1974)
5.33 E. Krätzig, H. Kurz: J. Electrochem. Soc. **124**, 131 (1977)
5.34 A.G. Chynoweth: Phys. Rev. **102**, 705 (1956)
5.35 A.M. Glass, D. von der Linde, T.J. Negran: Appl. Phys. Lett. **25**, 233 (1974)

5.36   O.F.Schirmer: J. Appl. Phys. **50**, 3404 (1979)
5.37   V.I.Belinicher, V.K.Malinovski, B.I.Sturman: Sov. Phys. – JETP **46**, 362 (1977)
5.38   *Standards of Piezoelectric Crystals*, Proc. IRE **37**, 1378 (1949);
       ibid **46**, 764 (1958)
5.39   V.M.Fridkin, R,M,Magomadov: JETP Lett. **30**, 686 (1979)
5.40   H.G.Festl, P.Hertel, E.Krätzig, R.von Baltz: Phys. Status Solidi B**113**, 157 (1982)
5.41   A.M.Glass, D.von der Linde, D.H.Auston, T.J.Negran: J. Electron. Mater. **4**, 915
       (1975)
5.42   R.Orlowski, E.Krätzig: Solid State Commun. **27**, 1351 (1978)
5.43   H.Heyszenau: Phys. Rev. B**18**, 1586 (1978)
5.44   R.von Baltz: Phys. Status Solidi B**89**, 419 (1978)
5.45   R.von Baltz, W.Kraut: Phys. Lett. A**79**, 364 (1980)
5.46   W.Kraut, R.von Baltz: Phys. Rev. B**19**, 1548 (1979)
5.47   R.von Baltz: Ferroelectrics **35**, 131 (1981)
5.48   H.Presting, R.von Baltz: Phys. Status Solidi B**112**, 559 (1982)
5.49   V.I.Belinicher, B.I.Sturman: Sov. Phys. – Usp. **23**, 199 (1980)
5.50   E.Krätzig: Ferroelectrics **21**, 635 (1978)
5.51   E.Krätzig, R.Orlowski, V.Doormann, M.Rosenkranz: Proc.-SPIE **164**, 33 (1978)
5.52   R.Orlowski, E.Krätzig, H.Kurz: Opt. Commun. **20**, 171 (1977)
5.53   D.von der Linde, A.M.Glass, K.F.Rodgers: Appl. Phys. Lett. **25**, 155 (1974)
5.54   D.von der Linde, A.M.Glass, K.F.Rodgers: J. Appl. Phys. **47**, 217 (1976)
5.55   H.Vormann, E.Krätzig: Solid State Commun. **49**, 843 (1984)
5.56   Ye Ming, E.Krätzig, R.Orlowski: Phys. Status Solidi A**92**, 221 (1985)
5.57   D.A.Bryan, R.Gerson, H.E.Tomaschke: Appl. Phys. Lett. **44**, 847 (1984)
5.58   N.V.Kukhtarev, V.B.Markov, S.G.Odulov, M.S.Soskin, V.L.Vinetskii: Ferroelec-
       trics **22**, 949 (1979)
5.59   V.L.Vinetskii, N.V.Kukhtarev, S.G.Odulov, M.S.Soskin: Sov. Phys. – Usp. **22**, 742
       (1979)
5.60   M.B.Klein, G.C.Valley: J. Appl. Phys. **57**, 4901 (1985)
5.61   R.Orlowski, E.Krätzig: Ferroelectrics **26**, 831 (1980)
5.62   G.A.Alphonse, R.C.Alig, D.L.Staebler, W.Phillips: RCA Rev. **36**, 213 (1975)
5.63   D.L.Staebler, J.J.Amodei: J. Appl. Phys. **43**, 1042 (1972)
5.64   D.W.Vahey: J. Appl. Phys. **46**, 3510 (1945)
5.65   J.J.Amodei, D.L.Staebler: Appl. Phys. Lett. **18**, 540 (1971)
5.66   F.B.Williams, W.J.Burke, D.L.Staebler: Appl. Phys. Lett. **28**, 224 (1976)
5.67   W.Bollmann, J.H.Stöhr: Phys. Status Solidi A**39**, 477 (1977)
5.68   W.Meyer, P.Würfel, R.Munser, G.Müller-Vogt: Phys. Status Solidi A**53**, 171 (1979)
5.69   R.G.Smith, D.B.Fraser, R.T.Denton, T.C.Rich: J. Appl. Phys. **39**, 4600 (1968)
5.70   O.W.Johnson, J.DeFord, J.W.Shaner: J. Appl. Phys. **44**, 3008 (1973)
5.71   W.Bollmann, K.Schlotthauer, O.J.Zogal: Krist. Tech. **11**, 1327 (1976)
5.72   J.L.Jackel, C.E.Rice: Appl. Phys. Lett. **41**, 508 (1982)
5.73   J.L.Jackel, A.M.Glass, G.E.Peterson, C.E.Rice, D.H.Olson, J.J.Veselka: J. Appl.
       Phys. **55**, 269 (1984)
5.74   J.P.Nisius, E.Krätzig: Solid State Commun. **53**, 743 (1985)
5.75   O.F.Schirmer, D.von der Linde: Appl. Phys. Lett. **33**, 35 (1978)
5.76   L.E.Halliburton, K.L.Sweeney, C.Y.Chen: Nucl. Instrum. Methods Phys. Res. B**1**,
       344 (1984)
5.77   See, e.g., [Ref. 5.7, pp. 56–75]
5.78   L.Arizmendi, J.M.Cabrera, F.Agulló-Lopez: J. Phys. C**17**, 515 (1984)
5.79   J.L.Ketchum, K.L.Sweeney, L.E.Halliburton, A.F.Armington: Phys. Lett. **94**A, 450
       (1983)
5.80   K.L.Sweeney, L.E.Halliburton: Appl. Phys. Lett. **43**, 336 (1983)
5.81   P.Nagels: "Experimental Hall Effect Data for a Small Polaron Semiconductor", in *The
       Hall Effect and Its Applications*, ed. by. C.L.Chen, C.R.Westgate (Plenum, New York
       1980) pp. 253–280

5.82   I.G.Austin, N.F.Mott: Adv. Phys. **18**, 51 (1969);
       P.Gerthsen, E.Kauer, H.G.Reik: "Halbleitung einiger Übergangsmetalloxide im
       Polaronenbild", in *Festkörperprobleme*, Vol. 5, ed. by F.Sauter (Vieweg, Berlin 1966)
       p. 1
5.83   D.M.Smyth: Ferroelectrics **50**, 93 (1983)
5.84   O.F.Schirmer, E.Salje: J. Phys. C**13**, L1067 (1980)
5.85   D. von der Linde, O.F.Schirmer, H.Kurz: Appl. Phys. **15**, 153 (1978)
5.86   O.F.Schirmer, S.Juppe, J.Koppitz: Cryst. Lattice Defects Amorph. Mater., in press
5.87   O.F.Schirmer: unpublished
5.88   P.Lerner, C.Degras, J.P.Dumas: J. Cryst. Growth **3**, 231 (1968);
       K.Nassau, M.E.Lines: J. Appl. Phys. **41**, 533 (1970)
5.89   See, e.g., O.F.Schirmer: J. de Phys. **41**, C6-479 (1980)
5.90   H.Kurz, D. von der Linde: Ferroelectrics **21**, 621 (1978)
5.91   L.A.Kappers, K.L.Sweeney, L.E.Halliburton, J.H.W.Liaw: Phys. Rev. **31**, 6792
       (1985)
5.92   V.A.Antonov, P.A.Arsenev, I.G.Linda, V.L.Farstendiker: Phys. Status Solidi A**28**,
       673 (1975)
5.93   E.M.Avakyan, K.G.Belabaev, L.A.Shuvalov: Sov. Phys. – Crystallogr. **28**, 676 (1983)
5.94   C.Y.Chen, K.L.Sweeney, L.E.Halliburton: Phys. Status Solidi A**81**, 253 (1984)
5.95   P.Günter, U.Flückiger, J.P.Huignard, F.Micheron: Ferroelectrics **13**, 297 (1976)
5.96   P.Günter, F.Micheron: Ferroelectrics **18**, 27 (1978)
5.97   P.Günter: Ferroelectrics **22**, 671 (1978)
5.98   P.Günter: Phys. Rep. **93**, 199 (1982)
5.99   E.Siegel, W.Urban, K.A.Müller, E.Wiesendanger: Phys. Lett. **53**A, 415 (1975)
5.100 E.Siegel: Ferroelectrics **13**, 385 (1976)
5.101 F.S.Chen: J. Appl. Phys. **38**, 3418 (1967)
5.102 S.R.King, T.S.Hartwick, A.B.Chase: Appl. Phys. Lett. **21**, 312 (1972)
5.103 D. von der Linde, A.M.Glass, K.T.Rodgers: Appl. Phys. Lett. **26**, 22 (1975)
5.104 L.A.Boatner, E.Krätzig, R.Orlowski: Ferroelectrics **27**, 247 (1980)
5.105 R.Orlowski, L.A.Boatner, E.Krätzig: Opt. Commun. **35**, 45 (1980)
5.106 D.Rytz, L.A.Boatner, E.Chatelain, U.T.Höchli, K.A.Müller: Helv. Phys. Acta **5**, 430
       (1978)
5.107 A.Agranat, Y.Yacoby: Phys. Rev. B**27**, 5712 (1983)
5.108 R.L.Townsend, J.T.LaMacchia: J. Appl. Phys. **41**, 5188 (1970)
5.109 F.Micheron, G.Bismuth: Appl. Phys. Lett. **20**, 79 (1972)
5.110 J.Feinberg, R.W.Hellwarth: Opt. Lett. **5**, 519 (1980)
5.111 W.T.H.Koch, R.Munser, W.Ruppel, P.Würfel: Solid State Commun. **17**, 847 (1975)
5.112 A.W.Hornig, R.C.Rempel, H.E.Weaver: J. Phys. Chem. Solids **10**, 1 (1959)
5.113 E.Siegel, K.A.Müller: Phys. Rev. B**20**, 3587 (1979)
5.114 E.Krätzig, F.Welz, R.Orlowski, V.Doormann, M.Rosenkranz: Solid State Commun.
       34, 817 (1980)
5.115 M.B.Klein: J. Opt. Soc. Am. B**3**, 293 (1986)
5.116 J.Feinberg, D.Heiman, A.R.Tanguay, R.W.Hellwarth: J. Appl. Phys. **51**, 1251 (1980)
5.117 N.V.Kukhtarev, E.Krätzig, H.C.Külch, R.A.Rupp: Appl. Phys. B**35**, 17 (1984)
5.118 J.J.Amodei, D.L.Staebler, A.W.Stephens: Appl. Phys. Lett. **18**, 507 (1971)
5.119 J.J.Amodei, D.L.Staebler: Appl. Phys. Lett. **18**, 540 (1971)
5.120 J.B.Thaxter: Appl. Phys. Lett. **15**, 210 (1969)
5.121 F.Micheron, G. Bismuth: Appl. Phys. Lett. **23**, 71 (1973)
5.122 J.B.Thaxter, M.Kestigian: Appl. Opt. **13**, 913 (1974)
5.123 A.V.Guinzberg, K.D.Kochev, Yu.S.Kusminov, T.R.Volk: Phys. Status Solidi A**29**,
       309 (1975)
5.124 J.P.Huignard, F.Micheron: Appl. Phys. Lett. **29**, 591 (1976)
5.125 F.Micheron, M.Peltier, J.P.Huignard: Opt. Commun. **18**, 216 (1976)
5.126 J.P.Huignard, J.P.Herriau: Appl. Opt. **17**, 2671 (1978)
5.127 S.L.Hou, R.B.Lauer, R.E.Aldrich: J. Appl. Phys. **44**, 2652 (1973)

5.128 A.Marrakchi, J.P.Huignard, P.Günter: Appl. Phys. **24**, 131 (1981)

5.129 T.G.Pencheva, S.J.Stepanov: Sov. Phys. – Solid State **24**, 687 (1982)

5.130 M.Peltier, F.Micheron: J. Appl. Phys. **48**, 3683 (1977)

5.131 A.M.Glass, A.M.Johnson, D.H.Olson, W.Simpson, A.A.Ballmann: Appl Phys. Lett. **44**, 948 (1984)

5.132 M.B.Klein: Opt. Lett. **9**, 350 (1984)

5.133 A.M.Glass, M.B.Klein, G.C.Valley: Electron. Lett. **21**, 221 (1985)

5.134 J.Schneider: "The Role of Point Defects in GaAs", in *Semi-Insulating III–V Materials*, ed. by S.Makram-Ebeid, B.Tuck (Shiva Publishing, Nantwich 1982) pp. 144–153

6. Photorefractive Measurements of Physical Parameters

Ruth A. Mullen

With 10 Figures

The photorefractive effect, whereby photoinduced charge migration between deep trapping centers brings about internal space-charge fields, the equipotential lines of which follow roughly the same contours as the incident intensity pattern, provides a convenient and sensitive optical probe of the material parameters. Some otherwise elusive information on the densities and cross sections of the trace impurities or defects which act as photorefractive trapping centers can be obtained from photorefractive measurments, as well as charge diffusion lengths, mobilities, and recombination (trapping) coefficients.

6.1 Background

Among the many difficult problems faced by those who grow and use photorefractive crystals is the accurate measurement of the relevant electrical parameters which result from the very low level of trace impurities and defects between which photorefractive charges migrate.

Conventional bulk photocurrent or photoconductivity measurements, such as deep-level transient spectroscopy (DLTS), which have been reasonably successful in determining some of the parameters in semi-insulating GaAs [6.1] suffer a distinct disadvantage when applied to more insulating materials such as $Bi_{12}SiO_{20}$ (BSO) and $BaTiO_3$. That is, since photocurrents are weak in insulating samples, the use of ultrasensitive detection electronics, thin samples, and/or large externally applied voltages becomes imperative. Nonohmic junction effects, surface currents, photovoltaic und pyroelectric effects, nonuniform field distributions throughout the crystal, and stray capacitances consequently may degrade the photocurrent measurements. Furthermore, DLTS measurements of photoexcitation cross sections (such as in [6.2], p. 259) are often available only at very low temperatures [6.3] or in arbitrary (rather than absolute) units. Absolute photorefractive absorption coefficient measurements can be made in a very straightforward manner with the photorefractive technique by simply taking care to measure absolute rather than relative optical intensities and correctly accounting for Fresnel reflection losses at the entrance surface of the crystal.

Destructive analysis techniques have been successfully used to measure the relative densities of various dopants in photorefractive $BaTiO_3$ [6.4, 5], but are unable to distinguish between different valence states (e.g., Fe^{3+} and Fe^{2+}) and are generally unable to measure absolute densities of impurities as small as those which are responsible for the photorefractive effect. Electron paramagnetic resonance (EPR) must be carefully calibrated with respect to a known standard in order to measure absolute densities of odd-valence state dopants (e.g., Fe^{3+}, as in photorefractive $BaTiO_3$ [6.4, 5]). No information on the densities of evenly charged valence state dopants (e.g., Fe^{2+}) is available from EPR. Furthermore, one cannot assume without separate photorefractive measurements that any spectroscopically measured species plays a significant role in the photorefractive effect.

In addition to measuring absolute rather than just relative trap densities and excitation cross sections, photorefractive measurements, unlike the spectroscopic and photocurrent techniques, are unambiguously sensitive only to the *bulk* (as opposed to *surface*) properties of the crystal, and have the additional advantage of measuring directly the parameter of interest for photorefractive studies. The amplitude and time response of small bulk photocurrents several millimeters from any electrodes or surfaces can be read out optically (by diffraction), eliminating both the need for highly sensitive electronic detection and difficulties with electrode or junction effects. The size feature of importance in photorefractive experiments, the holographic fringe spacing, is conveniently alterable over a wide range of physically important dimensions (from less than 0.1 μm to several micrometers) by adjusting the angle between two interfering writing beams. Questions such as the applicability of a particular mobility measurement to transport leading to photorefractivity or the participation of a specific dopant in photorefractivity do not arise.

6.1.1 Nature of the Charge Transport Mechanism

The charge transport equations governing the photorefractive effect have been thoroughly described in Chap. 3. These equations, which are based on a model in which charges move between traps via a three-step process of photoexcitation, drift or diffusion in the conduction band, and finally recombination into another empty trap, correctly describe most of the photorefractive experiments performed up until this chapter was prepared, and provide a solid framework for understanding the important parameters for a materials characterization. An alternative model [6.6, 7] in which charges "hop" between trapping centers under the influence of light provides a helpful intuitive picture of photorefractive charge dynamics and describes all the zero-field photorefractive experiments very well, but cannot describe the effects of an electric field on the time constants for the transient buildup and decay in BSO without the choice of a rather complicated probability distribution function [6.8]. Application of a dc electric field to a photorefractive crystal thus may prove to be a convenient means to discriminate between conventional band transport and hopping transport [6.8].

Interestingly, this failure of the hopping model to describe photorefractive transient measurements in the presence of externally applied fields seems to be at odds with bulk ac dark conductivity [6.9, 10] and photoconductivity [6.11] experiments in sillenites which indicate that the transport in these experiments is via hopping. The experimental evidence for hopping as opposed to band transport are reports [6.9–11] that the dark conductivity and photoconductivity of sillenites are observed to be increasing functions of frequency for a wide range of externally applied frequencies between 1 kHz and 10 MHz. A simple formal proof by *Pollak* [6.12] directly associates this frequency dependence with hopping as opposed to band transport.

6.1.2 Photorefractive Charge Transport via Hopping

Since the apparent disagreement between bulk ac photoconductivity and photorefractive time constant experiments in the presence of externally applied fields has not yet been resolved, since both models work equally well in the absence of externally applied fields, and since the applicability of photorefractive charge hopping to $BaTiO_3$ and $LiNbO_3$ has not yet been tested by the application of external fields, the essential elements of the hopping model will be reviewed and the formal equivalence between the two models in the absence of external fields will be outlined.

a) Discretization of Conduction Band Transport Equation

The formal similarity between hopping and conduction band charge transport becomes immediately clear upon discretization of the conduction band transport equations at some small volume element of the crystal indexed with the subscript n. The divergence of the electron density from a small volume element at position n is

$$(\nabla n)_n = \frac{n_{n+1} - n_{n-1}}{2l} \, ,$$

where n_i is the free electron density at position i and l is the separation between sites so that

$$(\nabla^2 n)_n = \frac{n_{n+1} + n_{n-1} - 2n}{l^2} \, .$$

The divergence of the electric field at position n is

$$(\nabla E)_n = -\frac{\phi_{n+1} + \phi_{n-1} - 2\phi_n}{l^2} \, ,$$

where ϕ_n is the electric potential at position n.

With the above three substitutions, some careful algebra, and the following three approximations:

- that the fringe spacings are large compared to the separation between unoccupied trapping centers ($\Lambda \gg l$),
- that the local electric potential differences between adjacent centers are small compared with $k_B T/q$, and
- $\delta n/\delta t \simeq 0$,

the continuity equation reduces to the form

$$\frac{dN_D^+}{dt} = \sum_m \frac{\mu k_B T}{l^2 q} n_n \exp\left(\frac{q\phi_{nm}}{2k_B T}\right) - \sum_m \frac{\mu k_B T}{l^2 q} n_m \exp\left(\frac{q\phi_{mn}}{2k_B T}\right), \tag{6.1}$$

where N_D^+ is the number of ionized (or vacant) donors, μ is the mobility of electrons in the conduction band, $k_B T$ is the product of Boltzmann's constant and temperature, and $\phi_{nm} = \phi_n - \phi_m$.

b) Hopping Theory

In the hopping picture, charges trapped in deep impurity or defect states can move between trapping centers only with the assistance of a photon. A trapped charge excited by a photon hops from site m to site n with a transition probability per second per unit intensity of D_{mn}. The asymmetry in the charge hopping direction is determined by the slope of the local electric potential contour according to the factor $\exp[q(\phi_m - \phi_n)/2k_B T]$, where ϕ_n is the electric potential at site n, and k_B is Boltzmann's constant.

The probability that a site n is occupied is W_n, the intensity of light is I_n, and the rate of change of occupation of site n is given by the hopping equation

$$\frac{dW_n}{dt} = -\sum_m D_{mn} W_n I_n \exp\left(\frac{q(\phi_m - \phi_n)}{2k_B T}\right)$$

$$+ \sum_n D_{nm} W_m I_m \exp\left(\frac{q(\phi_n - \phi_m)}{2k_B T}\right). \tag{6.2}$$

The first summation on the right-hand side, proportional to $W_n I_n$, accounts for the rate of charge hopping *out* of site n, while the second summation, proportional to the intensity and occupation probability at site m, accounts for the rate of charge hopping *into* site n from some other site m.

By comparing (6.1) with the above hopping equation, the following relation between the band drift mobility μ, the free electron density n_n at position n in the crystal, and the parameters of the hopping model becomes apparent:

$$\frac{\mu k_B T}{q} n_n = N_T D_{mn} W_n I_n l^2 \; ,$$

where $N_T = N_A(1 - N_A/N_D)$ is the effective density of mobile charges, N_A is the acceptor density, and N_D is the donor density.

This equation treats W_n whereas the charge transport equation for electrons traveling in the conduction band treats an averaged quantity, $\langle 1 - W_n \rangle_n$, called $N_D^+(x)$, the average density of ionized sites at the position x (along the grating wave vector) in the vicinity of n. Unlike the band transport equation, the hopping equation is equally valid for all charge transport processes, including band transport, tunneling, unscreened transport (when particles act discretely) and screened transport (when the individual character of particles is blurred and the flow of charge is more continuous, as in a liquid). Each of these processes would have a distinctly different form and temperature dependence for D_{mn}, which could be specified in (6.2). The band model could also accommodate these different transport processes, but only by defining an effective mobility, and not from first principles. Similarly, the hopping model can always be made to accommodate (though sometimes with considerable difficulty) a band transport mechanism by appropriate definition of D_{mn}.

c) Hopping in Barium Titanate (Short Range)

In barium titanate, it was found [6.6] that charge hopping distances are small relative to typical holographic fringe spacings, so that to a good approximation charges hop only between nearest neighbor sites. Also to a good approximation in barium titanate, the hopping probability was found to be symmetric so that the $D_{mn} = D_{nm}$. The hopping probability in barium titanate was thus written as

$$D_{mn} = \begin{cases} 1 & \text{for } m = n \pm 1 \; , \\ 0 & \text{otherwise} \; . \end{cases} \tag{6.3}$$

In this special case of nearest neighbor hopping, the infinite sum in (6.2) thus reduced to a sum over four terms.

The approximation in the hopping model that the local fluctuation in the density of charges is small ($W_n/W_0 \ll 1$) is similar to but less restrictive than the approximation made by *Kukhtarev* et al. [6.13] that the modulation in the optical intensity pattern was small ($m_{opt} \ll 1$).

Finally, by noting that $q(\phi_m - \phi_n)k_B T \ll 1$ in the hopping model, the exponent in (6.2) was expanded to first order in $\phi_{mn} = \phi_m - \phi_n$.

With these approximations and Poisson's relation between the complex electric potential $\phi(K)$ and the trapped charge distribution $\varrho(x, K) \propto \exp(iKx)$,

$$\phi(K) = \frac{\varrho(K)}{\varepsilon \varepsilon_0 K^2} \; ,$$

and the approximation that the separation between nearest neighbor sites is small compared to a fringe spacing, $d_{hop} \ll \Lambda$, both the steady-state and transient space-charge fields were predicted [6.6]. The steady-state prediction agreed with that of Kukhtarev for small m, small gain and small phase changes. The transient solution of the nearest-neighbor hopping model is a simple exponential decay whose intensity-normalized exponential decay rate Γ depends on the grating wave vector K as

$$\Gamma(K) = \frac{1}{I\tau_{di}} \left[1 + \left(\frac{K}{k_s}\right)^2 \right] , \qquad (6.4)$$

where τ_{di} is the dielectric relaxation time, K is the grating wave vector, and k_s is the space-charge screening wave vector. This solution agrees with Kukhtarev's prediction [6.14] only in the limit of $Kd_{hop} \ll 1$, and with no externally applied fields.

d) Hopping in the Sillenites (Long Range)

Charges hop much farther in n-BSO than in barium titanate [6.7]. In as-grown n-BSO from Crystal Technology, electrons are the dominant carriers and their range is longer than the Debye screening length ($r_e < l_s$), whereas in most BaTiO$_3$ crystals, holes are dominant and $r_h < l_s$, as illustrated in Fig. 6.1. Consequently, the short hopping distance approximation that charges hop only between nearest neighbors expressed in (6.3) must be replaced in the case of n-BSO by a longer range hopping probability $D(r)$, the form of which depends on the nature of the impurity, on the charge transport mechanism, and on other properties of the crystal.

A generalized form for exponential rates of intensity-normalized charge grating erasure decays in any material with short or long hopping distances is

$$\Gamma(K) = \left[1 + \left(\frac{k_s}{K}\right)^2 \right] \left(\sum D_{mn} \{ 1 - \exp [iK(x_m - x_n)] \} \right) .$$

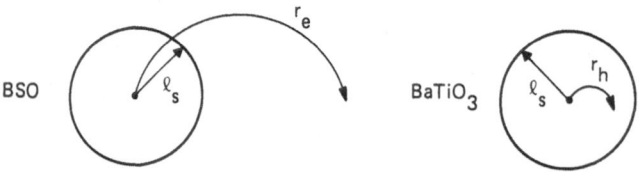

Fig. 6.1. In short transport length materials such as BaTiO$_3$, charges (generally holes) relocate between trapping centers which are within a sphere of radius equal to a Debye screening length ($r_h < l_s$). In long transport length materials such as BSO, charges (generally electrons) can diffuse a relatively long distance, far outside of a Debye sphere, before being retrapped ($r_e > l_s$)

The average over all sites n indicated by the angle brackets can be approximated by an integral

$$\left(\sum D_{mn}\{1 - \exp\left[iK(x_m - x_n)\right]\}\right) = N_0 \int d^3x\, D(x)\left[1 + \exp(iKx)\right] ,$$

where N_0 is the total number of occupied and unoccupied sites.

The solution of this integral depends on the form of $D(x)$. For the nearest neighbor hopping probability of (6.3) it gives the $\Gamma(K)$ form of (6.4) correct for small grating wave vectors in barium titanate.

Two longer-range hopping probabilities which were hypothesized for BSO both predicted $\Gamma(K)$ curves which agreed with experiments. A hopping probability of

$$D(r) = D_0 \frac{e^{-\kappa r}}{r}$$

(where $\kappa^2 = r_e^{-2} = d_{hop}^{-2} = 6/d_{rms}^2 = 4/d_{mean}^2$) leads to a wave vector dependence of intensity normalized erasure rates which agrees exactly with the prediction of *Kukhtarev* [6.14]:

$$\Gamma(K) = \frac{1}{I\tau} = \frac{1}{I\tau_{di}}\frac{1 + K^2/k_s^2}{1 + K^2/\kappa^2} . \tag{6.5}$$

A slightly longer-range hopping probability

$$D(r) = D_0 e^{-\kappa r} \tag{6.6}$$

leads to a different functional form for $\Gamma(K)$

$$\Gamma(K) = \frac{1}{I\tau_{di}}\frac{\kappa^2(K^2 + k_s^2)(K^2 + 2\kappa^2)}{2k_s^2(K^2 + \kappa^2)^2}$$

than is predicted by the band theory. In this case κ^2 is equal to $12/d_{rms}^2$ and $9/d_{mean}^2$.

These two functional forms are almost identical, only differing from one another very slightly for grating wave vectors in close proximity to κ. Experiments in this wave vector range in BSO [6.7, 15] were not precise enough to determine which hopping probability applies to the trapping centers in BSO. The simpler of these two possible functional forms for $\Gamma(K)$, (6.5), agrees exactly with the band transport prediction and was chosen for the analysis in [6.7, 15].

Although the nature of the charge transport mechanism definitely affects the numerical values of cross sections and mobilities, neither model at this level relies on a detailed knowledge of the nature of the trapping center or the microscopic details of how the charge gets from one center to another. In fact, the transient and steady-state photorefractive predictions in the absence of externally applied fields seem to be remarkably insensititive to these details. Application of an external field to BSO [6.8] removes some of this insensitivity. Nevertheless, the microscopic details of how the charges get from one trapping site to another are among the most physically interesting problems for future investigations.

6.1.3 Relevant Material Parameters

In the band transport model of photorefractivity [6.14, 16] there are at least eight parameters of the material that control the formation of refractive index gratings when the material is illuminated with an optical interference pattern:

ε the low frequency dielectric constant
n_b the background refractive index
r_{eff} the effective electro-optic coefficient
s the cross section for photoionization
N_D the number density of dopants
N_A the number density of acceptor that compensate for the number density of ionized dopants N_D^+ in the dark
μ the mobility
γ_R the coefficient for electron recombination at ionized traps N_D^+.

The first three of these parameters are well-known, intrinsic properties of photorefractive crystals which are not amenable to change in well-poled crystals and are not expected to vary significantly from one sample of a crystal to another.

The remaining five parameters depend on the impurity and/or the defect content of the crystal, and thus may be adjusted either by doping during crystal growth, oxidation and reduction treatments, radiation, or in-diffusion. For the description of photorefractive responses to cw lasers, only three parameters are required. The required parameters are $N_T = N_A(1 - N_A/N_D)$, r_e (or r_h) and sN_D/N_A. The separate parameters N_D and N_A are not distinguishable in these cw transient experiments, but instead are lumped together into an effective photorefractive density, or a "density of mobile charges" of $N_T = N_A(1 - N_A/N_D)$. Similarly, the parameters μ and τ_e are measurable only as a mobility-lifetime product, or charge range, since, in this "quasi-steady" time regime, the photoexcited electron density is known to be in constant equilibrium with the empty traps (accordingly, this time regime has also been referred to as the "adiabatic" [6.7] regime). That is, the electron trapping times τ_e are so short compared to the time scale of the buildup or decay of a grating under illumination with a cw laser that the transfer of charge from one site to another can be thought of as being instantaneous. It is only under conditions in which the duration of an intense illumination pulse is comparable to or shorter than a recombination time that the parameters μ and τ_e are thought to be separately important. The experimentally observable effect of changing the ratio of N_A/N_D from a very small value ($\ll 1$) to nearly unity by oxidation and reduction treatments is to cause photorefractive beam coupling measurements of

$N_T = N_A(1 - N_A/N_D)$ to drop from their normally large values of several times 10^{16} to around 5×10^{15} [6.17]. A photorefractive technique to measure N_D and N_A separately would be very useful but has not yet been devised.

6.2 Photorefractive Measurement Techniques

Photorefractive measurements of trap-related material parameters fall into three general categories: steady-state, "quasi-steady" or "adiabatic" transient, and short pulse. Steady-state measurements of the maximum amplitude of photorefractive space charge gratings are made after the grating has been completely formed, on a time scale long compared to the photorefractive response time, typically milliseconds to seconds. For materials in which one carrier dominates, these measurements can provide only the following two pieces of information: the effective trap density (or density of photorefractively mobile charges) N_T, and a lumped parameter that includes the effective electro-optic coefficient [6.6, 18–20].

Quasi-steady transient measurements are measurements of the time constants characteristic of creating or erasing photorefractive gratings with cw illumination. Since the time constants to build up (or erase) a photorefractive grating with a cw laser are long compared to the carrier trapping time, the mean carrier number density n is constant in the quasi-steady regime. Photorefractive response rates depend, for the most part, nearly linearly on intensity in this regime [6.6, 7]. As such, it is generally the intensity-normalized photorefractive response rates that are of interest. Measurements of the response times as a function of grating period in this regime were originally used to obtain two independent parameters in $BaTiO_3$: the trap density N_T and the dielectric relaxation time [6.6, 21], and to obtain three independent parameters in $Bi_{12}SiO_{20}$ [6.7, 15]: the trap density N_A, the $\mu\tau_e$ product (where $\tau_e = 1/\gamma_R N_A$) or the effective electron range $r_e = (\mu\tau_e k_B T/e)^{1/2}$ (where $k_B T$ is the Boltzmann constant times temperature and e is the charge on the electron), and a cross section for electron hopping sN_D/N_A or photorefractive absorption coefficient $s(N_D - N_A) = N_T s N_D/N_A$. The photorefractive quantum efficiency is given by the ratio of α_{pr} to the ordinary optical absorption, $Q = \alpha_{pr}/\alpha$. The use of this technique to measure the charge range in $BaTiO_3$ was delayed by the fact that along the c-axis in $BaTiO_3$, the charge range is comparable to but not much smaller than the smallest achievable holographic fringe spacing. The technique has been recently been used with a doubled, Q-switched YAG laser ($\lambda = 532$ nm) to measure the (longer) charge range perpendicular to the c-axis in $BaTiO_3$ and to place bounds on the charge range parallel to the c-axis [6.22]. Since then, improved experimental accuracy obtainable with a cw argon ion laser ($\lambda = 515$ nm) has decreased the uncertainty in both ranges [6.23]. The same three parameters are obtainable by measuring the functional dependence of grating erasure rates on the strengths of internal fields arising from externally applied fields, provided the internal field is accurately measured with either the electro-optic effect [6.8] or by careful calibration of the absolute diffraction efficiency as a function of applied field [6.24].

Short pulse measurements are those made with pulses less than or equal to the carrier recombination time and can provide a measurement of τ_e for materials in which the rate equation model of *Kukhtarev* [6.14] is valid. By combining short pulse measurements of τ_e with the quasi-steady transient measurements of N_A and $\mu\tau_e$ one should be able to obtain μ and γ_R independently.

6.3 The Steady State

In the steady state, wihout externally applied fields, for small charge modulation $|m| \ll 1$ and for a material in which there is only one type of charge carrier, the amplitude of the photorefractive space-charge field E_{sc} depends on only one photorefractive parameter, the density of photorefractively mobile charges, N_T, according to the formula [6.6, 13]

$$E_{sc} = \frac{K k_B T}{q} \frac{1}{1 + K^2/k_s^2} , \tag{6.7}$$

where $k_s = (N_T q^2 / k_B T \varepsilon \varepsilon_0)^{1/2}$ is the Debye screening wave vector.

By measuring either the diffraction efficiency η or the steady-state gain as a function of grating period for grating periods both larger than and smaller than the Debye screening wave vector, the point at which the grating reaches its maximum diffraction efficiency can be discerned. This peak in diffraction efficiency corresponds to the grating period at which the diffusion field $E_D = K k_B T/e$ is equal to the limiting space-charge field attainable in a crystal with N_T photorefractively mobile charges, $E_q = N_T q / K \varepsilon \varepsilon_0$. This maximum space-charge field occurs for grating fringe spacings exactly equal to a Debye space-charge screening length l_s. For larger fringe spacings ($\Lambda \gg l_s$), the space-charge field is simply equal to the diffusion field. As fringe spacings decrease, the diffusion field increases until Λ is as small as the Debye screening length l_s. For fringe spacings smaller than l_s, the space-charge grating becomes partially screened, reducing the space-charge field to below that which would be expected from purely diffusion effects.

Since (for small η and small beam coupling) both the diffraction efficiency and the gain coefficient of a photorefractive grating are proportional to the square of the internal space-charge field, and since, in a fractionally poled, two carrier crystal, the internal space-charge field E_{sc}' is reduced from the value in (6.7) by

$$E_{sc}' = F \bar{\sigma} E_{sc} , \tag{6.8}$$

a second piece of information can be extracted from careful measurements of absolute gain coefficients [6.25]; that is the product of the fractional poling F of the crystal with the normalized differential conductivity $\bar{\sigma}$ (see Chap. 7). Measurements of the amplitudes of steady-state gain at a single small grating wave vector in $LiNbO_3$ crystals with an independently measured range of Fe^{2+}/Fe^{3+} ratios enabled *Orlowski* and *Krätzig* [6.26] to determine the relative contribution of electrons and holes to photorefractive charge transport.

For a single carrier situation, the steady-state grating efficiency (and hence the gain coefficient) is proportional simply to the square of this space-charge field and is independent of the charge range $r_e = (\mu\tau_e k_B T/e)^{1/2}$, the photorefractive absorption coefficient, and the photoexcitation cross section, as the latter three parameters affect only the speed at which a grating forms, not the grating's steady-state amplitude. Steady-state experiments thus provide no information on the charge range, but provide good measurements only of the density N_T (for crystals dominated by a single carrier) and on the product of the fractional poling with the differential conductivity. To measure the charge range, the rates of grating buildup and decay must be monitored.

There are three experimental difficulties associated with steady-state measurements which are not of concern to the quasi-steady transient measurements. These are beam overlap effects, accuracy in path-length matching, and acoustic noise.

Inaccuracies of about a factor of two in steady-state measurements of N_T can result from failing to account correctly for the dependence of grating thickness on the angle between two intersecting writing beams which have diameters smaller than the crystal dimensions. While for very small angles or for counterpropagating beams, the thickness of the grating is equal to the length of the crystal, for intermediate angles the grating thicknesses t depends on the internal half-angle between the beams θ_{in} according to $t = D/\cos\theta_{in}$, where D is the diameter of the writing beams. This geometric effect tends to shift the peak in diffraction efficiency and gives an inaccurate measurement of N_T unless the proper correction factor can be applied. Experimental geometries such as those of *Klein* and *Valley* [6.25] in which one of the writing beams overfills the crystal while the other writing beam, which has a smaller diameter, probes the crystal, avoids this source of systematic error by arranging for the grating thickness to be equal to the thickness of the crystal. In this case, the thickness of the grating is always equal to the thickness of the crystal.

Care must be taken in these measurements to see that the path lengths of the writing beams are well-matched at all angles to within the coherence length of the laser. Acoustic noise in the laboratory must be effectively eliminated in order to accurately measure steady-state diffraction efficiencies. When a crystal has two charge carriers which both contribute significantly to photorefractivity, this steady-state technique becomes much less useful because the experimental data begin to depend on the diffusion (or hopping) lengths of the electrons and holes as well as on the relevant densities [6.27].

6.4 "Quasi-Steady" Transient Experiments

By adding a measurement of the charge range and the dielectric time constant to knowledge about N_T, one immediately obtains all the parameters necessary to predict all photorefractive effects obtainable with cw visible lasers. Measurement of the charge range requires an experimental setup which has sensitivity to the response times of grating buildup and decay under cw illumination. The density N_T and the dielectric time constant τ_{di} can be measured with the same apparatus.

6.4.1 Without Externally Applied Fields

In the absence of externally applied fields or moving gratings, the photorefractive charge transport equations [6.7, 14] predict that gratings are created and erased with exponential rates given by (6.5). This equation for the wave vector dependence of the intensity-normalized erase rates in photorefractive materials has several interesting features illustrated in Figs. 6.2, 3, which, once measured, lead to unambiguous measurements of three independent photorefractive parameters. From this expression, it is clear that the range of photorefractive response rates obtainable in a crystal (over a range of wave vectors) is set by two limits in the extreme cases of very large and very small fringe spacings.

For fringe spacings large compared to both the charge hopping distance and the Debye screening radius, the intensity-normalized grating erasure rate in photorefractive materials is independent of grating period and is equal to the intensity-normalized dielectric relaxation rate

$$\Gamma_0 = \lim_{K \to 0} \frac{1}{I\tau} = \frac{1}{I\tau_{di}} \, , \tag{6.9}$$

where $\tau_{di} = \varepsilon\varepsilon_0/n_0\mu e$. In this regime, the independence of response times from the period of a fringe spacing can be seen by modeling each fringe as an RC circuit. The capacitance C of a fringe decreases inversely with increasing fringe spacing Λ according to $C = \varepsilon\varepsilon_0 A/\Lambda$, while the resistance R increases in proportion to the fringe period as $R = \Lambda/ne\mu A$, where A is the area of the fringe. The fringe-spacing dependence of the RC product clearly cancels out with the simple result that for macroscopic fringe spacings (bigger than either the Debye screening length or the charge range), the grating erasure rates are independent of fringe spacing.

For fringe spacings small compared to both the Debye screening radius and the charge hopping distance, the importance of macroscopic collective motion of the charges fades and the intensity-normalized grating erasure rate is controlled by microscopic photoexcitation processes, which are independent of grating period. The intensity-normalized grating erasure rate in this limit of very large wave vectors and small fringe spacings is equal simply to the intensity-normalized charge generation rate, which can be called the energy-normalized

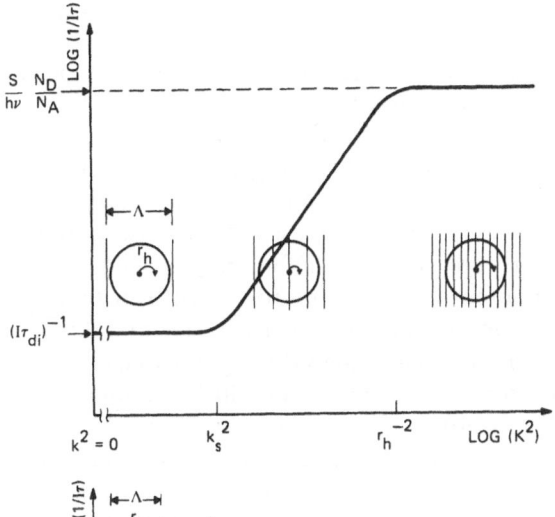

Fig. 6.2. In BaTiO$_3$ and LiNbO$_3$, the short range of the charge transport mechanism brings about intensity-normalized photorefractive response rates which are increasing functions of grating wave vector

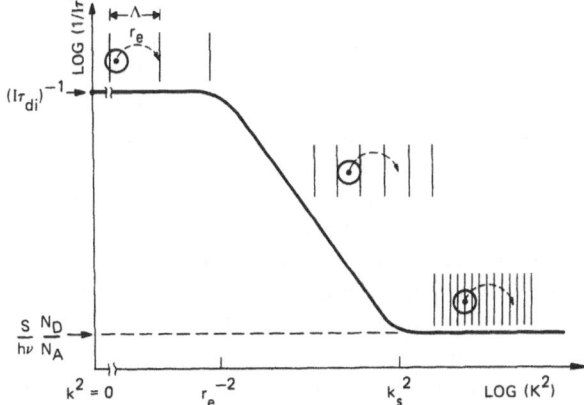

Fig. 6.3. In BSO and GaAs, the long range of the charge transport mechanism brings about intensity-normalized photorefractive response rates which are decreasing functions of the grating wave vector

cross section for charge transfer or "hopping cross section"

$$\Gamma_\infty = \lim_{K \to \infty} \frac{1}{I\tau} = \frac{s}{h\nu} \frac{N_D}{N_A} .$$ (6.10)

This charge transfer cross section is the photoionization cross section multiplied by the ratio of occupied to empty traps.

In materials such as BaTiO$_3$, in which the charge range is smaller than the Debye screening length ($r_h \ll l_s$), the photoexcitation cross section limits the maximum attainable speed (per unit intensity) of the photorefractive effect. Gratings in materials with short charge ranges cannot be erased any faster than the photoexcitation rate for charges, nor any slower than the intensity-normalized dielectric relaxation rate.

By contrast (see Fig. 6.3), in long-range materials such as BSO, the dielectric relaxation rate is faster than the photoexcitation rate. The dielectric relaxation rate in this case limits the fastest rate at which a grating can be erased while the photoexcitation rate sets the slowest rate at which a photorefractive grating can be erased.

For fringe spacings intermediate between the charge range and the Debye screening length, the intensity-normalized grating erasure rates depend either proportionally or inversely on K^2 according to whether the charge range is smaller or larger than the Debye screening length l_s (see Fig. 6.1). Crystals such as as-grown $BaTiO_3$ in which the charge range is smaller than the Debye screening length exhibit short-range effects; the erase rates increase with increasing grating wave vector. Since the charge ranges are smaller than the fringe spacings in this regime, an increasing number of photoexcitation events (or "hops") are required to erase (or buildup) larger fringes as compared to smaller fringes and the time constant associated with the erasure of widely-spaced fringes is longer.

Both GaAs and BSO are examples of materials in which the long-range nature of the Coulomb force causes gratings of larger fringe spacing to be erased more quickly than gratings of a smaller fringe spacing. Since, in these materials, for $K > \kappa$, charges migrate across a full fringe in a single hop, thin fringes are erased no faster than thick fringes. In fact, the screening effects act to make the intensity-normalized erase rates faster for the larger period fringes in the range $\kappa < K < k_s$.

It was this important distinction between "short-range" and "long-range" photorefractive materials that was raised by *Moharam* et al. [6.28, 29] with regard to photorefractive sensitivities and beam coupling with and without externally applied fields and internal ferroelectric fields. By now, analytical solutions of photorefractive charge transport theories [6.6, 7, 13, 14] can correctly describe materials of both types, and there are several ways within the framework of these theories to understand physically why the short-range photorefractive crystals have erase rates which are increasing functions of grating wave vector while the erase rates of long-range materials are decreasing functions of grating wave vector. In terms of the charge hopping picture, the key distinction between these two types of materials is that in the short-range materials (such as $BaTiO_3$), each charge hop occurs within a Debye screening sphere (within which all of the space-charge fields are screened), whereas in long-range materials (such as BSO), charges hop well outside the Debye sphere, with a long-range hopping probability $D(r) = D_0 \exp(-\kappa r)/r$.

6.4.2 Zero-Field Experimental Results

A few notable device implications of the physics of long-range versus short-range charge hopping (or transport) are illustrated in Fig. 6.4.

The response rate of a photorefractive device can be sensitively dependent on the experimental or device geometry. The response rate of BSO, for example, varies by more than three orders of magnitude depending on the angle between the writing beams. While at small grating wave vectors, BSO is five orders of magnitude faster than $BaTiO_3$, the erase rates of the two materials are much

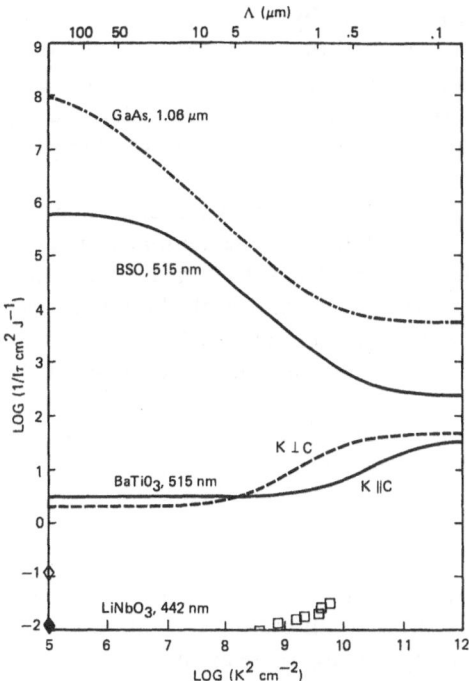

Fig. 6.4. Photorefractive response rates are known to vary by as many as 10 orders of magnitude in different materials. The curves for GaAs [6.18], BSO [6.7], and BaTiO$_3$ [6.23] are theoretical, using photorefractively measured material parameters (except for τ_{di} in GaAs, which has not yet been measured photorefractively). The boxes are measurements of simple exponential photorefractive response rates in a LiNbO$_3$ crystal which had been annealed at 400 °C [6.30]. The diamonds are the experimental results of bulk photocurrent measurements of the dielectric relaxation rates in two samples of LiNbO$_3$ [6.31]. The solid diamond is for a partially reduced crystal and the empty diamond is for an oxidized crystal

more comparable (within one order of magnitude) at the large grating wave vectors of importance in reflection grating geometries. Long transport length materials such as BSO will act as low pass filters (blurring sharp edges) when used under transient conditions while short transport length materials will act as high pass filters (edge enhancers).

Temporal modulation functions of photorefractive materials can be sensitively controlled by doping, oxidation and reduction treatments, and, in the case of BaTiO$_3$, by adjusting the crystal orientation, l_s, and r_h (or r_e). The densities of photorefractively mobile charges N_T can be adjusted in crystals of BaTiO$_3$ by oxidation and reduction treatments [6.17]. Another way to change the shapes of the temporal modulation functions would be to change either the mobility-lifetime product or the photoexcitation cross section s. These can both be expected to vary somewhat with temperature, at least over some temperature ranges. Since charge ranges in BaTiO$_3$ are known to vary by nearly an order of magnitude depending on crystal orientation [6.22, 23], single crystals of BaTiO$_3$ can be expected to have different temporal modulation transfer functions as the crystal is rotated with respect to the crystal's c-axis, with the crystal responding fastest for large wave vectors when the grating wave vector is oriented perpendicular to the c-axis. In [6.23], the erase rate curves were found to vary somewhat with the intensity and polarizations of the erase beams; the curves shown in Fig. 6.4 are for an erase intensity of 400 mW cm^{-2} and an ordinary erase beam polarization.

Materials such as GaAs and BSO in which the Debye screening radius k_s^{-1} is very different from the diffusion length or charge hopping range κ^{-1} exhibit a wider range of erase rates Γ_∞/Γ_0 as a function of grating wave vector than materials (such as BaTiO$_3$) in which these two quantities are comparable. In fact, the range of achievable erase rates is related to the ratio of the Debye screening to charge hopping wave vectors by $\Gamma_\infty/\Gamma_0 = \kappa^2/k_s^2$.

Experimental difficulties associated with the quasi-steady transient experiments have been dealt with extensively in [6.15]. These include the appearance of nonexponential decays at low erasing beam intensities, variations in signal-to-noise ratio at different grating wave vectors, the need to uniformly illuminate the crystal, the effect of absorption on the shape of the transient decay, the need for transparent electrodes, and optical activity. Since photorefractive decays are, for the most part, simple exponentials, no special care need be taken to see that the maximum obtainable grating amplitude is obtained for each different grating wave vector alignment, as the time constant associated with a simple exponential is independent of the initial amplitude of the grating. This eliminates the need for very thorough acoustical dampening and greatly reduces the importance of exactly matching path lengths for the various experimental geometries. The transient measurements can be made particularly insensitive to table vibrations by measuring the rates of grating erasure rather than grating formation rates. Since only one beam is present during grating erasure, there are no interference effects which are sensitive to acoustical vibrations. While well-matched path lengths will most definitely enhance the signal-to-noise ratio for the transient measurements, the time constants themselves are insensitive to any mismatching of path lengths. Setting up the experimental configurations at different grating wave vectors can be facilitated by the use of a narrow linewidth laser, but no special care need be taken to see that the path lengths for all angles are equally well-matched. The wavelength dependence of the densities and cross sections can be determined with the use of cw dye lasers.

Even without studying the whole range of grating wave vectors as a function of intensity, by making careful measurements at just two wave vectors ($K > k_s$ and $\kappa < K < k_s$), the photorefractive absorption $\alpha_{pr} = s(N_D - N_A)$ (which equals $\Gamma_\infty N_T$ in BSO) as a function of wavelength can be determined. The ratio of α_{pr} to the ordinary linear absorption α gives the photorefractive quantum efficiency $Q = \alpha_{pr}/\alpha$, for the fractional number of electrons caused to move (or hop) between trapping centers per photon absorbed by the crystal. Nonphotorefractive absorption (arising, for example, from band-to-band absorption or from the presence of extraneous impurities) is generally small in the best samples of these crystals but must be minimized for device applications in order to avoid wasting energy and unnecessarily heating the crystal.

a) Experimental Description

Experimental configurations which have been used previously for measuring the quantities N_T, sN_D/N_A, and $\mu\tau_e$ in Bi$_{12}$SiO$_{20}$ [6.7, 15] are shown in Figs. 6.5

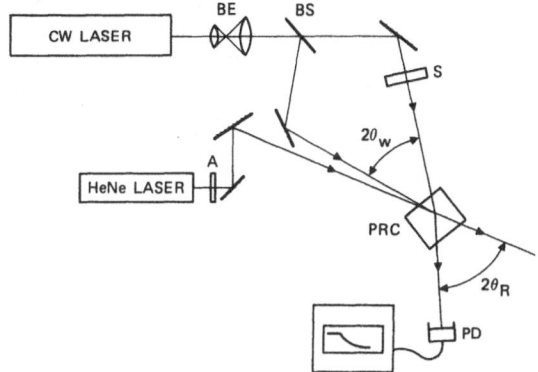

Fig. 6.5. A cw laser beam is expanded with beam expander BE and split into two beams with the beam splitter BS. The two writing beams interfere inside the photorefractive crystal PRC to write a grating of wave vector $K = 4\pi\sin\theta/\lambda$. The shutter S in one of the writing beams can be closed to turn off one of the writing beams and erase the photorefractive grating. A weak He-Ne laser continuously probes the grating, and the decay of the grating during erasure is captured by a photodetector and a transient recorder

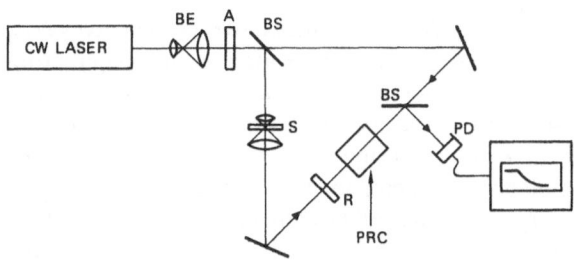

Fig. 6.6. The smallest fringe spacings of $\Lambda = \lambda/2n$ can be achieved when the grating is written with counterpropagating beams

and 6.6. To obtain sN_D/N_A and $\mu\tau_e$ the quasi-steady transient response must be monitored, as the steady-state behavior of photorefractive materials is insensitive to these parameters. In the apparatus shown in Fig. 6.5 the quasi-steady transient response is monitored by inserting a shutter in one of the writing beams, by using a fast detector on the diffracted signal and by capturing the transient decay of the diffracted signal with a transient recorder (such as a digital oscilloscope). The writing and erasing beams are expanded to illuminate the crystal uniformly, thus avoiding the buildup of bulk space-charge fields which have cylindrical symmetry along the beam paths. All three parameters can be obtained for BSO by adjusting the angle between the writing beams so that the fringe spacings vary from larger than the charge hopping distance or range $\mu\tau_e$ to smaller than the Debye screening radius. The measured data agree well with theory, as shown in Fig. 6.7.

Reflection gratings offer a convenient way to obtain fringe spacings a factor of five smaller than the Debye screening radius, as necessary to observe the asymptotic (large grating wave number K) limit of the intensity-normalized

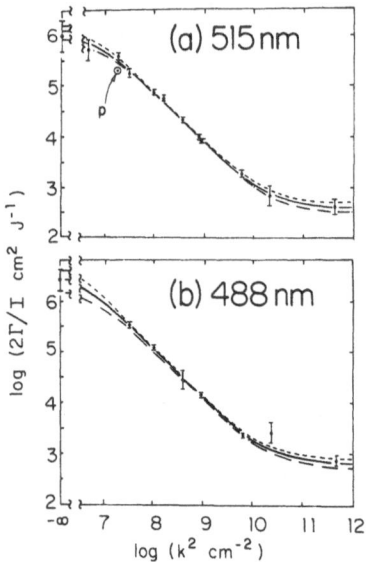

Fig. 6.7. Transient data for BSO. Experimental values for the intensity-normalized erasure rates $2\Gamma/I = 2/I\tau$ for the decay of the diffraction efficiency of a photorefractive grating, after [6.7, 15]. The solid curve is the best fit of (6.5) to the data, with 3 free parameters

photorefractive response rate $(\Gamma_\infty = sN_D/N_A)$, and to measure r_s (which gives N_T). In this geometry, shown in Fig. 6.6, the writing beams are counterpropagating and the fringe spacing inside the crystal is $\Lambda = \lambda/2n_3$; for $n_3 = 2.66$ (as in BaTiO$_3$) and $\lambda = 488$ nm one obtains $\Lambda = 92$ nm. If the crystal exhibits optical activity; as in the case of BSO, then a half-wave plate inserted into one of the beams assures that the writing beams are co-polarized and can consequently write a grating as they pass through the crystal. In practice this can be achieved by simply rotating the wave plate until a peak in the diffraction efficiency is observed. In this counterpropagating geometry, the helium-neon wavelength is too long to allow for Bragg matching and one of the writing beams must be used as the probe. The diffracted signal propagates backwards along the read-out beam and is monitored as a reflection from a beam splitter as shown.

In all of these configurations, the writing and erasing beams are expanded to uniformly illuminate the crystal, thus avoiding two possible complications. First, a well-expanded erase beam ensures a uniform erasure intensity in the region of the crystal being probed. Second, expanded beams ensure that the intensity contours transverse to the beam directions are uniform in the crystal, thus avoiding the occurrence of photorefractive self-focusing effects.

Fresnel reflections and their dependence on the angle of the erase beam with respect to the normal to the crystal surface must be accounted for in order to accurately measure absolute intensities inside the crystal.

At wavelengths for which $\alpha l \geq 1$, the intensity of the erase beam in the crystal is nonuniform due to absorption effects. The effect of this exponential decrease in erase beam intensity with distance into the crystal is to alter the shape of the temporal decay from a simple exponential to an exponential integral function. For large αl, the apparent initial time constant of the decay is lengthened somewhat; this effect can fortunately be quantified as in [6.7, 15] once the absorption of the crystal is measured. Then an appropriate correction factor on the intensity can be applied and the erase rates can be normalized by this average intensity inside the crystal.

Transient beam coupling between the probe and its first diffracted order can also alter the shape and apparent time constant of a decay under certain conditions. An analytical form for the temporal shape of an erasure decay in the presence of strong beam coupling has not yet been worked out. In practice it is not difficult to avoid beam coupling effects in BSO, which has negligibly small beam coupling in the absence of applied fields and moving gratings.

While transient beam coupling effects are negligible in the case of BSO, they can be large enough to affect the shape of a transient response in materials with larger electro-optic coefficients such as $BaTiO_3$. The effects of energy coupling in $BaTiO_3$ can be minimized by using thin crystals, orienting the grating wave vector to pick up the smallest r coefficient (r_{13}), consistently probing the grating with a laser at a wavelength where the crystal is relatively insensitive and by setting up an independent erase beam which is not Bragg-matched to scatter off of the existing grating. A convenient check to quickly assess the effects of beam coupling on the shape or time constant of an erasure (or writing) transient is to rotate the crystal by 180° so that the direction of the grating wave vector in the crystal is reversed. If no change is seen in the shape or the time constant of the decay, one can be sure that the effects of energy coupling are not distorting the shapes of the erasure decays.

Unfortunately, the applicability of this technique as a measurement of the charge range to materials with charge ranges smaller than about 100 nm is limited by band-to-band absorption effects which occur at short wavelengths. In $BaTiO_3$, for example, this technique can set limits on the charge range along the c-axis but cannot easily determine it directly with a high degree of accuracy, since in $BaTiO_3$ the charge range $r_h = (\mu\tau_e k_B T/e)^{1/2}$ is comparable to but not much smaller than the smallest achievable fringe spacing of 82 nm ($\lambda = 416$ nm, $\alpha \leq 1$ cm^{-1}, $n_3 = 2.55$). The accuracy of this technique in $BaTiO_3$ is also limited by uncertainties in measurements of ε parallel and perpendicular to the c-axis (Chap. 7). A photorefractive technique to accurately measure ε would be very helpful here.

b) Nonlinear Intensity Dependence

Unfortunately, the current photorefractive transport theories which are used to extract material parameters from these time-resolved photorefractive experiments all predict a linear dependence of grating erasure rates on intensity. However, a distinctly nonlinear (power law) dependence $1/\tau = I^x$ has been observed in $BaTiO_3$ in an experiment in which erasure intensities were varied over six order of magnitude [6.21]. Current theories provide no rigorous explanation for this power law dependence of photorefractive rates on intensity; the effect is thought to arise either from a distribution of impurities in the band gap or from band tailing. It is not known how these complications will affect the interpretation of experimental results, but it is interesting to note that the exponent $x=0.62$ is apparently close enough to unity that the density N_T of photorefractively mobile charges in at least one $BaTiO_3$ crystal measured using the transient technique is in excellent agreement (within 3%) with the density measured using steady-state beam coupling techniques (which appear to be unaffected by power-law complications) [6.6].

c) Nonexponential Decays

Complications arise when the observed decay is not a simple exponential, as is sometimes the case in BSO [6.15] and in $LiNbO_3$ [6.30]. Nonexponential transients can, in theory, result from several different effects, including beam coupling, transverse intensity nonuniformities, absorption, charge density modulations near unity, dark conductivity effects, and multiple species of charge trapping centers [6.15]. In BSO [6.15], nonexponential erasure decays such as are shown in Fig. 6.8 were observed when photorefractive gratings written with either 515 nm or 633 nm light were erased with 633 nm light from a weak (several milliwatts) helium-neon laser. Careful experimental design and the performance of several simple tests eliminated all of the above-mentioned possible causes of the nonexponential decays except for the possibility that they arose from the participation of several trapping centers in the photorefractive charge transport process.

While nonexponential decays observed in BSO qualitatively appeared to be a sum of two exponentials (as shown in Fig. 6.8), attempts to fit the data rigorously to a simple sum of two exponentials were not satisfactory enough to proceed with an analysis based on a simple model [6.32] of two types of trapping centers participating in the photorefraction.

In $LiNbO_3$ [6.30], nonexponential decays attributed to the participation of two types of trapping centers in the photorefractive charge migration process were observed and analysed as a sum of two exponentials. The effects of the participation of two types of trapping centers were assumed to be completely decoupled and the densities of these two centers were extracted from plots of the

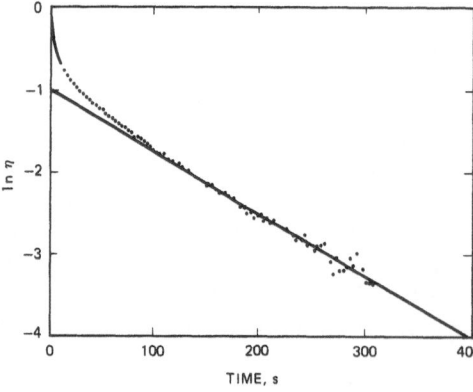

Fig. 6.8. The decay of a grating under erasure with red light from a He-Ne laser is not a simple single exponential

erase rates of the fast and slow components versus the grating wave vector. Corroborative steady-state beam coupling checks of these densities would help to determine the accuracy of the assumption that the two levels are completely decoupled.

6.4.3 With Externally Applied Fields

Stepanov et al. [6.33] were the first to observe the marked slowing down of the erasure rates of photorefractive gratings in BSO in the presence of an externally applied field. The use of the parametric dependence of erase rates on electric fields to measure crystal material parameters was, however, hampered until techniques were developed to monitor the often highly nonuniform and time-varying internal fields arising from externally applied fields.

Extreme spatial and temporal nonuniformities in the internal fields arising from externally applied fields in BSO are easily observable in a transmission experiment with the crystal between crossed polarizers. The same apparatus, once calibrated, can be used to measure internal fields directly. The electro-optic effect in such an experiment is simply used to probe the internal field in the particular small volume element of the crystal being studied [6.8]. Knowledge of the internal field in the small crystal region being probed can then be used to advantage in determining some of the quasi-steady parameters with measurements at a single (but carefully chosen) grating wave vector.

The band transport theory [6.14] predicts that the intensity-normalized grating response rate slows down in the presence of an externally applied field according to the expression

$$\Gamma(E) = \frac{1}{I\tau} = \frac{1}{I\tau_{di}} \frac{[1 + (K/k_e)^2][1 + (K/\kappa)^2] + (K\kappa/k_E k_s)^2}{[1 + (K/\kappa)^2]^2 + (K/k_E)^2} , \qquad (6.11)$$

where $k_E = (\mu\tau_e E_0)^{-1}$ is the inverse drift length.

An early experimental check of the validity of this expression in BSO is shown in Fig. 6.9. In this experiment, the electric field inside the crystal was determined by fitting the data to (6.11), and was found to be equal to about one-third the value of the ratio of the externally applied voltage to the electrode separation. This voltage dividing factor could be determined with confidence, as all other relevant photorefractive parameters in this crystal sample had been measured independently in a zero-field cw transient experiment, and as the data with this correction factor fit the prediction of the band transport model so well.

Without knowledge of a particular crystal's material parameters, one must measure the internal field arising from an externally applied field directly, using the electro-optic effect as in [6.8], or by measuring and calibrating absolute diffraction efficiencies, along the lines of [6.24]. Once the internal field arising from an externally applied field is known, the crystal parameters can be determined from the parametric dependence (6.11) of the erase rates on externally applied fields.

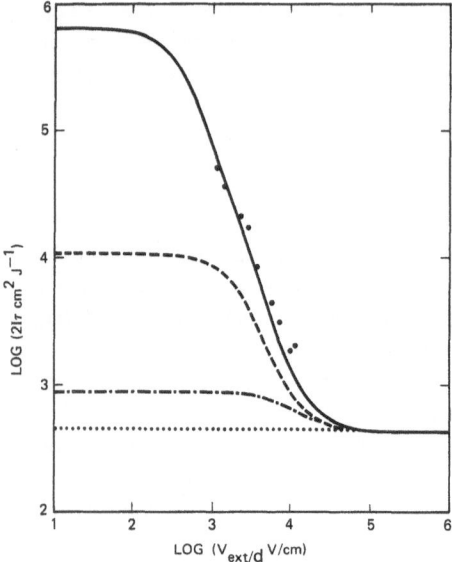

Fig. 6.9. Early experimental check of the validity of the band transport model. Dependence of photorefractive grating erasure rates on the magnitude of an externally applied field: ○: experiment, $K = 2.53 \times 10^3$ cm^{-1} ($\Lambda = 25$ μm). ——: fit of the data to (6.11) using independently measured photorefractive parameters of this crystal sample and the relation $E_{int} = V_{ext}/3d$, where d is the electrode separation. The three other curves are predictions of (6.11) for larger grating wave vectors. - - -: $K = 8.79 \times 10^4$ cm^{-1} ($\Lambda = 0.72$ μm); — ·: $K = k_s = 1.35 \times 10^5$ cm^{-1} ($\Lambda = 0.47$ μm); · · ·: $K = 6.39 \times 10^5$ cm^{-1} ($\Lambda = 98$ nm), the largest grating wave vector, obtainable with counterpropagating writing beams

The response rate of a small wave vector grating ($K < k_s$ and $K < \kappa$) under the influence of an externally applied field is given by a particularly simple expression

$$\Gamma_0(E) = \frac{1}{I\tau} = \frac{1}{I\tau_{di}} \frac{1 + (K\kappa/k_s k_E)^2}{1 + (K/k_E)^2} \; , \tag{6.12}$$

which has a zero-field limit equal to the intensity-normalized dielectric relaxation rate

$$\Gamma_0(E=0) = \lim_{k_E \to \infty} \frac{1}{I\tau} = \frac{1}{I\tau_{di}} \tag{6.13}$$

and a large field limit of

$$\Gamma_0(E=\infty) = \lim_{k_E \to 0} \frac{1}{I\tau} = \frac{1}{I\tau_{di}} \left(\frac{\kappa}{k_s}\right)^2 \; . \tag{6.14}$$

The intercept of this expression on a log-log plot of rates versus the internal electric field E_{in}^2 is

$$b = \frac{1}{I\tau_{di}} \frac{k_B T}{e} \left(\frac{\kappa}{k_s}\right)^2 \; .$$

Knowing E_0 from measurements of the electro-optic effect, one can get $\kappa_e^{-1} = r_e = (\mu\tau_e k_B T/e)^{1/2}$. The density N_T is then obtainable with information from the large applied field asymptote $\Gamma_{E=\infty}$ according to the expression

$$k_s^2 = \frac{\Gamma_{E=0}}{\Gamma_{E=\infty}} \kappa^2 \ . \tag{6.15}$$

Two problems associated with this technique, which limited the range of fields shown in the figure, are a low signal-to-noise ratio for the very weak diffraction efficiencies characteristic of small grating wave vectors in the absence of external fields, and electrical breakdown across the crystal surface in the high field limit. The first of these problems is shared with the zero-field quasi-steady techniques, and should be made less severe by using lock-in detection techniques. The second problem of surface breakdown can be minimized by using a crystal whose surfaces have been very carefully cleaned and polished, then performing the experiment in an atmosphere which has a very high threshold for electrical breakdown.

Alternatively, the signal-to-noise ratio problems at small grating wave vectors can be lessened somewhat by performing the experiment at a slightly larger grating wave vector and fitting the data to the more complicated expression, but at the expense of a weakening dependence of the response rates on E_0, as shown in Fig. 6.9. In the limit of the largest grating wave vectors, photorefractive response rates are independent of the strength of any externally applied fields.

6.5 Intense Short-Pulse Measurements

In intense short-pulse experiments, using a configuration such as shown in Fig. 6.10, a nonequilibrium distribution of free electrons (or holes) is created with an intense short (nanosecond or picosecond) pulse of light. The excited conduction band electrons (or valence band holes if holes are the dominant carrier) will separate from one another by diffusion even after the light pulse has ended to form a space-charge grating in a time on the order of the electron trapping time (also known as the electron recombination time) [6.34]. The amplitude of the grating can be probed continuously after the writing pulse has ended, and the time constant to reach the steady state can be measured. This time constant

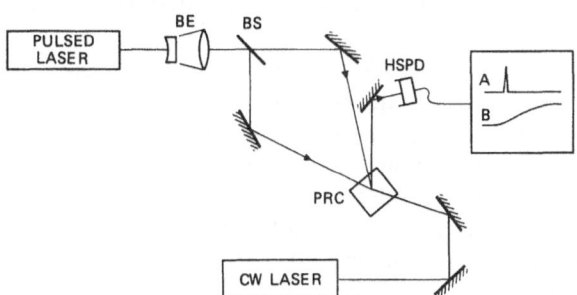

Fig. 6.10. Short pulse schematic to illustrate the use of a short intense pulse of light and high-speed detectors (or a mechanical delay apparatus) to resolve the electron trapping time

should be exactly equal to the trapping time, as long as the trapping time (τ_e or τ_h) is shorter than the time required for a charge to diffuse a grating period. An experiment to resolve the buildup or erasure of such a short-pulse grating thus provides a way in which the electron trapping time τ_e can be measured separately from the mobility. Knowledge of the $\mu\tau_e$ product from the quasi-steady transient technique described above then should give a measurement of the mobility separately from τ_e.

Doubled Nd:YAG lasers are conveniently near the optimal wavelength for photorefractive recording in materials such as BSO, BaTiO$_3$, and KTN. The 1.06 μm fundamental of a YAG is nearly optimal for studying GaAs. Hundreds of millijoules per pulse are available in 10–40 ns pulses so that large amplitude gratings can easily be written with single pulses (with due consideration of damage thresholds of photorefractive crystals).

As discussed above, the quantities N_A and $\mu\tau_e$ can be obtained from cw measurements. Therefore, since $\tau_e = 1/\gamma_R N_A$, once one has measured τ_e, one can then get both μ and γ_R.

One very direct way to resolve τ_e would be to write (or erase) a grating with a laser pulse that is short compared to the recombination time, the dielectric relaxation time and the diffusion time. In this limit, essentially a delta function illumination pulse, all of the photorefractive grating formation or erasure takes place on a time scale long compared to the pulse length. As shown in [6.34], the band transport theory predicts that the grating will be completely formed in a time equal to a few recombination times τ_e, provided τ_e is much shorter than the time required to diffuse a grating period. For BSO, if τ_e were as long as it was thought to be (several microseconds), it would be easily resolvable as the grating erasure or writing time constant in this sort of experiment. The recent results of this experiment [6.35] indicate that either τ_e is much shorter than was previously thought or μ is much larger. Experiments with grating periods of 21 μm and larger, large enough to reach the regime $\tau_e < \tau_D$, are probably impractical. For BaTiO$_3$, in which pulsed experiments have been done only in crystals in which holes are thought to dominate, the relevant time constant is the time required to trap a hole, τ_h. Attempts along these lines to resolve the hole trapping time τ_h in BaTiO$_3$ using doubled YAG lasers with pulse widths of 20 ns [6.36] and 8 ns [6.22] have indicated that τ_h in BaTiO$_3$ is shorter than one nanosecond [6.22]. More recently, experiments with a mode-locked, doubled Nd:YAG laser [6.37] with a 30 ps pulse duration indicate that τ_h in BaTiO$_3$ is probably less than 100 ps.

A reliable photorefractive mobility measurement opens up other areas of investigation, one of the most interesting being the dependence of mobility on temperature. Erase rate as a function of temperature has been measured [6.21, 38], but in BaTiO$_3$ the range of variation of temperature was limited by the requirement to avoid phase transitions. In BSO these measurements have been made over a much larger range of temperatures [6.38], but as yet at only one (unspecified) grating wave vector. The dependence of mobility on temperature indicates whether the mobility is trap-related or intrinsic, and could help to further discriminate between hopping and band transport.

6.6 Directions for Future Experiments

The techniques just described can be applied repeatedly to crystals with different doping, radiation, or treatment histories in order to learn how the different dopants or treatments affect photorefractive cross sections, diffusion lengths, densities, and mobilities. Comparison of theoretical cross sections with measurements of the photorefractive cross section as a function of wavelength in a material with a known photorefractive center could help to clarify the effects of the host lattice on the wave functions of charges trapped at deep impurity or defect centers. The effect of temperature on the photorefractive mobility and conductivity could be determined with this technique in order to help ascertain whether the charge transport is via band transport or hopping. The recent realization that $BaTiO_3$ occurs in both n-type and p-type forms [6.25] and that the relative dominance of one type of carrier over another can be altered after growth by oxidation and reduction promises to enable the close tailoring of these materials to specific applications requiring various degrees of coupling strengths and speeds of response. Effects expected to occur in the limits of high intensity (when the quasi-steady approximation cannot hold and the crystal is in a highly transient regime) which are just now beginning to be investigated [6.39–41] promise to be very interesting.

6.6.1 Temperature

While the room temperature quasi-steady transient measurements just described provide an excellent means to quantify the lumped parameters N_T, sN_D/N_A, and r_e (or r_h), they provide no further information, being notably insensitive to the scattering processes and other details that mediate the charge transfer from one site to another. Application of an electric field to BSO has indicated that the charge hopping probability has the same electric field dependence as would be expected from a conduction band charge transport mechanism. Seemingly at odds with this observation are the results of bulk ac dark conductivity and photoconductivity experiments that indicate that the charge transfer in BSO is accomplished via hopping in bulk photoconductivity experiments: an exponential rise in mobility with an increase in temperature and a photoconductivity that is an increasing function of the frequency of an applied electric field.

No analogous ac photorefractive experiments have been done to uncover the nature of photorefractive charge transport, although it is easy to see how the experimental techniques described in the previous section could facilitate such an investigation.

a) Temperature Dependence Intrinsic to Photorefractive Transport Equations

The band transport equations of *Kukhtarev* [6.14] indicate that, in the absence of a temperature dependence of the mobility or electrical conductivity, photorefractive response rates increase linearly with increasing temperature whenever the grating wave vector is larger than the Debye screening wave vector ($K > k_s$).

In BSO, this is only in the large wave vector limit $K > k_s > \kappa$ where

$$\Gamma_\infty(T) = \frac{\kappa^2}{I\tau_{di}} \frac{\varepsilon\varepsilon_0 k_B T}{N_T q^2} \ . \tag{6.16}$$

In BaTiO$_3$, a linear temperature dependence of erase rates should be observable over a wide range of experimentally obtainable grating wave vectors, including $k_s < K < \kappa$, where

$$\Gamma(T) = \frac{K^2}{I\tau_{di}} \frac{\varepsilon\varepsilon_0 k_B T}{N_T q^2} \ . \tag{6.17}$$

For grating wave vectors smaller than a Debye screening wave vector, no temperature dependence of Γ is called out explicitly in the band transport theory. Since the temperature dependence of Γ observed in BSO [6.38] was apparently observed for $K < \kappa$, it must directly reflect a temperature dependence in μ, σ, τ_{di}, n, or in some other lumped parameter of the crystal.

b) Temperature Dependence of Mobility

The temperature dependence of the dark mobilities and conductivities of BSO [6.9] and BGO [6.10] have been measured and found to be increasing functions of temperature over a wide range of conditions. Photocurrents in BGO were found to be either increasing or decreasing functions of temperature depending on the temperature range and wavelength of illumination [6.42]. Thus, it is clear that a more complicated temperature dependence of photorefractive erase rates than is explicitly included in the band transport theories can be expected.

c) Measurement of Activation Energies

The temperature dependence of photorefractive erase rates has been used in BaTiO$_3$ [6.17, 21] to determine the activation energy of photorefractive centers. This technique will beome increasingly useful as the crystal growth technology develops to alter the types of impurities acting as photorefractive trapping centers.

d) Importance to Applications

The practical importance of the effect of temperature on photorefractive response rates is inarguable when experiments indicate that the rate of response of BSO to red light (633 nm) increases by a factor of 16 [6.38] as the temperature is raised from room temperature to 300 °C and that the response rate to 515 nm illumination of BaTiO$_3$ varies by over two orders of magnitude from 20 °C to 125 °C [6.43].

6.6.2 Topics Not Yet Investigated

As mentioned above, it seems likely that measurements of the photorefractive effect with short intense pulses will produce effects that cannot be explained by analysis [6.34] of the rate equations given by *Kukhtarev* [6.14]. Two-photon ionization and the photorefractive effect have already been investigated [6.44, 45]. Excited state polarization and pyroelectric effects may be important [6.46–48]. The carrier number density may be large enough that the nonlinearities usually seen in semiconductors are important [6.49]. Some of these effects have been recently observed in short pulse experiments in GaAs [6.39–41]. If the grating formation time is fast enough it may be necessary to use the frequency dependent dielectric constant $\varepsilon(\omega)$ in place of the static dielectric constant to describe the formation of the space-charge field.

Separation of these effects from the standard photorefractive effect will undoubtedly open up many new and very rich areas for investigation.

Acknowledgements. Helpful discussions with Dr. R. W. Hellwarth, Dr. M. B. Klein, and Dr. G. C. Valley are gratefully acknowledged.

References

6.1 A. Chantre, G. Vincent, D. Bois: Phys. Rev. B**23**, 5335 (1970)
6.2 R. K. Willardson, C. Beer: *Semiconductors and Metals* (Academic, New York 1984)
6.3 D. V. Lang, R. A. Logan: J. Electron. Mater. **4**, 1053 (1975)
6.4 M. B. Klein: Proc. SPIE **519**, 12 (1984)
6.5 M. B. Klein, R. N. Schwartz: J. Opt. Soc. Am. B**3**, 293 (1986)
6.6 J. Feinberg, D. Heimen, Jr., A. R. Tanguay, R. W. Hellwarth: J. Appl. Phys. **51**, 1297 (1980)
6.7 R. A. Mullen, R. W. Hellwarth: J. Appl. Phys. **58**, 40 (1985)
6.8 J. M. C. Jonathan, R. W. Hellwarth, G. Roosen: IEEE J. QE-**22**, 1936 (1986)
6.9 V. P. Avramenko, A. Yu. Kudzin, G. Kh. Sokolyanskii: Sov. Phys. – Solid State **22**, 1839 (1980)
6.10 V. P. Avramenko, L. P. Klimenko, A. Yu. Kudzin, G. Kh. Sokolyanskii: Sov. Phys. – Solid State **19**, 702 (1977)
6.11 V. P. Avramenko, A. Yu. Kudzin, G. Kh. Sokolyanskii: Sov. Phys. – Solid State **26**, 290 (1984)
6.12 M. Pollak: *Some Aspects of Non-Steady-State Conduction in Bands and Hopping Processes* (Institute of Physics, London 1962) pp. 86–93
6.13 N. V. Kukhtarev, V. B. Markov, S. G. Odulov, M. S. Soskin, V. L. Vinetskii: Ferroelectrics **22**, 949, 961 (1979)
6.14 N. V. Kukhtarev: Sov. Tech. Phys. Lett. **2**, 438 (1976)
6.15 R. A. Mullen: "Time-Resolved Holographic Measurements of Bulk Space-Charge Gratings in Photorefractive $Bi_{12}SiO_{20}$"; Ph. D. Thesis, University of Southern California (1984)
6.16 G. C. Valley, M. B. Klein: Opt. Eng. **22**, 704 (1983)
6.17 S. Ducharme, J. Feinberg: J. Opt. Soc. Am. B**3**, 283 (1986)
6.18 M. B. Klein: Opt. Lett. **9**, 350 (1984)
6.19 A. Krumins, P. Günter: Appl. Phys. **19**, 153 (1979)

6.20 Y.H.Ja: Opt. Quantum Electron. **16**, 355 (1984)
6.21 S.Ducharme, J.Feinberg: J. Appl. Phys. **56**, 839 (1984)
6.22 Chyr-Pwu Tzou, T.Y.Chang, R.W.Hellwarth: Proc. SPIE **613**, 58 (1986)
6.23 T.Y.Chang: "Non-linear Optical Studies of Photorefrative Barium Titanate: Parameter Measurements and Phase Conjugation"; Ph. D. Thesis, University of Southern California (1986)
6.24 G.Pauliat, J.M.Cohen-Jonathan, M.Allain, J.C.Launay, G.Roosen: Opt. Commun. **59**, 266 (1986)
6.25 M.B.Klein, G.C.Valley: J. Appl. Phys. **57**, 4901 (1985)
6.26 R.Orlowski, E.Krätzig: Solid State Commun. **27**, 1351 (1978)
6.27 F.P.Strohkendl, J.M.C.Jonathan, R.W.Hellwarth: Opt. Lett. **11**, 312 (1986)
6.28 M.G.Moharam, T.K.Gaylord, R.Magnusson, L.Young: J. Appl. Phys. **50**, 5642 (1979)
6.29 M.G.Moharam, T.K.Gaylord, R.Magnusson, L.Young: Ferroelectrics **27**, 255 (1980)
6.30 K.Tyminski, R.C.Powell: J. Opt. Soc. Am. B**2**, 440 (1985)
6.31 E.Krätzig, H.Kurz: Opt. Acta **24**, 475 (1977)
6.32 G.C.Valley: Appl. Opt. **22**, 3160 (1983)
6.33 S.I.Stepanov, V.V.Kulikov, M.P.Petrov: Opt. Commun. **44**, 19 (1982)
6.34 G.C.Valley: IEEE J. QE-**19**, 1637 (1983)
6.35 G.Lesaux, G.Roosen, A.Brun: Opt. Commun. **56**, 374 (1986)
6.36 L.K.Lam, T.Y.Chang, J.Feinberg, R.W.Hellwarth: Opt. Lett. **10**, 475 (1981)
6.37 A.L.Smirl, G.C.Valley, R.A.Mullen, K.Bohnert, C.D.Mire, T.F.Boggess: Opt. Lett. **12** (1987)
6.38 M.A.Powell, C.R.Petts: Opt. Lett. **1**, 36 (1986)
6.39 G.C.Valley, A.L.Smirl, M.B.Klein, K.Bohnert, T.F.Boggess: Opt. Lett. **11**, 647 (1986)
6.40 A.L.Smirl, G.C.Valley, K.Bohnert, T.F.Boggess: "Picosecond Photorefractive and Free-Carrier Transient Energy Transfer in GaAs at 1 μm", to appear in IEEE J. QE, Feb. (1988)
6.41 G.C.Valley, A.L.Smirl: "Theory of Transient Energy Transfer in Gallium Arsenide", to appear in IEEE J. QE, Feb. (1988)
6.42 B.Kh.Kostyuk, A.Yu.Kudzin, G.Kh.Sokolyanskii: Sov. Phys. – Solid State **22**, 1429 (1980)
6.43 D.Rytz, M.B.Klein, R.A.Mullen, R.N.Schwartz, G.C.Valley, B.A.Wechsler: "High Efficiency, Fast Response in Photorefractive BaTiO$_3$ at 120 °C", submitted to Appl. Phys. Lett.
6.44 D. von der Linde, O.F.Schirmer, H.Kurz: Appl. Phys. **15**, 167 (1978)
6.45 D. von der Linde, A.M.Glass, K.F.Rogers: Appl. Phys. Lett. **26**, 22 (1975)
6.46 A.M.Glass, D.H.Auston: Opt. Commun. **5**, 45 (1972)
6.47 A.M.Glass, D. von der Linde, D.H.Auston, T.J.Negran: J. Electron. Mater. **4**, 915 (1975)
6.48 M.E.Lines, A.M.Glass: *Principles and Applications of Ferroelectrics and Related Materials* (Clarendon, Oxford 1977)
6.49 R.K.Jain, M.B.Klein: "Degenerate Four-Wave Mixing in Semiconductors", in *Optical Phase Conjugation*, ed. by R.Fisher (Academic, New York 1983) pp. 307–415

7. Photorefractive Properties of BaTiO$_3$

Marvin B. Klein

With 16 Figures

Barium titanate (BaTiO$_3$) was one of the first ferroelectric materials to be discovered, and also one of the first to be recognized as photorefractive [7.1]. The particular advantage of BaTiO$_3$ for photorefractive applications is the very large value of its electro-optic tensor component r_{42}, which in turn leads to large values of grating efficiency, beam coupling gain and four-wave mixing reflectivity. As an example, four-wave mixing reflectivities as large as 20 have been observed in BaTiO$_3$ [7.2], with no electric field applied to the crystal. Such large reflectivities are particularly desirable in phase conjugate resonator applicattions, where to date BaTiO$_3$ has been the material of choice.

The first observation of the photorefractive effect in BaTiO$_3$ was reported by *Townsend* and *LaMacchia* in 1970 [7.1]. This report came quite soon after the first reports of optical damage [7.3] and holographic storage [7.4] in LiNbO$_3$. In the work of *Townsend* and *LaMacchia* [7.1], simple sinusoidal holograms were written at 514.5 nm in melt-grown samples of BaTiO$_3$, and were read out at 632.8 nm. The diffraction efficiency and response time of the holograms were measured as a function of several experimental parameters, and a model was developed to explain some of the results. In 1972, *Micheron* and *Bismuth* [7.5] reported diffraction efficiency measurements on flux-grown samples doped with Fe and Ni. It was noted that Fe was more effictive than Ni, and an optimum Fe concentration was determined. Between 1972 and 1980, the photorefractive effect in BaTiO$_3$ received little or no attention, except for the studies of the bulk photovoltaic effect (using uniform illumination) by *Koch* et al. [7.6, 7]. The lack of progress on BaTiO$_3$ during this time interval was probably due to the sustained interest in LiNbO$_3$ for holographic memory applications and to the scarcity of BaTiO$_3$ samples with high optical quality.

In 1980, a renewal of interest in BaTiO$_3$ was generated by the work of *Feinberg* et al. [7.8] and *Krätzig* et al. [7.9], in which many of the favorable features of this material for real-time applications were demonstrated, and several important material parameters were measured. Since 1980, a large number of papers on the fundamental properties and applications of BaTiO$_3$ have been published. For a description of device applications, see *Topics in Applied Physics*, Vol. 62, Chaps. 4 and 5.

At present, the advantages of BaTiO$_3$ for phase conjugate resonators and other applications requiring large signal levels are widely recognized. At the same time, many of the physical, optical and dielectric properties of this material are not widely understood. The purpose of this chapter is to review the important properties of $BaTiO_3$, especially as they relate to the photorefractive effect. The photorefractive effect can be a powerful technique in itself for making physical measurements, and this feature will be emphasized. In Sect. 7.1, we discuss the crystal growth, lattice and domain structure of $BaTiO_3$, and describe the dielectric and electro-optic properties of this material. A description of defect and impurity centers in $BaTiO_3$ is given in Sect. 7.2, with special emphasis on the photorefractive centers. Section 7.3 reviews the band transport model of the photorefractive effect, and its application to the particular case of $BaTiO_3$. In Sect. 7.4 we review the photorefractive measurements of materials properties in $BaTiO_3$. Other characterization studies of photorefractive crystals are described in Sect. 7.5. Techniques for optimizing the photorefractive properties of $BaTiO_3$ are described in Sect. 7.6, and our conclusions are given in Sect. 7.7.

7.1 Basic Properties and Technology

7.1.1 Crystal Growth

The early experiments on ferroelectricity in $BaTiO_3$ were performed on ceramic samples. The first practical technique developed for the growth of single crystals was the KF-flux or Remeika technique [7.10]. In this technique, $BaTiO_3$ powder is dissolved in molten KF. The mixture is then cooled to $\sim 900\,°C$, the saturation temperature of the mixture. Upon further cooling, spontaneous crystallization occurs at the bottom of the crucible, producing single crystals in the form of platelet pairs attached to each other along one edge at an angle of 39°. The resulting crystals resemble butterflies, and the platelets are sometimes called "butterfly wings". This unusual morphology is due to (111) twinning, and is characteristic of KF-flux-grown crystals. While the Remeika crystals were relatively easy to grow and their quality was satisfactory for the early measurements on $BaTiO_3$, the use of these crystals for applications was limited by scattering due to flux inclusions and absorption due to substitution of K and F for the Ba and O ions in the lattice.

A major advance in $BaTiO_3$ crystal growth technology was the development of the top-seeded solution growth (TSSG) technique in the late 1960s [7.11]. In this technique, a solution of BaO and TiO_2 was used, with TiO_2 added in excess of the amount necessary for stoichiometry. The purpose of the excess TiO_2 was to lower the melting point of the solution from a value of $\sim 1620\,°C$ (in stoichiometric $BaTiO_3$) to $\sim 1400\,°C$ (with approximately 65 mol% TiO_2 in the melt). This eases the requirements on furnace and crucible, but also avoids the high temperature hexagonal phase of stoichiometric $BaTiO_3$, which is deleterious to the production of high quality crystals. The actual crystal growth process is performed in air, and involves the slow cooling of the solution by $\sim 40\,°C$, and the simultaneous pulling of the crystal from the melt. Single crystal boules grown by the TSSG technique are typically 2–3 cm in diamter, and polished, poled samples from these boules (typically $0.5 \times 0.5 \times 0.5$ cm^3) are generally clear and free of visible defects. The only significant disadvantage of the TSSG technique is that it is incongruent, and is thus characterized by a slow growth rate.

7.1.2 Lattice Structure

At the growth temperatures used for the TSSG technique, the crystals which nucleate are cubic, with point group $m3m$ (symmetry O_h). Upon cooling, BaTiO$_3$ transforms to a tetragonal, ferroelectric phase (point group $4mm$, symmetry C_{4v}) at $T_c \simeq 130\,°C$, where T_c is the Curie temperature. The exact value of T_c in a given crystal depends on the concentration of impurities and other defects. Values of T_c a few degrees above $130\,°C$ have been reported in pure, melt-grown crystals [7.12, 13], while values closer to $120\,°C$ are measured in flux-grown crystals [7.14, 15].

The tetragonal phase of BaTiO$_3$ is the stable one at room temperature, and is of primary interest for applications. At $\sim 9\,°C$ a transition to an orthorhombic phase (point group $mm2$) occurs. This phase transition is of some practical concern. Crystals cooled too rapidly through this phase transition are subject to cracking, and thus excessive cooling during shipment and handling must be avoided. At $-90\,°C$ a third and final transition to a rhombohedral phase (point group $3m$) occurs.

BaTiO$_3$ is a member of the perovskite family of ABO$_3$ compounds, which includes other well-known materials such as KNbO$_3$, KTaO$_3$, PbTiO$_3$ and SrTiO$_3$. The structure of BaTiO$_3$ (and other perovskites) in the cubic phase is a simple one (Fig. 7.1), with Ba^{2+} ions at the cube corners, Ti^{4+} ions at the body centers, and O^{2-} ions at the face centers. The structure can be considered as a rigid grouping of oxygen octahedra linked at their corners by shared oxygen ions. The Ti^{4+} ions thus lie at the center of each octahedron, while the Ba^{2+} ions lie outside the octahedra. Many of the dielectric and optical properties of BaTiO$_3$ are determined by the characteristics of the basic TiO$_6^{8-}$ octahedron. For example, the energy band structure of BaTiO$_3$ near the band edge is determined by titanium $3d$ orbitals (which are responsible for the low-lying levels of the conduction band) and oxygen $2p$ orbitals (which are responsible for the upper

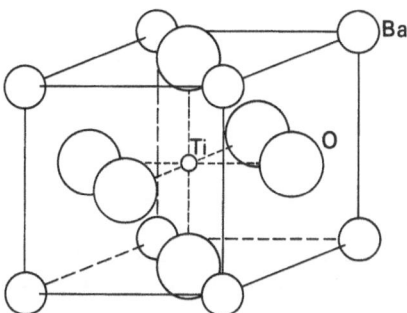

Fig. 7.1. BaTiO$_3$ unit cell. The size of the ions is reduced for clarity

levels of the valence band). Furthermore, the dominant transverse optical (TO) mode in BaTiO$_3$ (which largely determines the dielectric properties in the cubic phase) consists of a vibration of the Ti^{4+} ion with respect to the TiO$_6^{8-}$ octahedron.

The cubic-tetragonal phase transition near 130 °C is a displacive one, the most important change being a shift of the Ti^{4+} ion from the center of the octahedron toward an oxygen ion at one of the face centers of the cubic unit cell. This shift produces a spontaneous polarization in the direction of motion. The polar nature of BaTiO$_3$ leads to a variety of well-known properties, including the piezoelectric effect, the pyroelectric effect, the electro-optic effect, and second harmonic generation. The reversibility of the polarization in BaTiO$_3$ leads to its well-known ferroelectric properties.

7.1.3 Domains and Poling

In the tetragonal phase of BaTiO$_3$ the spontaneous polarization may be oriented along any of the six pseudodocubic $\langle 001 \rangle$ directions. Different regions of the crystal may polarize in each of these directions, each region of uniform polarization being referred to as a domain. In an as-grown crystal, the multidomain state that appears on cooling generally produces no net polarization. Therefore, such crystals show very small, if any, piezoelectric, pyroelectric and electro-optic effects. These effects depend on the direction and sign of the polarization, and are thus averaged out to zero in a multidomain crystal. For this reason, it is important to find a procedure (called poling) that leads to a single domain state [7.16].

The domain structure of BaTiO$_3$ in the tetragonal phase is shown schematically in Fig. 7.2. Two types of domains are found to exist: domains whose polarizations lie at 90° to each other, and domains with antiparallel polarizations. The former, called 90° domains, are readily seen when the crystal is placed between crossed polarizes, and may frequently be seen by the unaided eye. The 180° domains are more elusive, being observable only by etching or more elaborate techniques. Both types of domains must be eliminated in order to reduce scattering and produce useful electro-optic properties.

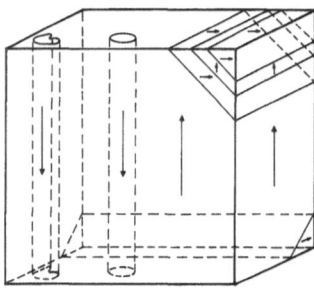

Fig. 7.2. Domain structure of BaTiO$_3$ in the tetragonal phase. The arrows represent the direction of polarization

The poling process always involves the nucleation of new domains and the movement of domain walls. In this picture, the single domain state is achieved when nucleation is complete and when all walls are induced to migrate to the boundaries of the crystal, where they disappear. There are two physical effects which tend to impede the movement of domain walls, and thus make poling more difficult: (1) imperfections (e.g. inclusions, dislocations, strains) tend to pin walls, and (2) charges at the walls associated with local changes in polarization are transported very inefficiently in insulating crystals. In fact, it is known that conducting samples of BaTiO$_3$ (produced permanently by reduction [7.17] or temporarily by illumination [7.18]) are much easier to pole than insulating samples.

A number of related techniques are used at present for the poling of melt-grown samples of BaTiO$_3$. They all rely on a particular combination of electric. field, heating, or an applied stress to produce a single domain sample. In all cases, as-grown samples are first oriented, cut and polished, with faces normal to the $\langle 100 \rangle$ family of axes in the cubic system. In the first technique, the 90° domains are removed at room temperature by applying uniaxial pressure alternately along two different directions. The resulting sample is left with only antiparallel 180° domains along the third direction. Following this mechanical poling process, an electric field is applied along the third direction at a temperature slightly below T_c. The sample is then cooled to room temperature, and the field is then removed. This step results in the removal of 180° domains, and produces a single domain crystal. It should be noted that the prolonged application of an electric field at temperatures near T_c can produce electrochemical damage in the form of cracks at the positive face of the crystal. This damage may be related to the large mobility of oxygen vacancies, which are positively charged and move to the negative electrode [7.19, 20]. This results in the liberation of free oxygen at the positive electrode, with the cracking resulting from the loss of stoichiometry or the buildup of pressure in pre-existing microcracks. The cracking could also be due to the lowering of the phase transition temperature (at the positive electrode) to a point below the temperature of the experiment. Regardless of the mechanism, the electrochemical damage can be controlled by reducing the electric field and cooling as fast as possible (except at the phase transition). However, these restrictions may not favorable for the elimination of all domains. Another mechanism for the electrochemical damage is the migration of metal ions into the crystal from the positive electrode (e.g. Ag$^+$ ions when silver paint is used for the contacts) into the bulk of the crystal.

A second poling technique involves cooling the sample through the cubic-tetragonal phase transition in the presence of an applied electric field [7.12]. This has the obvious advantage of eliminating the need for mechanical poling. However, there is a risk of cracking if the sample is cooled too rapidly through the phase transition. In addition, this technique may not eliminate all 180° domains [7.12, 13]. Recently *Muser* et al. [7.21] have improved upon this technique by first etching the sample to remove surface strains, which can pin domains. The sample is also held for an extended period of time at its highest temperature, in order that all charges associated with the original polydomain state can relax away.

A third poling technique involves the application of a large electric field to the sample at room temperature, with no previous mechanical poling. This technique is potentially the easisest one, and is commonly employed (with alternating fields) to study hysteresis loops in ferroelectrics [7.16]. However, it typically requires the application of large fields to overcome the sluggish motion of domain walls at room temperature.

After the completion of the poling process, it is important for many applications to determine the *degree of poling*. A useful figure of merit in this case is the *fractional poling factor F*, defined as the ratio of any second-order coefficient (e. g. electro-optic, nonlinear optic, piezoelectric, pyroelectric) in the sample in qeuestion to the same coefficient in a single domain sample. It is difficult to determine F from a direct measurement of a particular second-order coefficient, first because many of the measurement techniques are inaccurate (especially when the use of electrical contacts is required), and second because disagreement frequently exists as to the value of the coefficient in a single domain sample. One useful approach for determining (at least qualitatively) the degree of poling is direct observation of the domain structure. *Lines* and *Glass* [7.16] review a number of useful techniques for visualizing domains. One particularly promising technique for observation of 180° domains is etching of the c-faces of the sample [7.16]. Certain etchants (such as HF and HCl) act preferentially on the positive (or negative) c-faces. thus allowing direct visualization of the pattern of 180° domains on a c-face. This technique is more accurate for thin samples, in which 180° domains are more likely to extend completely to the opposite c-face.

7.1.4 Dielectric and Electro-optic Properties

The photorefractive performance of $BaTiO_3$ is significantly influenced by its dielectric and electro-optic properties. Specifically, the very large values of induced index change, beam coupling gain coefficient and degenerate four-wave mixing (DFWM) reflectivity are due to the large values of the electro-optic tensor components in $BaTiO_3$. Similarly, the relatively long time constants observed in $BaTiO_3$ are partially the result of the high values of dielectric constant in this material.

As indicated earlier, $BaTiO_3$ at room temperature is in a tetragonal phase, with uniaxial optical and dielectric behavior. The resulting values of dielectric constant are $\varepsilon_a = \varepsilon_b$, and ε_c. The magnitude and temperature dependence of ε_a and ε_c in the tetragonal phase are quite different [7.13], as shown in Fig. 7.3. In order to understand this behavior, we must first consider the behavior of $BaTiO_3$ on cooling through the cubic-tetragonal phase transition. At temperatures above T_c, the Ti^{4+} ion in each unit cell is in a cubic environment, and is highly polarizable along all three cubic axes, resulting in a large value of dielectric constant. Upon cooling toward the Curie temperature, the Ti^{4+} ion experiences

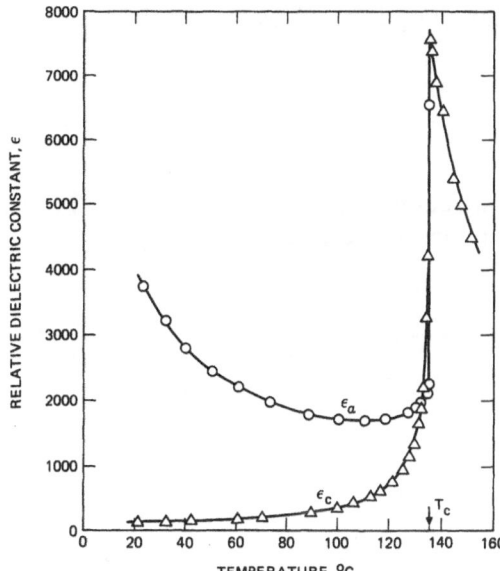

the formation of a double potential well along one particular cubic axis. At the phase transition, the depth of the double well has increased enough to "trap" the Ti^{4+} ion on one side or the other, leading to a macroscopic polarization, and a restricted polarizability for electric fields applied along the axis of polarization. However, for electric fields polarized along the a-axis (or b-axis), the Ti^{4+} ion remains free to vibrate and is highly polarizable. This leads to the large difference in the values of ε_a and ε_c as indicated in Fig. 7.3. Note also in the figure that ε_c is resonant at the cubic-tetragonal transition at $T_c \simeq 135\,°C$, consistent with the particular structural transformation at this temperature. On the other hand, the dielectric properties in the a-direction (or b-direction) are sensitive to the tetragonal-orthorhombic transition near $9\,°C$, where a second structural transformation occurs.

For a crystal of point group $4mm$, there are three independent components of the electro-optic tensor: $r_{113} = r_{223}$, r_{333}, and $r_{232} = r_{131}$ [7.22]. These coefficients relate the impermeability tensor $(1/n^2)_{ij}$ to the applied electric field E_k:

$$\Delta \left(\frac{1}{n^2}\right)_{ij} = \sum r_{ijk} E_k \tag{7.1}$$

An alternative (and physically more fundamental) description uses the field-induced crystal polarization as the driving term. In this case the linear effect is described by

$$\Delta \left(\frac{1}{n^2}\right)_{ij} = \sum f_{ijk} P_k , \tag{7.2}$$

where the f_{ijk} are the linear polarization-optic coefficients, and the quadratic effect is described by

$$\Delta\left(\frac{1}{n^2}\right)_{ij} = \sum g_{ijkl} P_k P_l \,, \tag{7.3}$$

where the g_{ijkl} are the quadratic polarization-optic coefficients. In BaTiO$_3$, the coefficients defined in the above relations are related by

$$r_{13} = f_{13} \varepsilon_0 \varepsilon_c \,,$$
$$r_{33} = f_{33} \varepsilon_0 \varepsilon_c \,, \tag{7.4}$$
$$r_{42} = f_{42} \varepsilon_0 \varepsilon_a \,,$$

$$r_{13} = 2 g_{12} \varepsilon_0 P_s \varepsilon_c \,,$$
$$r_{33} = 2 g_{11} \varepsilon_0 P_s \varepsilon_c \,, \tag{7.5}$$
$$r_{42} = g_{44} \varepsilon_0 P_s \varepsilon_a \,,$$

where P_s is the spontaneous polarization and ε_0 is the permittivity of free space [7.22]. In (7.4 and 7.5) we have used the contracted notation: $r_{113} = r_{13}$, $r_{333} = r_{33}$, and $r_{231} = r_{42}$. The relation to the quadratic coefficients (7.5) is quite useful, for several reasons. First, the explicit dependence on P_s is consistent with the fact that $r_{ij} = 0$ in the cubic phase. Second, the g-coefficients are nearly invariant with temperature, frequency and material for most oxide ferroelectrics [7.22]. Thus, the relations above are a form of Miller's rule [7.23], and show the explicit dependence of the electro-optic coefficients on the relative dielectric constant. The above relations clearly indicate why the coefficient r_{42} is so large in BaTiO$_3$: its magnitude is determined by the polarizability normal to the c-axis, and is thus proportional to ε_a. Thus, r_{42} is resonant at the tetragonal-orthorhombic transition [7.24] and r_{13} and r_{33} are resonant at the cubic-tetragonal phase transition [7.24].

The similarity in temperature dependence of r_{42} and ε_a in tetragonal BaTiO$_3$ has been noted by *Johnson* and *Weingart* [7.24], who plotted these two values together, as shown in Fig. 7.4. The enhancement of r_{42} near the tetragonal-orthorhombic phase transition leads to an increase in the photorefractive beam coupling gain Γ for the particular interaction geometry which exploits r_{42}. This enhancement in gain is sufficient to allow operation of a self-pumped phase conjugate resonator at 840 nm [7.25]. It was shown in this work that the phase conjugate resonator only exceeded threshold when the sample temperature was less than 18 °C.

There is considerable variation among the values of dielectric constant reported in the literature for BaTiO$_3$ [7.26]. These variations are related to the experimental conditions (especially the quality of the contacts), and to composition variations from sample to sample (e.g., flux-grown vs solution-grown samples). In the case of the electro-optic coefficients, the opposite

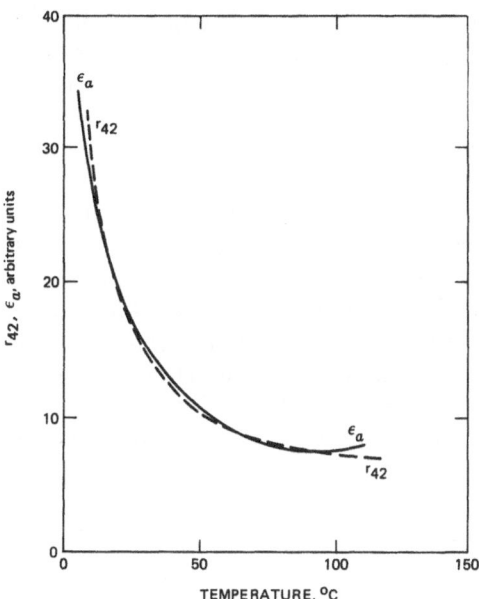

Fig. 7.4. Temperature dependence of the unclamped dielectric constant ε_a and the electro-optic coefficient r_{42} in single domain BaTiO₃. (After [7.24])

problem exists: there is a scarcity of measured values, with few measurements reported for any component [7.27]. Furthermore, most measurements have only been performed on flux-grown crystals. In Table 7.1 we list selected values of the dielectric constant and electro-optic tensor components in BaTiO₃. We have included values in Table 7.1 for both unclamped (constant stress) and clamped (constant strain) conditions. For an electric field applied uniformly throughout an unconstrained sample, the unclamped condition corresponds to applied frequencies below the bulk piezoelectric resonances of the sample, which can range between 1 MHz and 100 MHz. The clamped condition corresponds to applied frequencies above the piezoelectric resonances. At these high frequencies, all piezoelectric effects are frozen out, and thus do not contribute to the dielectric or electro-optic properties. In a mechanically constrained sample, the clamped coefficients are appropriate at all frequencies.

Table 7.1. Dielectric constants and electro-optic coefficients for single crystal BaTiO₃ at 20 °C. The dielectric constant data are taken from measurements on solution-grown samples. All values of electro-optic coefficient (except the recent measurement [7.28]) were obtained using flux-grown samples. Reference numbers are given in square brackets after each value

	ε_a	ε_c	r_{13} [pm/V]	r_{33} [pm/V]	r_{42} [pm/V]
Unclamped	3700 [7.12, 13]	135 [7.13]	24 [7.29] 19.5 [7.28]	80 [7.29] 97 [7.28]	1640 [7.24]
Clamped	2400 [7.12]	60 [7.13]	8 [7.30]	28 [7.30]	820 [7.31]

There has been some question over the years as to whether the dielectric constants and electro-optic coefficients used in models for the photorefractive effect should be the clamped or unclamped values. One might first expect that a sample uniformly illuminated throughout its volume by two intefering beams might be unclamped, while a sample illuminated in a portion of its volume might be considered as clamped (by the rigid mass of unilluminated crystal). However, in the illuminated portion, the strain or stress induced by the space charge electric field (via the piezoelectric effect) is spatially periodic along the direction of the space charge field, and no average strain or stress is induced. Thus, the crystal can be considered as unclamped in the direction of periodicity. In recent beam coupling experiments in $BaTiO_3$ [7.32, 33], the measured values of the electro-optic coefficient r_{13} in many samples were found to be larger than the literature value [7.30] for clamped conditions (Tables 7.1 and 7.3). This suggests that the appropriate coefficient for photorefractive measurements must be the unclamped value.

7.2 Band Structure and Defects

7.2.1 Intrinsic Band Structure

The electronic band structure of $BaTiO_3$ and other ABO_3 perovskite oxides has been extensively studied by *Michel-Calendini* et al. [7.34–40] and others [7.41] over the past few years. Two basic calculational approaches have been used. The NLCAO (nonorthogonal linear combination of atomic orbitals) approach has allowed the calculation of the band structure of $BaTiO_3$ throughout the Brillouin zone, using $2p$ orbitals centered on the oxygen ions and the $3d$ orbitals centered on the titanium ions in the unit cell [7.34, 35, 41]. The self-consistent field–multiple scattering–$X\alpha$ (SCF-MS-$X\alpha$) technique is used primarily for electronic structure calculations at the zone center. Its specific advantage is that it allows the calculation of energy levels due to substitutional impurities, as well as the intrinsic electronic structure. *Michel-Calendini* and co-workers have used the SCF-MS-$X\alpha$ technique to calculate the electronic structure of pure $BaTiO_3$ [7.36, 37], as well as $BaTiO_3$:Fe [7.37–39] and $BaTiO_3$:Co [7.40]; their work will be followed in the rest of this discussion.

On the basis of the SCF-MS-$X\alpha$ calculations carried out on TiO_6^{8-} clusters, the following picture emerges of the intrinsic electronic states in cubic (O_h) and tetragonal (C_{4v}) $BaTiO_3$. For both symmetries the top of the valence band is composed of oxygen $2p$ states, whereas the conduction bands are titanium $3d$ states. The states of the barium ion, on the other hand, lie well below the top of the valence band, and do not play an important role. In O_h symmetry the valence band edge can be decomposed into states with t_{1g} and t_{1u} symmetry, whereas the bottom of the conduction band has t_{2g} classification (Fig. 7.5a). In this picture, the optical band gap results from t_{1u}–t_{2g} transitions, while the thermal gap is due to t_{1g}–t_{2g} transitions. As the symmetry is lowered from O_h to C_{4v}, the cubic levels

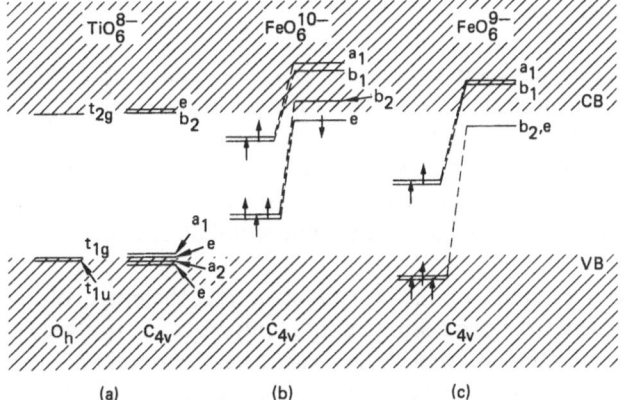

Fig. 7.5. Schematic representation of the energy levels of $M^{n+}O_6^{(12-n)-}$ clusters in BaTiO₃ : (a) TiO_6^{8-}, (b) FeO_6^{10-} (Fe^{2+}), and (c) FeO_6^{9-} (Fe^{3+}). In (b) and (c), the up-spin levels are located on the left side of each panel, and the down-spin levels are on the right. The levels in (a) and (c) have been calculated by the SCF–MS–Xα method [7.36, 39]. The levels in (b) are extrapolated from those of the $Fe^{2+}-V_0$ center [7.38]. The CB and VB correspond respectively to the conduction band minimum (t_{2g} Ti $3d$ states) and the valence band maximum (t_{1g} O $2p$ states of the TiO_6^{8-} cluster). Note the tendency of the higher valence states (e.g., Fe^{3+}) to lie lower in the band gap. Note also that Fe^{3+} can act either as a donor [occupied (a_1, b_1) states] or as an acceptor [empty (b_2, e) states]

split as well as shift. For example, in lower symmetry the levels (t_{1g}, t_{1u}) go over into ($a_1, e; a_2, e$) and (t_{2g}) into (b_2, e).

The optical gaps calculated by the Xα technique for undoped, tetragonal BaTiO₃ are 3.61 eV ($E \parallel \hat{c}$) and 3.56 eV ($E \perp \hat{c}$) [7.39]. The calculated values can be expected only approximately to predict the experimental values, for two reasons: (1) necessary approximations in the Xα approach limit its accuracy, and (2) all states in the conduction and valence bands are assumed to be discrete, so that transitions between them neglect broadening due to phonons, impurities, etc. The neglect of broadening should result in an overstimation of the band-gap energy.

The band edge properties of BaTiO₃ can also be elucidated through the study of optical absorption. In a series of absorption studies of solution-grown samples, *Wemple* [7.42] and co-workers determined that the absorption edge of tetragonal BaTiO₃ has a significant Urbach tail, due to indirect (phonon-assisted) transitions. The optical gap was defined as the photon energy at which the absorption coefficient was 5×10^3 cm^{-1}. The resultant values at room temperature are 3.27 eV ($E \perp \hat{c}$) and 3.38 eV ($E \parallel \hat{c}$). Other optical measurements have yielded values of 3.1 eV [7.43] and 3.3 eV [7.44]. Subsequent absorption studies [7.45] have shown that doping with Fe^{3+} ions leads to a decrease in the band-gap energy, in proportion to the Fe concentration. This is likely due to the introduction of O $2p$–Fe $3d$ charge transfer bands concentrated at energies just below the band gap energy.

7.2.2 Vacancies and Impurities

There are two major types of defects in crystalline BaTiO$_3$ which could lead to deep levels in the band gap (and thus participate in the photorefractive effect): vacancies and impurities. It is always expected that impurities will be present in the 10–100 ppm range, unless extraordinary precautions are taken in purifying starting materials and controlling the growth environment. Vacancies are also commonly present, if only to compensate for the charge imbalance introduced by the impurities.

Of the three elements which make up the BaTiO$_3$ structure, the most volatile is oxygen. During crystal growth or processing at high temperatures, this species is the most likely to evaporate, thus creating oxygen vacancies, denoted V_O. Since the oxygen ion in the BaTiO$_3$ lattice has a charge 2 −, an oxygen vacancy has a charge 2 +. Thus, an oxygen vacancy can trap one or two electrons, leading to donor levels in the BaTiO$_3$ band gap. The binding energy for the trapped electrons is thought to be quite small [7.44, 46–48], leading to shallow levels in the bandgap. *Berglund* and *Braun* [7.46] estimate an energy depth of 0.2 eV for an O vacancy with one trapped electron and 0.025 eV for an O vacancy with two trapped electrons. At room temperature these levels are expected to be thermally ionized, and thus not significantly populated.

Vacancies in BaTiO$_3$ may also be produced from the incorporation of excess TiO$_2$ during solution growth [7.47, 49]. Both Ba and O vacancies are formed by the reaction

$$TiO_2 \rightarrow V_{Ba} + Ti_{Ti} + 2O_O + V_O \;, \tag{7.6}$$

where the subscripts on the right-hand side refer to the site in the lattice. According to the commonly used phase diagram for the BaO-TiO$_2$ system [7.50], the solubility of TiO$_2$ in BaTiO$_3$ at the typical growth temperature of 1400 °C is ∼1 mol%, leading to a very large concentration of Ba and O vacancies. More recent measurements [7.47, 49] have shown that the solubility of TiO$_2$ in BaTiO$_3$ is at most 100 ppm. Thus, the maximum concentration of O and Ba vacancies due to TiO$_2$ incorporation is 100 ppm. When acceptor impurities are also present (see below), the additional O vacancies required for charge compensation will cause a corresponding reduction in the number of Ba vacancies, through the equilibrium relation [7.51]

$$[V_{Ba}] [V_O] = \text{const} \;.$$

Ba vacancies are deep acceptors [7.51], and could contribute to the photorefractive behavior of high purity crystals.

We now wish to consider the possible impurities in BaTiO$_3$ and their relation to the photorefractive effect. Three major groups of impurities have been shown to exist in BaTiO$_3$ [7.52–54]: (1) calcium and strontium, (2) aluminum and silicon, and (3) transition metals. Calcium and strontium are in the same family of the periodic table as barium. They almost always occur with barium in nature, and are difficult to separate from it. Since calcium and strontium are isovalent with barium and substitute for barium in the BaTiO$_3$ lattice, they do not introduce any levels in the BaTiO$_3$ band gap, and thus cannot participate in the photorefractive effect.

The sources of aluminum and silicon impurities are the furnace walls and heating elements. However, aluminum and silicon possess only one stable valence state (Al^{3+} and Si^{4+}), and thus these elements cannot support the intervalence transfer which is a key requirement of the photorefractive effect.

Transition metal impurities are ubiquitous in many oxide crystals, because of their abundance in nature, their chemical similarity to constituents of the compound in question (e.g., Ti in BaTiO$_3$), and possible contamination of starting powders from crucibles or metal utensils. The most likely transition metal impurities are Cr, Co, Ni, Fe and Cu, and they are all expected to be present at levels as high as 50 ppm in BaTiO$_3$. Fe is the most abundant of the transition metals, and could be expected at still higher concentrations. It is believed that transition metal impurity ions (as well as Al^{3+} and Si^{4+}) substitute for Ti^{4+} in the BaTiO$_3$ lattice, due to the close match between their ionic radius and that of Ti^{4+}. In the case of Fe impurities in BaTiO$_3$, this has been proven in a number of separate studies [7.53, 55]. Due to the low binding energies of their $3d$ electrons, each of the transition metals can exist in several valence states, ranging typically between $+1$ and $+4$.

7.2.3 Charge Balance

When an impurity ion (with valence $+3$ or less) substitutes for Ti^{4+} in BaTiO$_3$, a charge imbalance is created. In the simplest case, the compensation for this imbalance is achieved through the creation of oxygen vacancies:

$$[M^{3+}] = 2[V_O^{2+}]_I \, , \tag{7.7}$$

where $[M^{3+}]$ is the concentration of a trivalent metal ion and $[V_O^{2+}]_I$ is the *impurity-related* concentration of oxygen vacancies. Note that the charge of the metal ion is expressed in relation to the isolated metal atom; the charge in the crystal is -1. In the simple model above, no free carriers are required for compensation, and the samples are expected to be highly insulating.

The above model does not account for the fact that the oxygen vacancy concentration is also influenced by the growth or processing environment of the crystal. At a given temperature there may be an excess or a deficiency of oxygen, compared with the amount required to produce $[V_O^{2+}]_I$. In either of these limits, free carriers are produced to maintain the charge balance:

$$[M^{3+}] + n = 2[V_O^{2+}] + p \, , \tag{7.8}$$

where n is the concentration of electrons, and p is the concentration of holes. The growth environment of commercial BaTiO$_3$ (grown by the top-seeded solution growth technique) [7.11] is oxidizing (i.e., there is excess oxygen), leading to a reduction in the number of vacancies below the value $[V_O^{2+}]_I$, and the creation of free holes. Thus, in as-grown crystals the transition metals are generally acceptor impurities, and the samples are p-type in the dark [7.47, 56–58]. Conversely, samples processed at low oxygen partial pressures requires the production of free electrons for charge balance, and are thus n-type in the dark [7.47, 56–58].

In considering charge balance we must also account for the fact that transition metal dopants or impurities (say Fe) can change valence state, while other metal impurities (say Al) cannot. We may thus write

$$[Al^{3+}] + 2[Fe^{2+}] + [Fe^{3+}] + n = 2[V_O^{2+}] + p \ . \tag{7.9}$$

The presence of several valence states in the transition metals allows the charge balance to be maintained over a range of oxygen partial pressures without the creation of large numbers of free carriers. In this pressure range the major impact is the variation of the relative amounts of Fe^{2+} and Fe^{3+}, due to changes in $[V_O^{2+}]$. In p-type samples, the free carrier concentration at room temperature is further reduced by trapping into acceptor levels, since these levels are thought to lie deep in the band gap [7.47, 56, 57, 59]. This leads to high resistivity values in samples grown or processed in air. By contrast, heavily reduced samples of $BaTiO_3$ are semiconducting (n-type) even at room temperature [7.47, 56, 60, 61], since the induced donor levels are shallow [7.47, 56] and are thus incapable of trapping electrons.

In addition to isolated metal impurities and fully ionized oxygen vacancies indicated in (7.9), we must also consider the possibility that association of ionic defects or trapping of free charges may lead to new species. Two specific cases must be considered. First, at high temperatures during crystal growth or processing, oxygen vacancies can associate with metal ions, due to Coulomb attraction. Assuming that this association occurs among nearest neighbors, the resulting structure in $BaTiO_3$ would be a $(M^{n+}O_5)(M^{n+}O_5)^{(10-n)-}$ cluster. This complex is called a $M^{n+}-V_O$ center. Such centers are known to exist in perovskites, but no consensus exists as to their importance. The concentration of these centers can be determined by electron paramagnetic resonance (EPR). According to recent EPR measurements of Fe^{3+} in melt-grown samples [7.52], the population of $Fe^{3+}-V_O$ centers is much less than that of isolated Fe^{3+}.

A second possible form of association is the trapping of one or two free electrons at oxygen vacancies. As indicated earlier, the resultant states are thought to be quite shallow [7.46–48], and are thus only likely to be populated significantly at low temperatures, or in heavily reduced samples.

7.2.4 Energy Levels of Fe^{2+} and Fe^{3+}

The electronic structure of a $3d$ ion embedded in $BaTiO_3$ depends on the oxidation state, the electronic ground term (spin configuration) of the impurity, and the symmetry of the local electrostatic crystalline field. *Michel-Calendini* and co-workers [7.36–39] have used the spin-unrestricted SCF-MS-Xα technique to calculate the molecular orbital (energy level) diagram of $FeO_6^{(12-n)-}$ (Fe^{n+}) clusters with O_h and C_{4v} symmetries. These calculations provide information on the location of the empty and occupied Fe impurity states in the TiO_6^{8-} band gap. The pertinent energy level diagrams for the FeO_6^{10-} (Fe^{2+}) and FeO_6^{9-} (Fe^{3+}) clusters are shown in Fig. 7.5b and c [7.36, 38, 39]. In order to gain a better understanding of the photorefractive mechanisms in BaTiO₃, we wish to relate the absolute levels given in Fig. 7.5 with the relevant ionization energies. The connection is easily seen from the following equation for the Fe^{2+}/Fe^{3+} system:

$$\{Fe^{3+}, (VB)^m, L^n\} \rightarrow \{Fe^{2+}, (VB)^{m-1}, L^{n+1}\} \ .$$

This equation describes the change in formal charge state which occurs during the conversion of Fe^{3+} ($e_\uparrow^2 b_{2\uparrow} a_{1\uparrow} b_{1\uparrow}$) to Fe^{2+} ($e_\uparrow^2 b_{2\uparrow} a_{1\uparrow} b_{1\uparrow} e_\downarrow$) by transferring an electron from the valence band (the fully occupied valence band has m electrons) to an $e_\downarrow$ impurity state (hole emission). Here L refers to the one-electron states $b_{1\uparrow}$, $e_\downarrow$, etc.; $\downarrow$ and $\uparrow$ refer to up- and down-spin electron spin states; and VB for the valence band. From this it follows that the energy of the ionization level above the valence band is given by

$$E(Fe^{2+}/Fe^{3+}) = E_T\{Fe^{2+}, (VB)^{m-1}, L^{n+1}\} - E_T\{Fe^{3+}, (VB)^m, L^n\} \ ,$$

where $E_T\{\cdots\}$ represents the total energy of the lowest multiplet of the corresponding charge state.

Calculated values of the ionization energies of transition metal impurities in BaTiO₃ are not available. Instead, we can gain experimental information on the energy levels from the thermogravimetric data of *Hagemann* and co-workers [7.44, 54]. The resulting ionization energy level diagram for BaTiO₃:Fe is given in Fig. 7.6. Also included are the levels due to oxygen vacancies with one or two trapped electrons [7.44, 46]. The notation in Fig. 7.6 is now consistent with that used by the semiconductor community. In this picture, the ionization energy

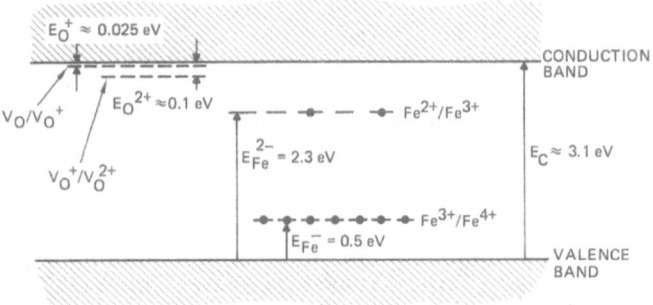

Fig. 7.6. Energy levels of oxygen vacancies [7.44, 46] and Fe^{n+} ions [7.44, 54] in BaTiO₃. The band gap energy of 3.1 eV is the value adopted by *Hagemann* [7.44]

of Fe^{2+} (leading to Fe^{3+} and a free electron) is $3.1-2.3=0.8$ eV. Similarly, the energy required to convert Fe^{4+} to Fe^{3+} through the generation of a free hole is 0.5 eV. In Fig. 7.6 we have assumed that the Fermi level is just below the Fe^{2+}/Fe^{3+} level, leading to an admixture of Fe^{2+} and Fe^{3+} (with $[Fe^{3+}] > [Fe^{2+}]$), and a negligible population of Fe^{4+}. Note that the energy positions determined by this and other approaches based on thermal activation may not be equal to the energies for *optical* transitions. Note also that the Fe energy level structure in Fig. 7.6 is clearly a simplification of the calculated structure given in Fig. 7.5. Nevertheless, it is adequate to describe most thermal and gravimetric measurements on $BaTiO_3$. We shall see below that it is also adequate to describe most photorefractive measurements.

In the semiconductor framework of Fig. 7.6, the populations of Fe^{2+}, Fe^{3+} and Fe^{4+} can be related to the Fermi energy E_F by

$$\frac{[Fe^{4+}]n}{[Fe^{3+}]} = 2N_C \exp\left[-\frac{(E_C - E_{Fe}^-)}{k_B T}\right] \qquad (7.10)$$

and

$$\frac{[Fe^{3+}]n}{[Fe^{2+}]} = \frac{1}{2} N_C \exp\left[-\frac{(E_C - E_{Fe}^{2-})}{k_B T}\right], \qquad (7.11)$$

where N_C is the density of states in the conduction band and E_C is the energy of the conduction band edge. Equation (7.10) is simply the law of mass action for

$$Fe^{3+} \leftrightarrows Fe^{4+} + e^- , \qquad (7.12)$$

while (7.11) is the corresponding relation for

$$Fe^{2+} \leftrightarrows Fe^{3+} + e^- . \qquad (7.13)$$

7.3 Band Transport Model

7.3.1 Energy Level Model

In order to develop a model for grating formation in a photorefractive material, it is first necessary to postulate a system of centers in the band gap which can generate carriers by photoionization, and trap carriers by recombination. Chapters 3 and 4 of this volume describe several energy level models which have been postulated in the past. The most frequently used model assumes a single photorefractive species which can exist in two ionization states. This model was originally proposed by *Chen* [7.62] and *Peterson* et al. [7.63] for $LiNbO_3$ with Fe dopants or impurities. In the particular case of $BaTiO_3$ (with Fe impurities), this model is sufficient to explain steady-state beam coupling measurements in as-

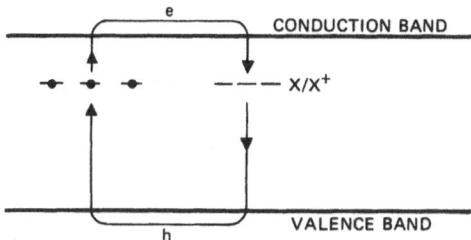

CONDUCTION BAND

Fig. 7.7. Energy level model for the photorefractive effect. Electrons are photoionized from filled states X and recombine at empty states X^+; holes are photoionized from level X^+ and recombine at X

grown crystals [7.32, 33], as well as certain features of their transient response [7.64]. A two-species model in which O vacancies are also considered to participate in charge creation and trapping has also been proposed for BaTiO₃ [7.33]. This model differs from the single species model with regard to the performance of reduced crystals (Sect. 7.6).

The energy level model used for our ensuing discussion is shown in Fig. 7.7. We assume that a single species X, which can exist in two valence states (X and X^+), is responsible for the photorefractive properties of BaTiO₃. In the case of iron-doped LiNbO₃, X corresponds to Fe^{2+} and X^+ corresponds to Fe^{3+}. In commercial samples of BaTiO₃, we also assume the Fe^{2+} and Fe^{3+} are dominant [7.52], since Fe impurities (primarily in the form of Fe^{3+} ions) are consistently present at large concentrations. However, it should be noted that the presence of Fe^{2+} has not been proven directly, and in fact the photorefractive species in as-grown samples could be Fe^{3+} (level X) and Fe^{4+} (level $X+$). The advantages of the Fe^{3+}/Fe^{4+} model are (1) the resulting location of the Fermi level below midgap is consistent with the p-type behavior of as-grown samples, and (2) the low hole ionization energy for Fe^{4+} is consistent with the observation of hole-dominated photoconductivity to 1.1 μm. In any case, the conclusions of the model described below are substantially unchanged for both Fe^{2+}/Fe^{3+} and Fe^{3+}/Fe^{4+}.

The representation of X/X^+ as a discrete, unique state is clearly a simplification of the Fe^{2+} and Fe^{3+} level structure given in Figs. 7.5b and c. As indicated above, this simplification does not appear to affect the utility of the model, except perhaps in the interpretation of certain transient measurements [7.64, 65], to be discussed in Sect. 7.4.2.

In our energy level model, we denote the concentration of X by N, and the concentration of X^+ by N^+. Other states, which are optically inactive, provide overall charge compensation within the crystal (Sect. 7.2.3). It is important to note that electrons or holes (or both) can contribute to the charge transport in BaTiO₃. For electron transport, state X is a donor, or "filled" state, and state X^+ is an ionized donor, or "empty" state. For hole transport, state X^+ is an acceptor or "filled" state, and state X is an ionized acceptor, or "empty" state. We further note that the sign of the space charge field is opposite for the two charge carriers. This changes the direction of beam coupling, and allows a measurement of the dominant photocarrier.

7.3.2 Transport and Rate Coefficients

In association with the energy level model and notation given above, we must also define rate coefficients for the individual levels, and transport coefficients for electrons and holes. The important coefficients are (1) the photoionization cross section, (2) the recombination rate coefficient, (3) the dark generation rate, and (4) electron and hole mobility. In the single trap species model, the photoionization cross section for level X (thus creating an electron) is s_e, while the cross section for hole creation from level X^+ is s_h. The recombination rate coefficient for electrons at centers X^+ is γ_e, and the corresponding coefficient for holes at centers X is γ_h. The dark creation rate for electrons (from centers X) is β_e, and the corresponding coefficient for holes (from centers X^+) is β_h. Finally, the mobility for electrons (or holes) is μ_e (or μ_h). It must also be kept in mind that the photoionization cross sections and mobilities for BaTiO$_3$ at room temperature are anisotropic; the cross sections depend on the polarization orientation of the incident radiation with respect to the c-axis (Sect. 7.5.1), and the mobilities depend on the direction of transport with respect to the c-axis (see below).

The photoionization cross sections for states X and X^+ are related to the total absorption coefficient α by

$$\alpha = \alpha_P + \alpha_{NP} , \quad \text{or} \tag{7.14}$$

$$\alpha = N s_e + N^+ s_h + \alpha_{NP} , \tag{7.15}$$

where α_P is the photorefractive contribution and α_{NP} is the non-photorefractive contribution. In as-grown samples of BaTiO$_3$, $N \ll N^+$ [7.52] and thus

$$\alpha_P \simeq N^+ s_h . \tag{7.16}$$

If we define the quantum efficiency as $\Phi = \alpha_P/\alpha$, then

$$\alpha = N^+ s_h/\Phi . \tag{7.17}$$

The recombination rate coefficients are related to the carrier recombination time for electrons (τ_{Re}) and holes (τ_{Rh}) by

$$\tau_{Re} = \frac{1}{\gamma_e N^+} , \quad \text{and} \tag{7.18}$$

$$\tau_{Rh} = \frac{1}{\gamma_h N} . \tag{7.19}$$

The mobility for holes and electrons is related to the conductivity σ by

$$\sigma = \sigma_e + \sigma_h = n e \mu_e + p e \mu_h , \tag{7.20}$$

where σ_e, σ_h are the electron and hole conductivities. The accurate measurement of mobility in insulating oxides is particularly difficult. In semiconductors, the Hall effect and the four-point probe technique can be used to measure the carrier concentration and conductivity, respectively, yielding accurate values of Hall mobility, via (7.20). In as-grown BaTiO₃, with a typical resistivity value of 10^{10}–10^{12} Ωcm, Hall measurements are not possible, and conductivity measurements are more difficult. In order to obtain meaningful Hall data for BaTiO₃, workers have been forced to measure reduced samples with much lower resistivities. Another approach to the measurement of mobility is the drift technique, in which carriers are injected from one surface of a plane-parallel sample, and their transit time across the sample is measured. This approach is limited by the small signals involved and the reliance on an accurate model of charge trapping and transport in the sample.

In Table 7.2 we have collected literature values for mobility in BaTiO₃. In most cases, the mobilities were determined by Hall measurements on reduced samples, or drift time measurements on as-grown samples. In a few cases, the measured thermal emf was used to determine the free carrier concentration. There is a large spread in the data, with measured values ranging from 3×10^{-4} to 1 cm²/Vs. Note also that although the mobility in BaTiO₃ at room temperature is known to be highly anisotropic [7.60], little attention has been paid to this property. One important exception is the photorefractive measurement by *Tzou* et al. [7.76], to be discussed later.

Table 7.2. Room temperature mobility measurements in BaTiO₃

Reference	Value [cm²/Vs]	Identification	Type of experiment	Sample
Electron mobility				
[7.66]	2	μ_e	Hall	KF flux (reduced)
[7.61]	0.1	μ_e	Hall	KF flux (reduced)
[7.67]	0.5	μ_e	Hall, drift	Ceramic
[7.68][a]	0.07	μ_e	Thermal emf	Ceramic
[7.60]	0.13	$\mu_{e\parallel}$	Hall	TiO₂ solution (reduced)
	1.0	$\mu_{e\parallel}$		
[7.69]	5×10^{-4}	μ_e	Hall	KF flux (reduced)
[7.70]	0.2	μ_e	Thermal emf	KF flux (reduced)
[7.71]	0.5	μ_e	Hall	Ceramic
	0.6	μ_e	Thermal emf	
[7.72]	0.6	μ_e	Hall	Verneuil (polydomain, reduced)
[7.73]	$(3\pm) \times 10^{-3}$	μ_e	Drift time	TiO₂ solution (as-grown)
Hole mobility				
[7.74]	$(3\pm1) \times 10^{-4}$	μ_h	Drift time	KF flux (as-grown)
[7.75]	0.02–0.03	$\mu_{h\parallel}$	Drift time	TiO₂ solution (as-grown)
[7.76]	>0.5	$\mu_{h\parallel}$	Photorefractive	TiO₂ solution (as-grown)

[a] Extrapolated from measurement at 1100 °C.

7.3.3 Grating Formation

The band transport model [7.77–80] is commonly used to describe grating formation in a photorefractive material. Under the influence of periodic illumination (Fig. 7.8), electrons (or holes) are optically excited from filled donor (or acceptor) sites to the conduction (or valence) band, where they migrate to dark regions in the crystal by drift or diffusion before recombining into an empty trap. The transported charges result in an ionic space-charge grating that is, in general, out of phase with the incident irradiance. The periodic space charge is balanced by a periodic space-charge electric field in accordance with Poisson's equation. This space-charge field modulates the refractive index through the electro-optic effect. If no electric field is applied to the crystal (as is generally the case in experiments with $BaTiO_3$), then diffusion alone leads to a phase shift of $\pi/2$ between the incident irradiance and the space-charge field. This shifted grating plays an important role in many device applications of $BaTiO_3$.

A mathematical description of the grating formation process for a single charge carrier was given in its most complete form by *Vinetskii, Kukhtarev*, and co-workers [7.77–79] (see also [7.80]). For the case where both charge carriers play an important role, the early work of *Orlowski* and *Krätzig* [7.81] has been developed and expanded by *Valley* [7.82] and by *Strohkendl* et al. [7.83], leading to solutions in the steady state and the transient regime. See Chap. 3 for a review of the band transport model. An important result of the steady-state two-carrier model is that the space-charge field (with no applied field and negligible

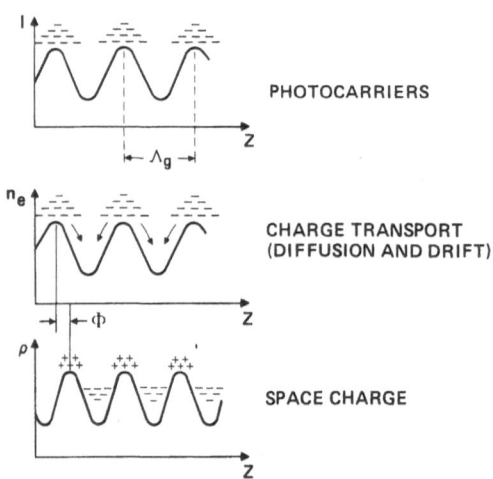

Fig. 7.8. Gratings in a photorefractive material. The periodic irradiance pattern results from the interference of two waves in the material

photovoltaic effect) is purely imaginary, and can be written as

$$\delta E = im E_{sc} = im \, \frac{k_B T}{e} \, \frac{K}{1 + K^2/K_s^2} \, \bar{\sigma}(K) \; , \tag{7.21}$$

where m is the fractional modulation, K is the grating wave number, and K_s is the Debye screening wave vector, given by

$$K_s = \left(\frac{e^2 N_E}{\varepsilon \varepsilon_0 k_B T} \right)^{1/2} . \tag{7.22}$$

In (7.22) ε is the relative dielectric constant and N_E is the effective number of empty traps, defined as

$$N_E = \frac{N N^+}{N + N^+} . \tag{7.23}$$

The factor $\bar{\sigma}(K)$ in (7.21) accounts for the relative contribution of electrons and holes to the photorefractive performance of the sample. It is given by [7.83]

$$\bar{\sigma}(K) = \frac{C-1}{C+1} \; , \quad \text{in which} \tag{7.24}$$

$$C = \frac{s_h N^+ (K^2 + K_e^2)}{s_e N (K^2 + K_h^2)} \; . \tag{7.25}$$

The quantities K_e^{-1} and K_h^{-1} are the diffusion lengths (also known as the hopping range) for electrons and holes:

$$K_e^{-1} = \left(\frac{k_B T \mu_e}{e \gamma_e N^+} \right)^{1/2} \; , \quad K_h^{-1} = \left(\frac{k_B T \mu_h}{e \gamma_h N} \right)^{1/2} . \tag{7.26}$$

For small values of K ($K \ll K_e, K_h$), $\bar{\sigma}$ is independent of K, and can be written in terms of the photoconductivity of each carrier[1],

$$\bar{\sigma} = \frac{\sigma_h - \sigma_e}{\sigma_h + \sigma_e} = \frac{\mu_h p - \mu_e n}{\mu_h p + \mu_e n} \; . \tag{7.27}$$

When the photoconductivity for holes is dominant, $\bar{\sigma} = +1$, whereas $\bar{\sigma} = -1$ when electron photoconductivity dominates. In a beam coupling experiment, this change in sign corresponds to a change in coupling direction. For the case where the photoconductivity values for the two carriers are nearly equal (and in the absence of an external or photovoltaic field), the net space-charge field (and

---

[1] Note also that for large values of K ($K \gg K_e, K_h$), $\bar{\sigma}$ is again independent of K and can be written in terms of the *absorption coefficient* for each carrier [7.82]: $\bar{\sigma} = (\alpha_h - \alpha_e)/(\alpha_h + \alpha_e)$.

thus the beam coupling gain) is small, being limited only by grating-dependent correction terms [7.82, 83]. In such crystals, the space-charge field (and thus the beam coupling gain) can actually change sign with grating period [7.32] (Sect. 7.4.3).

It is useful to determine the trap populations at the compensation point, using our two-level model. We start by writing the compensation condition

$$\sigma_h = pe\mu_h = ne\mu_e = \sigma_e , \quad \text{or} \tag{7.28}$$

$$\frac{p}{n} = \frac{\mu_e}{\mu_h} . \tag{7.29}$$

The free carrier concentrations n and p can be related to the trap concentrations N and N^+ by the rate equations

$$s_e I N = \gamma_e n N^+ \quad \text{and} \tag{7.30}$$

$$s_h I N^+ = \gamma_h p N , \tag{7.31}$$

where I is the irradiance. These two equations may be combined to yield

$$\frac{N^+}{N} = \left(\frac{\gamma_h s_e p}{\gamma_e s_h n}\right)^{1/2} . \tag{7.32}$$

Finally, by combining with (7.29), we find the condition at the compensation point as

$$\frac{N^+}{N} = \left(\frac{\gamma_h s_e \mu_e}{\gamma_e s_h \mu_h}\right)^{1/2} . \tag{7.33}$$

Note that we are concerned here with the compensation of the *photoconductivity* – dark conductivity is neglected. In the dark, the rate equations (7.30 and 31) would be replaced by

$$\beta_e N = \gamma_e n N^+ \quad \text{and} \tag{7.34}$$

$$\beta_h N^+ = \gamma_h p N , \tag{7.35}$$

leading to a different condition for compensation. Note also that the values of free carrier density and irradiance in the above discussion are spatially averaged values.

It may be seen from the above discussion that if one can adjust the ratio of trap concentrations to be equal to the value given in (7.33), the photorefractive gain (in the absence of an applied or photovoltaic field) can be reduced to a very small value. This could be important for applications in which the photorefractive effect is undesirable. Furthermore, the measurement of N^+/N at the compensation point gives a value of the ratio

$$R = \frac{\gamma_h s_e \mu_e}{\gamma_e s_h \mu_h} \;, \tag{7.36}$$

and thus provides information regarding transport coefficients. Techniques for varying $N^+/N = [\text{Fe}^{3+}]/[\text{Fe}^{2+}]$ are discussed in Sect. 7.6.

The transient response of a photorefractive grating when both electrons and holes contribute to the transport of charge has been analyzed by *Valley* [7.82] and *Strohkendl* et al. [7.83] (see Chap. 3). In the absence of an applied or photovoltaic field, the erasure rate for our single-species model is given by

$$\gamma = \gamma_{die} \frac{1 + (K/K_s)^2}{1 + (K/K_e)^2} + \gamma_{dih} \frac{1 + (K/K_s)^2}{1 + (K/K_h)^2} \;, \tag{7.37}$$

where γ_{die} and γ_{dih} are the dielectric relaxation rates for electrons and holes:

$$\gamma_{die} = \frac{\sigma_e}{\varepsilon\varepsilon_0} = \frac{ne\mu_e}{\varepsilon\varepsilon_0} \;, \qquad \gamma_{dih} = \frac{\sigma_h}{\varepsilon\varepsilon_0} = \frac{pe\mu_h}{\varepsilon\varepsilon_0} \;. \tag{7.38}$$

The dielectric relaxation rates contain contributions from dark carriers and photocarriers. By substituting the rate equation solutions for the photocarrier concentrations n and p (7.30 and 31) into (7.38), we may write

$$\gamma_{die} = \frac{\sigma_{de}}{\varepsilon\varepsilon_0} + \left(\frac{e\alpha_{Pe}I}{\varepsilon\varepsilon_0}\right)\mu_e \tau_{Re} \;, \tag{7.39}$$

and

$$\gamma_{dih} = \frac{\sigma_{dh}}{\varepsilon\varepsilon_0} + \left(\frac{e\alpha_{Ph}I}{\varepsilon\varepsilon_0}\right)\mu_h \tau_{Rh} \;, \tag{7.40}$$

where σ_{de}, σ_{dh} are the dark conductivities for electrons and holes, and α_{Pe}, α_{Ph} are the photorefractive absorption coefficients for electrons and holes. Note that the dielectric relaxation rates vary linearly with intensity in this model. It should also be noted that the dielectric relaxation rates and the diffusion rates are anisotropic, through their dependence on μ and ε. The application of (7.37) to the specific case of BaTiO$_3$ will be discussed in Sect. 7.4.2.

7.4 Physical Measurements Using the Photorefractive Effect

Aside from the many device applications of photorefractive materials, the photorefractive effect may also be used to great advantage for measuring fundamental materials properties. The specific advantage of the photorefractive effect for such measurements is that it can be used to generate internal electric fields without the use of contacts. This allows the measurement of electrical and electro-optic properties without concern for the quality and characterization of the contacts.

In our discussion below, we will describe techniques for steady-state and transient photorefractive measurements in $BaTiO_3$. A more detailed description of techniques in the transient (and short pulse) regime is provided in Chap. 6. We will then discuss the specific measurements which have been made in $BaTiO_3$.

7.4.1 Steady-State Measurements

One useful technique which provides information on the magnitude and sign of the steady-state space-charge field (and thus a number of important photo-refractive parameters) is the measurement of the beam coupling gain Γ, given by

$$\Gamma = 2\pi n^3 \frac{Fr_{ang}E_{sc}}{\lambda} , \tag{7.41}$$

where n is the refractive index, F is the fractional poling factor, and r_{ang} is the appropriate combination of electro-optic tensor components and angular and polarization factors for a fully poled crystal. If we substitute for E_{sc} in (7.41), we obtain

$$\Gamma = \frac{2\pi n^3}{\lambda} \frac{k_B T}{e} r_{eff} \frac{K}{1 + K^2/K_s^2} , \tag{7.42}$$

or

$$\Gamma = \frac{(2\pi)^2 n^3}{\lambda} \frac{k_B T}{e} r_{eff} \frac{\Lambda}{\Lambda^2 + \Lambda_0^2} , \tag{7.43}$$

where

$$r_{eff} = F\bar{\sigma} r_{ang} , \quad \text{and} \tag{7.44}$$

$$\Lambda_0 = l_s = \frac{2\pi}{K_s} = \left(4\pi^2 \varepsilon \varepsilon_0 \frac{k_B T}{e^2 N_E} \right)^{1/2} \tag{7.45}$$

is the Debye screening length. In including $\bar{\sigma}$ as part of the effective electro-optic coefficient, we follow the approach of the previous studies of steady-state beam coupling in $BaTiO_3$ [7.33, 52], which neglected the K-variation of $\bar{\sigma}$. This assumption is only strictly true for $K \ll K_e, K_h$, or when one photocarrier is dominant ($\bar{\sigma} = \pm 1$).

In (7.42) [or (7.43)] there are two material parameters which can be measured directly in a beam coupling experiment: r_{eff} and l_s (or K_s). We may then obtain $F\bar{\sigma}$ from r_{eff} provided r_{ang} is known, and $N_E = NN^+/(N + N^+)$ from l_s (or K_s), provided ε is known. The most direct means of determining these parameters is to fit the function given by (7.42) or (7.43) to the data of Γ versus K (or Λ), with r_{eff} and K_s (or l_s) as variables. An alternative approach [7.32] is to recast (7.42) as

$$\frac{K}{\Gamma}=\frac{e\lambda}{2\pi n^3 k_B T r_{\text{eff}}}\left(1+\frac{K^2}{K_s^2}\right) .$$

(7.46)

Thus, if r_{eff} is independent of K, the experimental data plotted as K/Γ vs K^2 should lie on a straight line, whose intercept yields r_{eff}, and whose value of slope/intercept yields K_s. In the more general case (accounting for the K-dependence of σ), the data may still retain an approximate straight-line behavior, but the interpretation is more complicated [7.83].

The photorefractive gain is commonly measured using the geometry shown in Fig. 7.9. A first step in all measurements is to orient the c-axis of the crystal to provide gain for the signal beam I_S. The knowledge of this orientation, along with an independent electrical measurement of the polarity of the c-axis, yields the sign of the dominant charge carriers [7.8, 81]. For the beam notation shown in Fig. 7.9 and the crystal orientation which gives gain for the signal beam, the transmission of the signal beam is given by

$$\frac{I_S(L)}{I_S(0)}=\frac{[I_S(0)+I_R(0)]\exp[(\Gamma-\alpha)L]}{I_R(0)+I_S(0)\exp(\Gamma L)} .$$

(7.47)

For negligible depletion of the pump wave [$I_S(0)\exp(\Gamma L)\ll I_R(0)$], (7.47) reduces to

$$I_S(L)=I_S(0)\exp[(\Gamma-\alpha)L] .$$

(7.48)

The quantity typically measured in the laboratory is the effective gain [7.84] γ_0, defined by

$$\gamma_0=\frac{I_S(L)\text{ with reference wave}}{I_S(L)\text{ without reference wave}} .$$

(7.49)

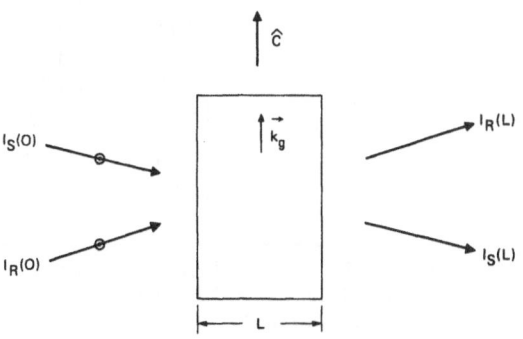

Fig. 7.9. Crystal orientation and beam notation for beam coupling measurements. Both beams are s-polarized

When depletion of the reference wave can be neglected, $\gamma_0 = \exp(\Gamma L)$. Thus, the measurement of γ_0 provides a direct means of determining Γ.

The measurement of beam coupling gain as a function of grating period has been used by several authors [7.8, 32, 33] to measure the sign of the dominant charge carrier, the Debye length and the effective electro-optic coefficient r_{eff} in as-grown samples of $BaTiO_3$. The effective number of empty traps is then obtained from the Debye length, using (7.45) and a given value of dielectric constant. The factor $F\bar{\sigma}$ is obtained from r_{eff}, using (7.44) and a given value of the electro-optic coefficient. Finally, the factor $\bar{\sigma}$ can be obtained from $F\bar{\sigma}$, using an estimated value of the fractional poling factor F.

As indicated earlier, all measurements to date [7.8, 32, 33] have ignored the K-dependence of $\bar{\sigma}$, an assumption which is only accurate in samples with small values of the diffusion length for both carriers. *Strohkendl* et al. [7.83] have shown that this assumption can lead to values of N_E which can differ by a factor of 3 from those obtained from a fit to the data using the full K-dependence of $\bar{\sigma}$.

The steady state beam coupling data for $BaTiO_3$ is collected in Table 7.3. We have also included in Table 7.3 values of Debye length and effective trap density determined from measurements of response time as a function of grating period [7.64, 86].

The measured values of N_E given in Table 7.3 vary between 0.5×10^{16} cm^{-3} and 8.7×10^{16} cm^{-3}. We noted earlier that the measured value of N_E depends on the chosen value of dielectric constant. Uncertainties in this parameter lead

Table 7.3. Measurements of Debye length and effective electro-optic coefficient in $BaTiO_3$

Crystal	Sign of dominant photo-carrier	Debye length $\Lambda_0 = l_s$ [μm]	Effective no. of empty traps N_E [$\times 10^{16}$ cm^{-3}]	Effective EO coefficient r_{eff} [pm/V]	$F\bar{\sigma} = \dfrac{r_{\text{eff}}}{r_{\text{ang}}}$	$\bar{\sigma}$
G/L[a]	+	0.46	4.4[f]	+ 7.3	+0.30[i]	+0.33[k]
R1[a]	+	0.34	8.1[f]	+ 9.5	+0.40[i]	+0.44[k]
R2[a]	−	0.57	2.8[f]	− 7.9	−0.33[i]	−0.37[k]
GB3[a]	+	0.39	6.1[f]	+ 9.7	+0.41[i]	+0.46[k]
K2[a]	+	0.32	8.7[f]	+12.0	+0.50[i]	+0.56[k]
SC[a]	+	0.47	4.2[f]	+ 8.3	+0.35[i]	+0.39[k]
GB4[a]	+	0.55	3.0[f]	+ 4.2	+0.18[i]	+0.20[k]
ROCKY[b]		1.17	0.6[g]	2.0	0.14[j]	0.14[l]
MMD[b]		1.28	0.5[g]	3.4	0.24[j]	0.24[l]
HOP[b,c]	+	0.66	1.9[g]	+ 5.0	+0.36[j]	+0.36[l]
CAT[b,d]	+	0.36	6.2[g,b] 6.0[g,d]	+ 9.8	+0.70[j]	+0.70[l]
SWISS[b]	+	0.44	4.2[g]	+12.5	+0.89[j]	+0.89[l]
DOYLE[b]	+	0.45	4.1[g]	+13.4	+0.96[j]	+0.96[l]
-----[e]		2.8	2.9[h]			

[a] [7.32], $\lambda = 442$ nm. [d] [7.64], $\lambda = 515$ nm. [g] $\varepsilon = \varepsilon_{33} = 150$. [j] $r_{\text{ang}} = r_{13} = 14$ pm/V.
[b] [7.33], $\lambda = 515$ nm. [e] [7.86], $\lambda = 497$ nm. [h] $\varepsilon = \varepsilon_{11} = 4200$. [k] $F = 0.9$.
[c] [7.8], $\lambda = 515$ nm. [f] $\varepsilon = \varepsilon_{33} = 168$. [i] $r_{\text{ang}} = r_{13} = 24$ pm/V. [l] $F = 1.0$.

to comparable uncertainties in the value of N_E. It should also be noted that in as-grown samples of BaTiO$_3$, $N \ll N^+$ [7.52], thus yielding [through (7.23)] $N_E \simeq N = [\text{Fe}^{2+}]$.

The values of $\bar{\sigma}$ given in Table 7.3 are dependent on the assumed values for r_{ang} and F. Unfortunately, there is disagreement as to the values of electro-optic coefficients for a perfectly poled (single domain) crystal, and the fractional poling in a given sample. In most beam coupling measurements, the grating normal is along $\hat{c}$ and the optical polarization is along $\hat{a}$ (or $\hat{b}$), thus coupling to the electro-optic tensor component r_{13}. This avoids the complicating effects of beam fanning [7.87]. Until very recently, the only reported measurement of the unclamped r_{13} in a single domain sample was the value $r_{13} = 24$ pm/V (see Table 7.1). However, the accuracy of this value is open to question, due to interference from nearby piezoelectric resonances. Recently, Ducharme et al. [7.28] have reported a value $r_{13} = 19.5$ pm/V at 1 kHz. This smaller value leads to larger values of $F\bar{\sigma}$, although there is still a variation of a factor of ~ 7 among the measured samples, indicating a significant variation in the admixture of electron and hole conductivity (and/or the fractional poling).

A number of approaches can be used to determine the fractional poling F in a given sample [7.16]. *Ducharme* and *Feinberg* [7.33] have measured r_{eff} after poling at two different voltages and obtained the same value, thus indicating that the sample was free of 180° domains ($F = 1$). On the basis of etching studies, *Klein* and *Schwartz* [7.52] concluded that $F \gtrsim 0.9$ in their samples.

It should be noted that there is at least one report of a BaTiO$_3$ sample with anomalous beam coupling behavior [7.32]: the magnitude of the steady-state gain was very small, and the sign of the gain changed as a function of grating period (Fig. 7.10). It was theorized in [7.32]: "that the photoconductivity was nearly compensated in this sample, leading to the appearance of a weaker effect of unknown origin". This is consistent with the later calculations of *Valley* [7.82] and *Strohkendl* et al. [7.83], who have shown that the change in sign can be explained by the weak dependence of $\bar{\sigma}$ on Λ (or K), in the limit when the electron and hole diffusion lengths cannot be neglected in comparison with the grating period.

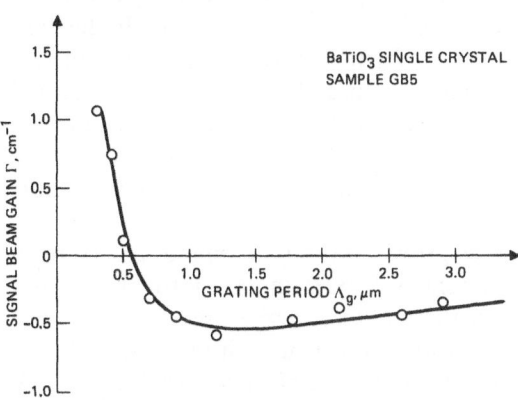

Fig. 7.10. Beam coupling gain as a function of grating period for the anomalous crystal GB5. The beams were *p*-polarized in this experiment, to exploit the larger electro-optic coefficient r_{33}

7.4.2 Transient Measurements

An expression for the erase rate when both electrons and holes contribute to the photoconductivity was given in Sect. 7.3.3 (7.37). When $K \ll K_e, K_h$, (7.37) reduces to

$$\gamma = \gamma_{di} \left[1 + \left(\frac{K}{K_s} \right)^2 \right] , \quad \text{where} \tag{7.50}$$

$$\gamma_{di} = \gamma_{die} + \gamma_{dih} . \tag{7.51}$$

Thus, a measurement of γ vs K at a fixed intensity yields values for K_s and γ_{di} [7.8, 64]. One may then obtain N_E from K_s, using (7.22). This measurement complements the separate determination of K_s from steady-state beam coupling measurements.

By comparison with (7.39 and 40), the dielectric relaxation rate may be written as [7.64]

$$\gamma_{di} = \gamma_{dark} + \gamma_{light} , \tag{7.52}$$

where γ_{light} is a function of intensity which is zero when $I = 0$. Thus, a measurement of γ_{di} as a function of intensity yields γ_{dark} from the low intensity limit, as well as the intensity dependence of γ_{light}. Assuming that the dark relaxation rate is impurity activated from a single center (with thermal activation energy E_0), then

$$\gamma_{dark} \sim \exp(-E_0/k_B T) , \tag{7.53}$$

According to our two-level model, see (7.39 and 40), we expect that $\gamma_{light} \sim I$. This simple linear dependence is not generally observed in transient experiments (see below).

As discussed in Sect. 7.4.2, the measurement of decay rate vs intensity (in the limit of small K) yields γ_{di} as a function of intensity. This in turn allows the measurement of the intensity dependence of γ_{light} in (7.52). In general, the intensity variation of γ_{light} has been found to be of the form $\gamma_{light} \sim I^x$. In the limit of zero intensity, the dark activation energy E_0 and the dark resistivity can also be measured. In Table 7.4 we list measured values of the exponent x, thermal activation energy and dark resistivity, determined from measurements of decay rate in $BaTiO_3$. We have also included data on the intensity variation of the photoconductivity, as well as the intensity variation of the write energy in a short pulse experiment.

It is clear from Table 7.4 that as-grown $BaTiO_3$ is characterized by a sublinear variation of erase rate with intensity. This is important for short-pulse applications; with nanosecond pulses, the measured write time is an order of magnitude longer than that predicted from a linear intensity variation [7.88]. The sublinear response is in disagreement with the simple model leading to (7.39

Table 7.4. Intensity dependence of decay rate, thermal activation energy and dark resistivity in as-grown BaTiO$_3$ at room temperature

Reference	Exponent x ($\gamma_{\text{light}} \sim I^x$)	Thermal activation energy E_0 [$\gamma_{\text{dark}} \sim \exp(-E_0/k_{\text{B}}T)$] [eV]	Dark resistivity [Ω cm]
[7.1]	0.5–1.0	0.7	
[7.8]	0.8		
[7.9]	0.6–0.9[a]		
[7.64, 88]	0.8[b]		
[7.65]	0.5		
[7.64]	0.62	0.98	10^{11}
	0.71 (40 °C)		
[7.33]	0.67 (oxidized)	0.97 (oxidized)	
	1.0 (reduced)		

[a] Exponent x determined from photoconductivity measurement.
[b] Exponent x determined from scaling of erase energy in short pulse regime.

and 40). This behavior suggests that more shallow levels may play an intermediate role in the trapping of electrons or holes [7.89]. Note also that the intensity variation becomes linear in reduced samples. This is consistent with the shallow position of the donor levels (V_0^+ or Fe^{2+}), which precludes the presence of more shallow intermediate levels.

In a recent transient and short pulse study of as-grown BaTiO$_3$, *Tzou* et al. [7.76] have measured the ratio $\mu_{\text{h}\parallel}/\mu_{\text{h}\perp}$ and the recombination time τ_{Rh} for holes. The key to this experiment was the use of the grating obtained from counterpropagating beam directions to access larger experimental values of grating wave vector K. The importance of this limit is that K is now comparable to K_{h} in the parallel and perpendicular directions, and thus K_{h} can be obtained from a fit to the data. From the measured values of K_{h}, the values $\mu_{\text{h}\parallel}\tau_{\text{Rh}} = 5 \times 10^{-10}$ cm^2/V, and $\mu_{\text{h}\perp}/\mu_{\text{h}\parallel} = 18 \pm 7$ were obtained, assuming $\varepsilon_\perp/\varepsilon_\parallel = \varepsilon_a/\varepsilon_c = 25$. In order to obtain individual values of $\mu_{\text{h}\parallel}$ and $\mu_{\text{h}\perp}$, the recombination time must be measured. This can be obtained from a time-resolved measurement of the grating formation in a short-pulse experiment [7.90]. Using 8 ns pulses, it was found that τ_{Rh} was less than 1 ns. This leads to the result $\mu_{\text{h}\parallel} > 0.5$ cm^2/Vs, which is consistent with estimates made earlier by *Valley* [7.90]. The measured value of hole mobility is larger than most reported values of μ_{h} in BaTiO$_3$ (see Table 7.2). It is possible that the sample measured contained a significant electron contribution to the photorefractive effect, so that the measured mobilities would also represent an admixture of the electron and hole mobility. Nonetheless, this is the first photorefractive measurement of the mobility in BaTiO$_3$; further measurements should help reduce the uncertainty in the correct value for photorefractive calculations.

7.5 Other Measurements in Photorefractive Crystals

While the physical and optical properties of $BaTiO_3$ have been studied for many years, only recently have attempts been made to study samples with known photorefractive properties, in order to understand the physical origins of the photorefractive effect. In this section we will describe such measurements on commercial as-grown samples, and the reasoning which leads to the conclusion that Fe is the dominant photorefractive species in $BaTiO_3$ [7.52].

7.5.1 Optical Absorption Coefficient Measurements

In Sect. 7.3.2 it was noted that the absorption coefficient for as-grown $BaTiO_3$ can be written as $\alpha = N^+ s_h/\Phi$. Thus, the measurement of α in a number of crystals yields a relative measurement of $N^+ \simeq [Fe^{3+}]$, provided that the quantum efficiency is constant. In Fig. 7.11 we show the measured absorption coefficients

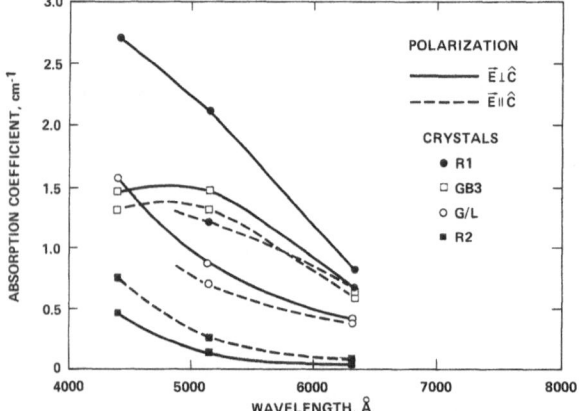

Fig. 7.11. Spectral variation of the absorption coefficient in the visible for four $BaTiO_3$ samples

Table 7.5. Absorption coefficient at 442 nm (for $E \perp \hat{c}$), iron impurity concentration, and Fe^{3+} concentration (by EPR) in as-grown single domain $BaTiO_3$

Crystal	Absorption coefficient [cm⁻¹]	Iron impurity concentration [ppm]	Fe³⁺ concentration by EPR [10¹⁸ cm⁻³]
G/L	1.6	120	2.0
R1	2.7	147	
R2	0.5	72	1.9
GB3	1.5	55	
BW1	3.5	51	4.7
1334	2.4	138	5.6

For Fe, 100 ppm $\simeq 5 \times 10^{18}$ cm⁻³.

as a function of wavelength for four different BaTiO$_3$ crystals [7.52]. We see that the absorption coefficient is anisotropic, as expected from the uniaxial symmetry of BaTiO$_3$ at room temperature. Note also that three of the four crystals show a broad, featureless "tail" extending from the fundamental absorption edge near 3500 Å. One sample (GB3) does show a more prominent peak at 5000 Å, indicating that the impurity content is different in this crystal. Crystal R2 is also anomalous in the sense that that the absorption anisotropy is reversed in sign from the other crystals. This effect may be related to the electron photoconductivity in this crystal. In Table 7.5 we present the value of α at 4416 Å for eight samples (with $E \perp \hat{c}$), in order to allow as relative determination of the density of filled traps.

7.5.2 Impurity Identity and Concentrations

Secondary ion mass spectroscopy (SIMS) has been used to determine the impurities in two different crystalline BaTiO$_3$ samples [7.52]. These measurements, while only qualitative, give a sensitive indication of all impurities. The elements observed were H, Li, Na, K, C, Al, Si, Ca, Sr, Ni, Mn, Cr and Fe. Of the elements listed, the first seven do not exist commonly in more than one valence state, and thus cannot support the intervalence transfer required for the photorefractive species. Ca and Sr substitute readily for Ba in the BaTiO$_3$ structure; however, they are isovalent with Ba, and thus produce no energy level in the BaTiO$_3$ band gap. The remaining species are transition metals and could contribute to the photorefractive effect.

In later measurements [7.52], six different photorefractive crystals (R1, R2, G/L, GB3, 1334 and BW1) were analyzed for impurities by spark emission spectroscopy. A semiquantitative determination of all metals was carried out, along with a quantitative measurement of transition metals. All samples contained Al and Si at concentration levels in the 50–500 ppm range. As indicated above, these species are not expected to participate in the photorefractive effect. Among the transition metals, iron was consistently the most abundant impurity, and the only element occurring consistently at concentrations greater than 50 ppm. The predominance of Fe impurities has previously been noted in BaTiO$_3$ [7.54]. The values of iron concentration for each crystal are included in Table 7.5.

7.5.3 EPR Measurements

EPR spectroscopy is a powerful technique for determining the presence and concentration of a variety of paramagnetic point defects and impurities. Furthermore, in the case of impurities, EPR can be used to identify the valence state. The disadvantage of EPR is that it is not equally sensitive to impurities in all possible valence states [7.52]. In the case of BaTiO$_3$, experiments are typically performed at room temperature, in order to avoid the tetragonal-orthorhombic phase transition near 9 °C. Thus, only those species with long spin-lattice

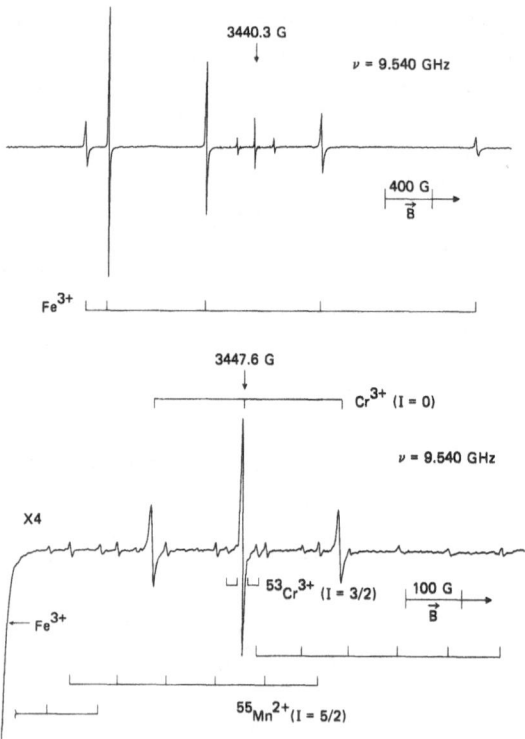

Fig. 7.12. EPR spectrum observed with the steady magnetic field directed along the [010] axis [$B \perp \hat{c}$ and in the (100) plane]

Fig. 7.13. Expanded EPR spectrum about the $g \simeq 2$ spectral region at the same magnetic field orientation as in Fig. 7.12

relaxation times (and thus negligible line broadening) can be observed. For the important case of Fe impurities, we note that Fe^{3+} is observable at room temperature, but Fe^{2+} and Fe^{4+} are not.

In [7.52], the EPR spectra of four different crystals of $BaTiO_3$ were measured at room temperature. In Figs. 7.12 and 7.13 we show representative spectra from crystal R2. The spectra were taken with the steady magnetic field directed along the [010] axis [$B \perp \hat{c}$ and in the (100) plane]. The spectrum shown in Fig. 7.12 exhibits five intense resonances, which is characteristic of a $^6S_{5/2}$ magnetic ion (Fe^{3+}) in a noncubic electrostatic crystalline field [7.91]. One of the unique features of this spectrum as well as spectra at other angles is the sharpness of the lines. This reflects the high microscopic quality (i.e., reduced number of structural imperfections and reduced strain) of the crystals studied. In Fig. 7.13 we show an expanded spectrum about the $g \simeq 2$ spectral region at the same magnetic field orientation. We have indicated in this figure the resonance transitions which were assigned to the even isotopes ($I = 0$) 50, 52 and 54 and the odd isotope 53 ($I = 3/2$, 9.55% abundance) of Cr^{3+}, along with the odd isotope ($I = 5/2$) 55 of Mn^{2+}. Note that every spectral feature in Fig. 7.13 has been identified.

The concentration of Fe^{3+} in several different $BaTiO_3$ crystals was determined by EPR [7.52] and the results are included in Table 7.5. By comparison, the concentrations of Cr^{3+} and Mn^{2+} were found to be at least two orders of magnitude lower than that of Fe^{3+}. No signals were observed in any of the samples from singly ionized oxygen vacancies. The detection limit for this measurement was $\sim 10^{16}$ cm^{-3}.

The angular dependence of the EPR spectra of crystal R2 was studied in detail. From the angular behavior in the plane perpendicular to the polar axis $\hat{c}$ of the five major transitions of the Fe^{3+} ion, it was established that the sites containing the paramagnetic species have a fourfold axis of symmetry (C_{4v}). It was further concluded that this symmetry resulted from a tetragonally distorted crystalline structure, rather than from the influence of a neighboring oxygen vacancy. The local site giving rise to the EPR spectrum of Fe^{3+} with C_{4v} symmetry in tetragonal BaTiO₃ is assumed to be associated with a tetragonally distorted FeO_6^{9-} cluster. This reasoning is based on the fact that if the local C_{4v} paramagnetic site were due to local compensation at a nearest-neighbor oxygen site, i.e., FeO_5^{7-} cluster ($Fe^{3+} - V_O$), then one would expect to observe a strong axial EPR spectrum, which is not consistent with the observations of [7.52]. Further, for the case of an $Fe^{3+} - V_O$ center, it is expected that at certain orientations of the magnetic field, three sets of axial resonances should be observed, each set having a unique tetragonal axis. This arises because there are three equivalent oxygen sites in the O_h FeO_6^{9-} unit from which an oxygen vacancy can be created to form the C_{4v} FeO_5^{7-} cluster. Again this was not observed, which supports the view that the iron resonances are associated with tetragonally distorted FeO_6^{9-} octahedra.

7.5.4 Correlation of Measured Parameters

The EPR and spectroscopic measurements described above indicate that Fe is the most abundant photoactive impurity in commercial samples of BaTiO₃. The importance of Fe for the photorefractive properties of BaTiO₃ can be determined [7.52] by seeking a correlation between the total iron concentration (from emission spectroscopy), the concentration of Fe^{3+} (by EPR), the relative concentration of filled traps (from the absorption coefficient) and the concentration of empty traps (from photorefractive beam coupling). In Fig. 7.14 we plot the parameters given above for each of four crystals. We see a clear correlation of the data, in the sense that all measured parameters increase monotonically with total Fe concentration. Ideally, a straight line behavior should be observed for each of the three plotted parameters, but the accuracy here is limited by experimental uncertainties, as well as possible variations in the oxidation/reduction state or quantum efficiency among our crystals.

To summarize the measurements of [7.52], we have

$$N = [Fe^{2+}] \simeq (2\text{–}9) \times 10^{16} \text{ cm}^{-3} \ ,$$

$$N^+ = [Fe^{3+}] \simeq [Fe] \simeq (2\text{–}8) \times 10^{18} \text{ cm}^{-3} \ ,$$

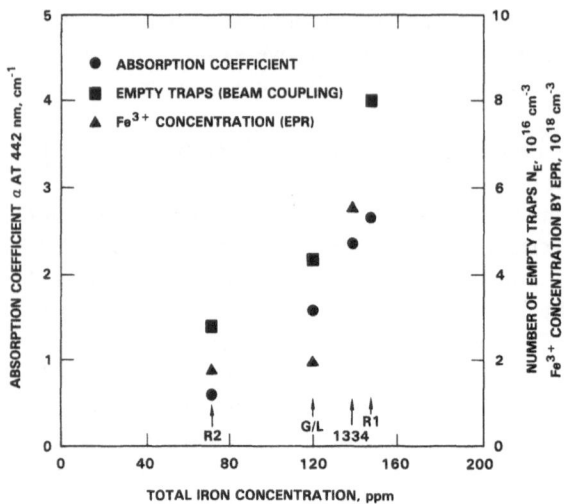

Fig. 7.14. Absorption coefficient, effective concentration of empty traps and Fe^{3+} concentration vs total iron concentration for four samples of $BaTiO_3$. The samples are identified at the bottom of the plot

and

$$\frac{[Fe^{3+}]}{[Fe^{2+}]} \simeq 40-120 \ .$$

Note that the large ratio of $[Fe^{3+}]$ to $[Fe^{2+}]$ is consistent with the oxidizing growth environment of $BaTiO_3$ when using the top-seeded solution growth technique [7.11]. It is also consistent with data on $BaTiO_3$: Fe ceramics [7.53, 54] processed in a similar atmosphere, for which Fe^{3+} is determined to be the dominant species.

A plot of electron and hole conductivity vs $[Fe^{2+}]/[Fe^{3+}]$ ratio provides a useful means for understanding the relative contribution of holes and electrons to the photorefractive effect in as-grown $BaTiO_3$ [7.52]. Following the approach of *Orlowski* and *Krätzig* (for $LiNbO_3$) [7.81], we define normalized electron and hole conductivities as

$$\bar{\sigma}_e = \frac{\sigma_e}{\sigma_e + \sigma_h} \quad \text{and} \tag{7.54}$$

$$\bar{\sigma}_h = \frac{\sigma_h}{\sigma_e + \sigma_h} \ . \tag{7.55}$$

In the limit of small diffusion lengths these parameters are related to the normalized differential conductivity [defined in (7.27)] by

$$\bar{\sigma} = \bar{\sigma}_h - \bar{\sigma}_e \ . \tag{7.56}$$

From the analysis of Sect. 7.3.3, we may write the normalized conductivities as

$$\bar{\sigma}_e = \frac{x^2 R}{1 + x^2 R} \tag{7.57}$$

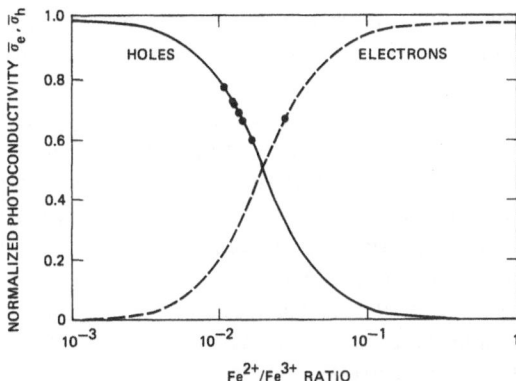

Fig. 7.15. Relative contribution of
electrons and holes to the photoconductivity in BaTiO$_3$, assuming
$R = 2500$. The experimental points
are taken from the beam coupling
data, using $\bar{\sigma} = \bar{\sigma}_h - \bar{\sigma}_e$

and

$$\bar{\sigma}_h = \frac{1}{1 + x^2 R} \,, \tag{7.58}$$

where $x = N/N^+ = [\text{Fe}^{2+}]/[\text{Fe}^{3+}]$, and R is defined in (7.36). We can thus plot the
normalized conductivity for BaTiO$_3$ if we know the value of the ratio R for the
Fe^{2+} and Fe^{3+} states.

For BaTiO$_3$, little information is available regarding the values of recombination rate coefficient and photoionization cross section for electrons and holes.
Measured values of the electron and hole mobilities for BaTiO$_3$ (given in Table
7.2) differ widely. While the existing data on photorefractive crystals provide no
direct information regarding these parameters, the condition $R \gg 1$ is most
consistent with the data. This may be explained with reference to Fig. 7.15, where
we have plotted the normalized conductivity for BaTiO$_3$ [using (7.57
and 58)], with $R = 2500$. We have also included data points for each of the crystals studied in [7.52], using the values of $\bar{\sigma}$ determined from the beam coupling
measurements (Table 7.3). We see that all crystals lie in the transition region
between hole and electron photoconductivity. With the assumed value of ratio
R, the compensation point corresponds to $x = R^{-1/2} = [\text{Fe}^{2+}]/[\text{Fe}^{3+}] = 0.02$. If
we had assumed a much larger or smaller value for the ratio R, the resulting
values of $x = [\text{Fe}^{2+}]/[\text{Fe}^{3+}]$ (as read from Fig. 7.15) would be inconsistent with
the values obtained using the individual measured values of $[\text{Fe}^{2+}]$ and $[\text{Fe}^{3+}]$.
Finally, we note that crystal R2, with dominant electron photoconductivity, is
still characterized by the condition $x \ll 1$, thus justifying the earlier approximation that $N_E \simeq N = [\text{Fe}^{2+}]$ in this crystal.

7.6 Optimization of Photorefractive Properties

As indicated earlier, the major advantage of BaTiO$_3$ for device applications is
the very large value of its electro-optic tensor component r_{42}. The major limitation of this material (in its as-grown form) is its relatively long response time.

Typical values of the response time at an intensity of 1 Wcm^{-2} are 0.1–1.0 s [7.1, 8, 64, 65], while for many applications response times less than 1 ms are desirable.

By analogy with earlier studies of $LiNbO_3$ [7.79, 85, 92], it has been suggested that chemical reduction should lead to an improvement in the photorefractive properties of as-grown $BaTiO_3$ [7.33, 52]. As discussed earlier, a reducing atmosphere is one with a low partial pressure of oxygen. When an as-grown sample is heated in a reducing atmosphere, the concentration of oxygen vacancies increases, leading to a reduction in the valence state of iron impurities, along with an increase in the population of free electrons (7.9). The major experimental question is whether a significant conversion to Fe^{2+} can be created without a significant increase in the free electron concentration (and thus the conductivity) in the sample.

7.6.1 Influence of Oxidation/Reduction

In order to understand the influence of reduction on the photorefractive properties of $BaTiO_3$, we have used the single-species, two-carrier model [7.82, 83] to calculate the absolute value of the beam coupling gain Γ (for a 0.7 μm grating period) and the photorefractive response time τ (also for $\Lambda = 0.7$ μm) as a function of the reduction ratio $N/N^+ = [Fe^{2+}]/[Fe^{3+}]$. The values of Γ and τ are calculated from (7.42) and (7.37), respectively, using $\tau = \gamma^{-1}$. In our calculation, we assumed $R = 2500$ (corresponding to a compensation point $N/N^+ = 2 \times 10^{-2}$), $\lambda = 500$ nm, a total dopant/impurity concentration $(N + N^+)$ of 5×10^{18} cm^{-3}, and $r_{ang} = r_{33} = 80$ pm/V. In order to plot the response time, the curve was scaled

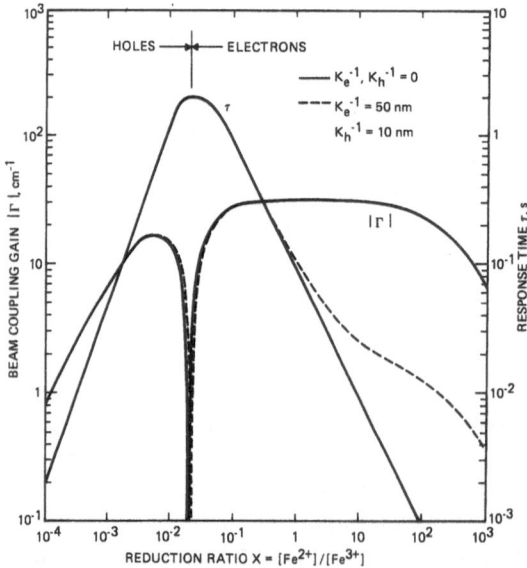

Fig. 7.16. Calculated response time and absolute value of beam coupling gain vs reduction ratio for $\Lambda = 0.7$ μm in $BaTiO_3$

to a value of 1.1 s at $N/N^+ = 8 \times 10^{-3}$ and $I = 0.25 \, \mathrm{W \, cm^{-2}}$, corresponding to the measured value in our experiment discussed below.

The results of our calculation are given in Fig. 7.16. We see that the gain changes sign at the compensation point, and reaches a broad maximum at $N/N^+ = 1$, where the effective trap density is the largest. On the other hand, the response time is largest (least favorable) at the compensation point, and decreases rapidly for larger or smaller values of reduction ratio. This suggests that in a reduced crystal with $[Fe^{2+}]/[Fe^{3+}] = 1\text{–}10$, the gain can be optimized, *and* the response time can be reduced to a value on the order 1–10 ms. Note that the inclusion of nonzero diffusion lengths in the calculation leads to a less rapid decrease in the response time at large values of reduction ratio. The influence on the gain is only noticeable at the compensation point.

The plots in Fig. 7.16 suggest an explanation for the observation by *Ducharme* and *Feinberg* [7.33] that the gain changed sign and increased in magnitude in a reduced crystal, but the response time (at $\sim 1 \, \mathrm{W \, cm^{-2}}$) did not change appreciably. We assume first that their as-grown sample was nearly compensated, with a typical value of reduction ratio $x \simeq 10^{-2}$. If the reduction treatment only increased $[Fe^{2+}]/[Fe^{3+}]$ by a factor of 4–6, then our single-species model indicates that the gain would increase, with only a slight reduction in response time.

The above calculations are based on the assumption that a single species (iron impurities or dopants) contributes to the photorefractive effect. This is entirely reasonable in as-grown crystals, in which we expect all oxygen vacancies to be fully ionized. However, in reduced crystals the increased concentration of free electrons will increase the rate of recombination with doubly ionized vacancies, leading to an increase in singly ionized vacancies, which are absorbing and could contribute to the photorefractive effect.

Two approaches have been used to calculate the photorefractive properties of a material with two separate optically active species [7.33, 82]. In both cases, each of the two species (labeled donor, with total concentration N_D, and acceptor, with total concentration N_A) can exist in two forms: filled and unfilled, or neutral and ionized. The model of *Ducharme* and *Feinberg* [7.33] assumes that the minority species is fully ionized and is present only for charge compensation. This model is equivalent to the single species model of *Kukhtarev* [7.77–79] applied separately to the regions $N_D > N_A$ and $N_D < N_A$. The model predicts that the number of photorefractive charges (equal to the effective density of empty traps N_E in our notation) should go to zero at the compensation point $N_D = N_A$, whereas the single species model predicts a maximum at this point. The model of *Valley* [7.82] assumes that the donor species contributes only to electron transport, while the acceptor species contributes only to hole transport. This model reduces to the basic model of *Kukhtarev* [7.77–79] when either of the two species is fully ionized. It also predicts that the density of empty traps should go to zero at the compensation point.

The best means of determining which of the above models applies to reduced BaTiO₃ is to measure the response time and Debye length over a wide range of oxidizing and reducing conditions. The measurements of *Ducharme* and *Feinberg* [7.33], while more limited in the range of oxygen partial pressure, do indicate a clear minimum in the number of photorefractive charges at or near the compensation point, which supports the importance of a two-species model.

We have recently studied the photorefractive and electrical properties of a reduced crystal of $BaTiO_3$, in which the oxygen partial pressure was varied over 12 orders of magnitude [7.93, 94]. The crystal (labeled G/L) had been previously characterized in its as-grown state [7.52], yielding $[Fe^{2+}] \simeq 4.4 \times 10^{16} \, cm^{-3}$ and $[Fe^{3+}] \simeq (4 \pm 2) \times 10^{18} \, cm^{-3}$. The beam coupling response time (see below) at an intensity of $0.25 \, W \, cm^{-2}$ and a grating period of $0.7 \, \mu m$ was $1.1 \, s$.

The crystal was subjected to three heat treatments. In each case the sample was heated to $1000 \, °C$, held for $\sim 40 \, h$, and then cooled to room temperature. A mixed atmosphere of CO and CO_2 was maintained throughout each experiment, with gas mixtures chosen to provide oxygen partial pressures in the range 10^{-4}–10^{-12} atm at $1000 \, °C$. Following each heat treatment, photorefractive beam coupling measurements were performed to determine the sign of the photo-carriers and the rise or decay time of the signal beam when the pump beam was chopped. Our results are given in Table 7.6. The response time measured via beam coupling is not generally equal to the fundamental material response time [7.95]. Nevertheless, our response time data should indicate the qualitative behavior of the material response time as a function of oxygen partial pressure.

The data presented in Table 7.6 are consistent with the model described earlier. As the oxygen partial pressure is decreased, the response time first increases and then decreases, as the compensation point is passed. At the same time, the dominant photocarrier changes from holes to electrons. Overall, a decrease in response time by a factor of ~ 10 was achieved. However, under the most reducing conditions, the resistivity of the crystal decreased dramatically. Although the crystal was free of $90°$ domains, electrical poling was only partially successful, due to the limited poling field which could be applied. We note also that we did not observe any change in T_c as a result of the heat treatment. This invariance in T_c differs from the results of *Ducharme* and *Feinberg* [7.33], who observed a decrease in T_c of $6 \, °C$ when an as-grown crystal was treated at $650 \, °C$ in 10^{-6} atm of oxygen.

Table 7.6. Resistivity, sign of photocarriers and beam coupling response time of $BaTiO_3$ crystal (sample G/L) subjected to several reduction treatments

Treatment CO/(CO+CO₂)	P_{O_2} at 1000 °C [atm]	Resistivity [Ω cm]	Sign of photocarriers	Response time [s]
as-grown	0.2	$\sim 10^{12}$	+	1.05
0.0	3×10^{-5}	$\sim 10^{12}$	+	1.75
0.001	8×10^{-9}	$\sim 10^{12}$	−	0.90
0.10	6×10^{-13}	10^6	−	0.1

From our response time measurements (Table 7.6) and our theoretical plot of τ versus $[Fe^{2+}]/[Fe^{3+}]$ (Fig. 7.16), we can estimate the ratio $[Fe^{2+}]/[Fe^{3+}]$ as a function of oxygen partial pressure. We find that the induced change in $[Fe^{2+}]/[Fe^{3+}]$ is relatively small, varying by less than a factor of 50 over a variation of nearly 12 orders of magnitude in oxygen partial pressure. The relative stability of Fe^{3+}, i.e., its resistance to conversion to Fe^{2+}, is a known property of BaTiO₃. In particular, in [7.53] negligible conversion of Fe^{3+} to Fe^{2+} was observed at oxygen partial pressures as low as 10^{-22} atm. Only in a pure H_2 environment (10^{-26} atm oxygen) was conversion of $\sim 50\%$ achieved. However, under these heavy reduction conditions, the crystals are highly conductive and opaque. We believe that the stability of Fe^{3+} in BaTiO₃ is a result of the shallow position of the Fe^{2+} levels (Fig. 7.5b). Any attempted conversion to Fe^{2+} is inhibited by thermal ionization of this species, leading to conversion back to Fe^{3+}, along with the generation of free electrons:

$$Fe^{2+} \rightarrow Fe^{3+} + e^- \ . \tag{7.58}$$

The equilibrium densities of Fe^{2+}, Fe^{3+} and n determined from this reaction are related in (7.13). We have also pointed out [7.52] that OH^- and H^+ ions tend to stabilize Fe^{3+}. This emphasizes the importance of avoiding the introduction of hydrogen during growth, and processing with other reducing agents, such as CO/CO_2.

There are a number of other approaches to inducing electron photoconductivity in BaTiO₃, while maintaining high resistivity. One approach would be to dope the crystals with transition metals other than iron. Here we must consider two cases: acceptor dopants (metals with valence states lower than 4+), and donor dopants (metals with valence states greater than 4+). For acceptor dopants, we require that the energy levels of the reduced species be substantially deeper than those of Fe^{2+}. Possible candidates are Cr or Mn, which are known to be more easily reduced than Fe in BaTiO₃ ceramics [7.54]. In fact, Mn is known to have deeper levels in the BaTiO₃ band gap than Fe [7.44]. However, even in this case, the ability to produce insulating, n-type behavior may be limited by the shallow nature of the compensating oxygen vacancies.

A second approach is donor doping, which offers the advantage of electron conductivity without the need for a reducing heat treatment. In donor-doped ceramics, charge compensation is by free electrons and metal ion vacancies, with the latter being favored at higher dopant concentrations [7.51, 96]. In single crystals, the required dopant concentration for metal ion compensation has not been determined. Metal ion-compensated samples are expected to be insulating, since the metal ion (acceptor) levels are deep in the band gap [7.51], and few oxygen vacancies are present.

7.7 Conclusions

Since the "rediscovery" of $BaTiO_3$ in 1980, this material has become the material of choice in applications requiring large gain and/or four-wave mixing reflectivity. At the same time, $BaTiO_3$ has developed a reputation for being slow, and thus unsuited for certain applications. However, all evidence suggests that the slow speed of $BaTiO_3$ is a specific property of this material in its as-grown state, and is not a fundamental property. The calculations from the previous section show that reduced samples of $BaTiO_3$ should have response times below 1 ms, with little or no decrease in beam coupling gain. Of course the challenge is to obtain reduced samples without sacrificing the high resistivity obtained in as-grown samples.

In order to further characterize commercial samples of $BaTiO_3$, it would be useful to subject a given sample to heat treatments over a wide range of oxygen partial pressures (as done by *Kurz* et al. [7.85] for $LiNbO_3$), and correlate the changes in EPR signals and absorption coefficient with changes in the photorefractive behavior. Mössbauer or low temperature EPR experiments should also be performed in order to measure the concentration of Fe^{2+} and Fe^{4+} ions. These measurements would serve to test the energy level models discussed earlier, and confirm the identification of Fe as the photorefractive species.

Acknowledgements. I would like to acknowledge helpful discussions with R.N.Schwartz, G.C.Valley, B.A.Wechsler, D.Rytz, R.Pastor, R.A.Mullen, S.Ducharme, and J.Feinberg.

References

7.1   R.L.Townsend, J.T.LaMacchia: J. Appl. Phys. **41**, 5188 (1970)
7.2   J.Feinberg, R.W.Hellwarth: Opt. Lett. **5**, 519 (1980)
7.3   A.Ashkin, G.D.Boyd, J.M.Dziedzic, R.G.Smith, A.A.Ballman, J.L.Levinstein, K.Nassau: Appl. Phys. Lett. **9**, 72 (1966)
7.4   F.S.Chen, J.T.LaMacchia, D.B.Fraser: Appl. Phys. Lett. **13**, 223 (1968)
7.5   F.Micheron, G.Bismuth: J. de Phys. **33**, Suppl. 4, Colloq. 2, 149 (1972)
7.6   W.T.H.Koch, R. Munser, W.Ruppel, P.Wurfel: Solid State Commun. **17**, 847 (1975)
7.7   W.T.H.Koch, R.Munser, W.Ruppel, P.Wurfel: Ferroelectrics **13**, 305 (1976)
7.8   J.Feinberg, D.Heiman, A.R.Tanguay, Jr., R.W.Hellwarth: J. Appl. Phys. **51**, 1297 (1980)
7.9   E.Krätzig, F.Welz, R.Orlowski, V.Doormann, M.Rosenkranz: Solid State Commun. **34**, 817 (1980)
7.10   J.P.Remeika: J. Am. Ceram. Soc. **76**, 940 (1954)
7.11   V.Belrus, J.Kalinajs, A.Linz, R.C.Folweiler: Mater. Res. Bull. **6**, 899 (1971)
7.12   S.H.Wemple, M.DiDomenico, Jr., I.Camlibel: J. Phys. Chem. Solids **29**, 1797 (1968)
7.13   I.Camlibel, M.DiDomenico, S.H.Wemple: J. Phys. Chem. Solids **31**, 1417 (1970)
7.14   W.J.Merz: Phys. Rev. **91**, 513 (1953)
7.15   F.Jona, G.Shirane: *Ferroelectric Crystals* (Macmillan, New York 1962)
7.16   M.E.Lines, A.M.Glass: *Principles and Applications of Ferroelectrics and Related Materials* (Clarendon, Oxford 1977)

7.17 M. DiDomenico, Jr., S.H. Wemple: Phys. Rev. **155**, 539 (1967)
7.18 V.M. Fridkin, A.A. Grekov, N.A. Kosonogov, T.R. Volk: Ferroelectrics **4**, 169 (1972)
7.19 J. Rodel, G. Tomandl: J. Mater. Sci. **19**, 3515 (1984)
7.20 R.C. Pastor: unpublished
7.21 H.E. Muser, W. Kuhn, J. Albers: Phys. Status Solidi A**49**, 51 (1978)
7.22 S.H. Wemple: "Electrooptical and Nonlinear Optical Properties of Crystals", in *Applied Solid State Science*, ed. by R. Wolfe, Vol. 3 (Academic, New York 1972), pp. 263–383
7.23 R.C. Miller: Appl. Phys. Lett. **5**, 17 (1964)
7.24 A.R. Johnston, J.M. Weingart: J. Opt. Soc. Am. **55**, 828 (1965)
7.25 M. Cronin-Golomb: Paper ThT3, CLEO '85, Baltimore, Md
7.26 W.R. Cook, H. Jaffe: "Piezoelectric, Electrostrictive and Dielectric Constants, and Electromechanical Coupling Factors of Piezoelectric Crystals", in *Landolt-Börnstein*, Group III, Vol. 11, ed. by K.-H. Hellwege, A.M. Hellwege, (Springer, Berlin, Heidelberg 1979) pp. 287–470
7.27 W.R. Cook, R.F.S. Hearmon, H. Jaffe, D.F. Nelson: "Piezooptic and Electrooptic Constants", in *Landolt-Börnstein*, Group III, Vol. 11, ed. by K.H. Hellwege, A.M. Hellwege (Springer, Berlin, Heidelberg 1979) pp. 495–670
7.28 S. Ducharme, J. Feinberg, R.R. Neurgaonkar: IEEE J. QE-**23**, 2116 (1987)
7.29 I.P. Kaminow: Appl. Phys. Lett. **7**, 123 (1965); Erratum, Appl. Phys. Lett. **8**, 54 (1966)
7.30 I.P. Kaminow: Appl. Phys. Lett. **8**, 305 (1966)
7.31 A.R. Johnston: Appl. Phys. Lett. **7**, 195 (1965)
7.32 M.B. Klein, G.C. Valley: J. Appl. Phys. **57**, 4901 (1985)
7.33 S. Ducharme, J. Feinberg: J. Opt. Soc. Am. B**3**, 283 (1986)
7.34 F.M. Michel-Calendini, G. Mesnard: J. Phys. C**6**, 1709 (1973)
7.35 P. Pertosa, F.M. Michel-Calendini: Phys. Rev. B**17**, 2011 (1978)
7.36 F.M. Michel-Calendini, H. Chermette, J. Weber: J. Phys. C**13**, 1427 (1980)
7.37 F.M. Michel-Calendini: Ferroelectrics **37**, 499 (1981)
7.38 P. Moretti, F.M. Michel-Calendini: Ferroelectrics **55**, 219 (1984)
7.39 F.M. Michel-Calendini, L. Hafid, G. Godefroy, H. Chermette: Solid State Commun. **54**, 951 (1985)
7.40 F.M. Michel-Calendini, P. Moretti: Phys. Rev. B**27**, 763 (1983)
7.41 L.F. Mattheiss: Phys. Rev. B**6**, 4718 (1972)
7.42 S.H. Wemple: Phys. Rev. B**2**, 2679 (1970)
7.43 C. Gahwiller: Phys. Kondens. Mater. **6**, 269 (1967)
7.44 H.-J. Hagemann: Ph. D. Thesis, Rheinisch-Westfälische Technische Hochschule Aachen (1980)
7.45 G. Godefroy, C. Dumas, P. Lompre, A. Perrot: Ferroelectrics **37**, 725 (1981)
7.46 C.N. Berglund, H.J. Braun: Phys. Rev. **164**, 790 (1967)
7.47 N.H. Chan, R.K. Sharma, D.M. Smyth: J. Am. Ceram. Soc. **64**, 556 (1981)
7.48 G.V. Lewis, C.R.A. Catlow: J. Phys. Chem. Solids **47**, 89 (1986)
7.49 R.K. Sharma, N.H. Chan, D.M. Smyth: J. Am. Ceram. Soc. **64**, 448 (1981)
7.50 D.E. Rase, Rustum Roy: J. Am. Ceram. Soc. **38**, 102 (1955)
7.51 J. Daniels, K.H. Hardtl: Philips Res. Rep. **31**, 489 (1976)
7.52 M.B. Klein, R.N. Schwartz: J. Opt. Soc. Am. B**3**, 293 (1986)
7.53 H.J. Hagemann, A. Hero, U. Gonser: Phys. Status Solidi A**61**, 63 (1980)
7.54 H.J. Hagemann, D. Hennings: J. Am. Ceram. Soc. **64**, 590 (1981)
7.55 E. Siegel, K.A. Muller: Phys. Rev. B**20**, 3587 (1979)
7.56 S.H. Wemple: "Electrical Contact to *n*- and *p*-Type Oxdies", in *Ohmic Contacts to Semiconductors*, ed. by B. Schwartz (Electrochemical Society, New York 1968)
7.57 N.H. Chan, R.K. Sharma, D.M. Smyth: J. Am. Chem. Soc. **65**, 167 (1982)
7.58 N.G. Eror, D.M. Smyth: J. Solid State Chem. **24**, 235 (1978)
7.59 J. Daniels: Philips Res. Rep. **31**, 505 (1976)
7.60 C.N. Berglund, W.S. Baer: Phys. Rev. **157**, 358 (1967)
7.61 S. Ikegami, I. Ueda: J. Phys. Soc. Jpn. **19**, 159 (1964)
7.62 F.S. Chen: J. Appl. Phys. **40**, 3389 (1969)

7.63 G.E.Peterson, A.M.Glass, T.J.Negran: Appl. Phys. Lett. **19**, 130 (1971)
7.64 S.Ducharme, J.Feinberg: J. Appl. Phys. **56**, 839 (1984)
7.65 D.Rak, I.Ledoux, J.P.Huignard: Opt. Commun. **49**, 302 (1984)
7.66 K.Kawabe, Y.Inuishi: Jpn. J. Appl. Phys. **2**, 590 (1963)
7.67 P.Gerthsen, R.Groth, K.H.Hardtl: Phys. Status Solidi **11**, 303 (1965)
7.68 H.Veith: Z. Angew. Phys. **20**, 16 (1965)
7.69 D.L.Ridpath, D.A.Wright: J. Mater. Sci. **5**, 487 (1970)
7.70 E.V.Bursian, Y.G.Girshberg, E.N.Starov: Sov. Phys.-Solid State **14**, 872 (1972)
7.71 P.Gerthsen, K.H.Hardtl, A.Csillag: Phys. Status Solidi A **13**, 127 (1972)
7.72 A.M.J.H.Seuter: Philips Res. Rep., Suppl. 3 (1974)
7.73 J.P.Boyeaux, F.M.Michel-Calendini: J. Phys. C **12**, 545 (1979)
7.74 G.A.Cox, R.H.Tredgold: Phys. Lett. **11**, 22 (1964)
7.75 L.Benguigui: Solid State Commun. **7**, 1245 (1969)
7.76 C.-P.Tzou, T.Y.Chang, R.W.Hellwarth: Proc. SPIE **613**, 58 (1986)
7.77 V.L.Vinetskii, N.V.Kukhtarev: Sov. Phys.–Solid State **16**, 2414 (1975)
7.78 N.V.Kukhtarev: Sov. Tech. Phys. Lett. **2**, 438 (1976)
7.79 N.V.Kukhtarev, V.B.Markov, S.G.Odulov, M.S.Soskin, V.L.Vinetskii: Ferroelectrics **22**, 949, 961 (1979)
7.80 G.C.Valley, M.B.Klein: Opt. Eng. **22**, 704 (1983)
7.81 R.Orlowski, E.Krätzig: Solid State Commun. **27**, 1351 (1978)
7.82 G.C.Valley: J. Appl. Phys. **59**, 3363 (1986)
7.83 F.P.Strohkendl, J.M.C. Jonathan, R.W.Hellwarth: Opt. Lett. **11**, 312 (1986)
7.84 J.P.Huignard, A.Marrakchi: Opt. Commun. **38**, 249 (1981)
7.85 H.Kurz, E.Krätzig, W.Keune, H.Engelmann, U.Gonser, B.Dischler, A.Rauber: Appl. Phys. **12**, 355 (1977)
7.86 N.V.Kukhtarev, E.Krätzig, H.C.Kulich, R.A.Rupp, J.Albers: Appl. Phys. B **35**, 17 (1984)
7.87 J.Feinberg: J. Opt. Soc. Am. **72**, 46 (1982)
7.88 L.K.Lam, T.Y.Chang, J.Feinberg, R.W.Hellwarth: Opt. Lett. **6**, 475 (1981)
7.89 A.Rose: *Concepts in Photoconductivity and Allied Problems* (Robert Krieger, New York 1978)
7.90 G.C.Valley: IEEE J. QE-**19**, 1637 (1983)
7.91 A.W.Hornig, R.C.Rempel, H.E.Weaver: J. Phys. Chem. Solids **10**, 1 (1959)
7.92 D.L.Staebler and W.Phillips: Appl. Opt. **13**, 788 (1974)
7.93 M.B.Klein: Paper FQ1, CLEO '86, San Francisco, Ca
7.94 B.A.Wechsler, M.B.Klein, D.Rytz: Proc. SPIE **681**, 91 (1986)
7.95 J.M.Heaton, L.Solymar: Opt. Acta **32**, 397 (1985)
7.96 H.M.Chan, M.P.Harmer, D.M.Smyth: J. Am. Ceram. Soc. **69**, 507 (1986)

Additional References

Motes,A., Kim,J.J.: "Beam Coupling in Photorefractive BaTiO$_3$ Crystals", Opt. Lett. **12**, 199 (1987)

Smirl,A., Valley,G.C., Mullen,R.A., Bohnert,K., Mire,C.D., Boggess,T.F.: "Picosecond Photorefractive Effect in BaTiO$_3$", Opt. Lett. **12**, 501 (1987)

Rytz,D., Klein,M.B., Mullen,R.A., Schwartz,R.N., Valley,G.C., Wechsler,B.A.: "High Efficiency, Fast Response in Photorefractive BaTiO$_3$ at 120 °C", submitted to Appl. Phys. Lett., 1988

Wechsler,B.A., Klein,M.B.: "Thermodynamic Point Defect Model of BaTiO$_3$ and Application to the Photorefractive Effect", submitted to J. Opt. Soc. Am. B, 1988

8. The Photorefractive Effect in Semiconductors

Alastair M. Glass and Jefferson Strait

With 20 Figures

The majority of experimental work on the fundamentals and applications of the photorefractive effect has been carried out with ferroelectric oxides. These materials have the strongest known electro-optic effects, which in turn lead to the largest photorefractive index changes. However, the magnitude of the electro-optic effect alone is not sufficient to characterize a material for all photo-refractive applications. It was realized many years ago [8.1] that for maximum sensitivity (index change per absorbed photon density) it is necessary to maximize the ratio of the electro-optic coefficient to the dielectric constant *and* to have a photocarrier drift or diffusion length comparable to the fringe spacing of the phase hologram. For these reasons, efficient photoconductors such as bismuth silicon oxide (BSO) can be as useful as or even more useful than ferroelectrics for optical processing applications, even though the electro-optic coefficient of BSO is relatively small.

It is perhaps surprising that many years passed after the initial demonstration of photorefraction in BSO [8.2] before any attention was paid to the electro-optic semiconductors which are in common use for other opto-electronic applications. Optically induced index changes had been observed in CdS [8.3] soon after the initial discovery of "optical damage" in $LiNbO_3$. Although these observations were made before a full understanding of the photorefractive effect was obtained, the effects in CdS were found not to have electro-optic origins.

In this chapter various characteristics of semiconductors which make them useful as photorefractive materials will be discussed. The underlying theory has already been discussed elsewhere in this volume and the reader is referred to those sections for details. At the present time, only a few experiments using semiconductors have been described in the literature [8.4–8] so major emphasis has been placed on those materials studied; namely, InP and GaAs. Other semiconductors are discussed from the point of view of their potential for photorefractive applications where appropriate, but the underlying principles remain the same.

8.1 Electro-optic Properties

A comparison of the electro-optic properties of semiconductor materials and several oxides is shown in Table 8.1. In this table the electro-optic coefficients r_{ij} are given for constant stress conditions at wavelengths well below the band gap

Table 8.1. Optical and electro-optical properties of some photorefractive materials. The data have been largely taken from [8.9, 10]. The electron and hole mobilities are given for undoped semiconductors. For most of the oxides these values are not well established

Material	Symmetry	E_g[eV]	r_{ij}[µm]	n_i[µm]	ε	$n_i^3 r_{ij}$ [pm/V]	$n_i^3 r_{ij}/\varepsilon_j$ [pm/V]	μ(300 K) [cm²/Vs] Electrons	Holes
III–V Semiconductors									
InP	4̄3 m	1.35	$r_{41}=$ 1.45 (1.06)	3.29 (1.06)	12.6	52	4.1	4600	150
GaAs	4̄3 m	1.42	$r_{41}=$ 1.2 (1.08)	3.5 (1.02)	13.2	43	3.3	8500	400
GaP	4̄3 m	2.26	$r_{41}=$ 1.07 (0.56)	3.45 (0.54)	12	44	3.7	110	75
II–VI Semiconductors									
CdTe	4̄3 m	1.56	$r_4=$ 6.8 (3.39)	2.82 (1.3)	9.4	152	16	1050	100
ZnS	4̄3 m	3.68	$r_{41}=$ 1.2 (0.4)	2.47 (0.45)	16	18	1.1	165	5
ZnSe	4̄3 m	2.68	$r_{41}=$ 2.0 (0.55)	2.66 (0.55)	9.1	38	4.1		
ZnTe	4̄3 m	2.27	$r_{41}=$ 4.45 (0.59)	3.1 (0.57)	10.1	133	13		
CdSe	6 mm	1.70	$r_{33}=$ 4.3 (3.39)	2.54 (1.15)	10.65	70	6.6	800	
CdS	6 mm	2.47	$r_{33}=$ 4 (0.59)	2.48 (0.63)	10.33	61	5.9		
Oxides									
LiNbO₃	3 m	3.2	$r_{33}=$ 32.2 (0.63)	2.27 (0.70)	32	320	11		
Bi₁₂SiO₂₀	23		$r_{41}=$ 5 (0.63)	2.54	47	82	1.8	1	
BaTiO₃	4 mm	~3.3	$\left\{\begin{array}{l}r_{51}=1640 (0.55)\\ r_{33}= 28\end{array}\right.$	2.40 (0.63)	$\left\{\begin{array}{l}3600\\ 135\end{array}\right.$	$\left\{\begin{array}{l}11300\\ 387\end{array}\right.$	$\left\{\begin{array}{l}4.9\\ 2.9\end{array}\right.$		
KNbO₃	4 mm	3.2	$r_{33}=$ 64	2.23	55	690	14		
Sr₀.₇₅Ba₀.₂₅Nb₂O₆	4 mm	3.2	$r_{33}=1340$ (0.63)	2.30 (0.63)	3400	16300	4.8		

of the materials [8.9]. The materials figures of merit relevant for maximum index change $(n_i^3 r_{ij})$ and for maximum sensitivity $(n_i^3 r_{ij}/\varepsilon_j)$ are also shown for comparison, where ε_i is the relative dielectric constant and n_i is the refractive index. Other relevant parameters such as optical band gap E_g, carrier mobility, and crystal symmetry are listed [8.10].

It is seen from the table that the ferroelectric materials have by far the largest figures of merit for maximum index change, which is necessary for steady-state applications such as phase conjugate mirrors which require high reflectivity or diffraction efficiency. However, for large sensitivity CdTe has the highest figure of merit listed in Table 8.1 – slightly greater then $KNbO_3$, but three times that of $BaTiO_3$. InP and GaAs have figures of merit about twice that of BSO. The relatively large value of $n_i^3 r_{ij}/\varepsilon_j$ for $LiNbO_3$ is offset by the fact that the free carrier drift and diffusion lengths are extremely small even in large applied fields and, in practice, the sensitivity of $LiNbO_3$ as well below that of all the other materials listed. In all of these other materials at sufficiently high applied electric fields, the drift length can be comparable to the grating period, as required for optimum sensitivity.

It is evident from Table 8.1 that the semiconductors provide a wide variety of choice for the wavelength of operation. GaAs, InP and CdTe are convenient materials for use at wavelengths in the near infrared, which is of importance for optical communications. Materials with band gaps below 1.35 eV have not been listed in the table since little is known about their electro-optic coefficients and nothing is known about their photorefractive behavior. The room temperature resistivity decreases rapidly with decreasing band gap. Low resistivity may be useful for short pulse, high repetition rate applications which require rapid relaxation times. However, practical sources become more scarce as the wavelength is increased. For longer wavelength operation it is feasible to use wider band-gap materials having donor (or acceptor) states at the requisite energy below (or above) the band edges, but no experiments have yet been performed at wavelengths longer than 1.32 μm.

8.2 Defects

Because of the importance of defects in semiconductors for electronic and opto-electronic applications, there is already a substantial body of knowledge in the literature concerning the controlled incorporation of electrically and optically active centers in these materials. This is particularly true for GaAs and InP. The electronic band gaps of these two materials are 1.42 eV and 1.35 eV, respectively, at 300 K, and photorefractive applications have been limited to wavelengths longer than this (resonant photorefractive effects will be discussed later).

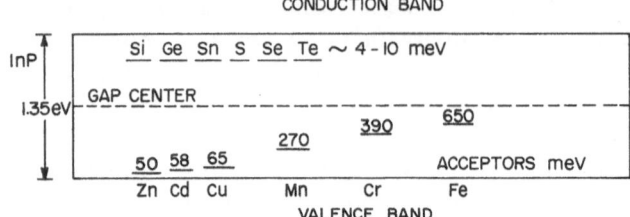

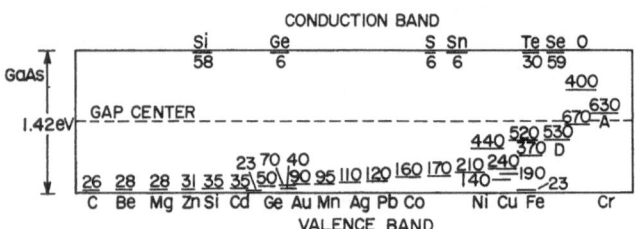

Fig. 8.1. Energy levels of donor and acceptor impurities within the InP [8.11] and GaAs [8.10] band gaps

8.2.1 Shallow Impurities

A variety of impurities have been studied which act as shallow donors or acceptors in GaAs and InP, thereby controlling the electrical behavior at room temperature. Several examples of these are shown in Fig. 8.1. Shallow impurities only alter the optical properties of the host crystal with a few tens of milli-electron volts below the fundamental absorption edge, and in the far infrared, corresponding to transitions from the valence to conduction bands. At room temperature these centers are to a large extent ionized, giving rise to electrical conductivity of the crystals. For instance, with only 10^{13} ionized donors per cubic centimeter in GaAs the conductivity is 10^{-2} $(\Omega\,\mathrm{cm})^{-1}$ and the calculated dielectric relaxation time $\varepsilon\varepsilon_0/\sigma$ is ~ 200 ps. Since it is extremely difficult to control impurities of these concentrations (indeed the impurity content of undoped crystals is typically greater than 10^{14} cm^{-3}) shallow impurity levels are only likely to be of interest in photorefractive applications for controlling the degree of compensation of deep levels. At low temperatures, however, shallow levels could play a more significant role as metastable states or for applications in the infrared. For instance, using 10 μm radiation, direct photoionization of levels 0.12 eV below the conduction band (or above the valence band) is possible. To date no such applications have been demonstrated.

8.2.2 Deep Levels

The deep impurity states shown in Fig. 8.1 affect both the optical and electrical characteristics of semiconductors which are of practical significance for photorefractive applications in the near infrared region of the spectrum. A more extensive list of deep levels in GaAs is given in [8.12]. By introducing a sufficiently high density of these levels, the conductivity due to shallow states is compensated and the Fermi level is located near the mid-gap region. In GaAs a chromium impurity ion concentration of about 10^{16} cm^{-3} can give a crystal resistivity as high as 6×10^8 Ωcm and a dielectric relaxation time of ~ 1 ms. Similar effects are achieved with "intrinsic" EL2 centers in GaAs, Fe impurities in InP, and a variety of other transition metal impurities. The deep levels greatly reduce the free carrier lifetime of photoexcited carriers as the concentration of centers is increased. In addition, photoionization of deep levels gives rise to optical absorption from the band edge down to wavelengths corresponding to near mid-gap energies (~ 0.7 eV in InP and GaAs). Careful control of deep level concentrations therefore affords considerable versatility for photorefractive recording in these materials: i.e. variation of material response time, absorption depth, trap density, wavelength of operation and photocarrier mobility. To the present time, three centers have been studied in GaAs and InP and these will be discussed individually in more detail below. A complication associated with the characterization of deep levels is the fact that excitation of holes from the valence band to empty deep levels occurs in the same spectral region as the excitation of electrons from occupied deep levels to the conduction band. Because of the higher electron mobility compared to holes, electronic excitation can be dominant electrically, even though the absorption coefficient of this process may be smaller. The relative importance of the two processes depends on the degree of compensation by shallower donor or acceptor states. Photorefractive studies lend themselves well to the elucidation of the optical and electrical behavior. For instance, the sign of the dominant photocarrier can be readily determined from studies of the direction of energy transfer in a two-wave mixing experiment [8.13–15]. This can be studied as a function of wavelength. Of particular practical importance, since semi-insulating GaAs and InP substrates are used for a number of electronic and opto-electronic devices, is the fact that the photorefractive effect offers a contactless method for mapping the optical and electrical characteristics of substrate wafers with optical imaging techniques. By simple measurements the local resistivity, absorption, sign of the photocarriers, and drift or diffusion length can be readily achieved.

a) GaAs:EL2

Many electrical and spectroscopic studies have been devoted to studies of EL2 centers in GaAs. A number of studies and reviews can be found in [8.16]. Even though full understanding of this center has not yet been achieved, a number of characteristics are generally agreed upon. These centers are observed in undoped GaAs as well as doped crystals. The EL2 centers have donor characteristics: They are electrically neutral when occupied by electrons. The EL2 is generally identified as an antisite defect, involving arsenic on gallium sites, but there is some evidence that a second defect may be associated with the antisite defect forming a complex.

The optical absorption band and photoconductivity spectrum [8.17] of the EL2 center is shown in Fig. 8.2. The bump observed at 1.2 eV is attributed to localized intra-center absorption. The absorption coefficient of EL2 is quite low for wavelengths longer than 1 eV, which limits the use of this defect at longer wavelengths. Following excitation, the EL2 center undergoes Frank-Condon relaxation with the lattice of about 130 meV, so that the activation energy for electron excitation and hole excitation are 0.86 eV and 0.85 eV, respectively.

At low temperatures the EL2 center is bleached by photoionizing radiation, so this center is not appropriate for conventional photorefractive recording below about 140 K.

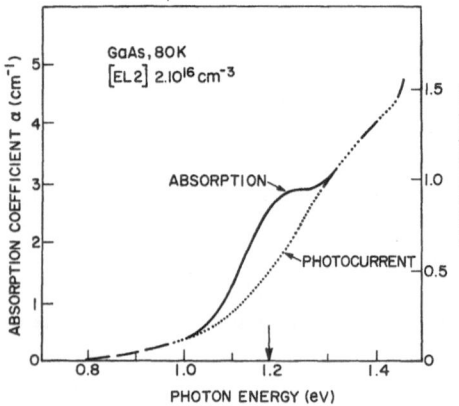

Fig. 8.2. Optical absorption and photoconductivity spectra of EL2 centers in GaAs [8.17]

b) GaAs:Cr

GaAs doped with about 10^{16} cm^{-3} chromium ions is normally semi-insulating. Resistivities as high as 6×10^8 Ωcm have been reported. The EL2 centers also present in chromium-doped crystals are fully ionized by the chromium ions. Thus the Fermi level is pinned at the Cr^{2+}/Cr^{3+} levels for [Cr] > [EL2^0], i.e.,

$$Cr^{3+} + EL2^0 \rightarrow EL2^+ + Cr^{2+} .$$

Chromium-doped GaAs also exhibits bipolar photoconductivity following the reactions

$$Cr^{3+} + h\nu \rightarrow Cr^{2+} + h^+ \quad \text{and}$$

$$Cr^{2+} + h\nu \rightarrow (Cr^{2+})^* \rightarrow Cr^{3+} + e^- ,$$

where the intermediate $(Cr^{2+})^*$ state is resonant with the GaAs conduction band. The Cr^{3+} state is situated 0.79 eV above the valence band. The extrinsic

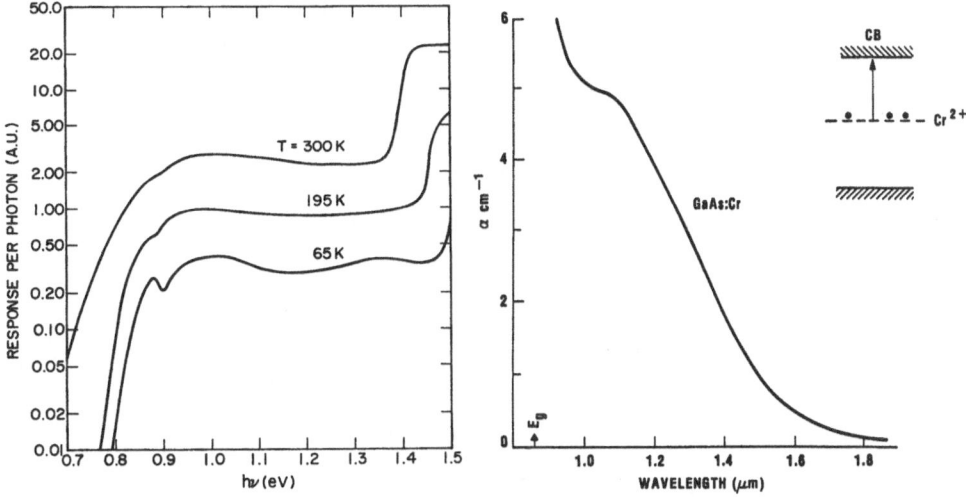

Fig. 8.3. The extrinsic photoconductivity in semi-insulating GaAs : Cr [8.18]

Fig. 8.4. The optical absorption spectrum (300 K) of GaAs : Cr ($\sim 10^{16}$ cm^{-3})

photoconductive response [8.18] for a semi-insulating GaAs:Cr is shown in Fig. 8.3 and an absorption spectrum is shown in Fig. 8.4. The optimum cross section for electron excitation From Cr^{2+} is at about 0.9 eV whereas that for holes continues to rise up to 1.1 eV. The cross section for holes is greater than that for electrons for all energies above 0.9 eV, so that for weakly compensated samples hole transport dominates the photoconductivity. The extrinsic absorption should allow photorefractive response from the GaAs band edge out to 0.7 eV (1.7 μm). However, at energies above 1.0 eV the large absorption cross section of chromium means that either short interaction lengths or low trap concentrations must be tolerated.

c) InP:Fe

At present, no stoichiometry-related deep level defects like EL2 centers have been observed in InP and all semi-insulating material is made by doping with transition metal ions in the melt – principally Fe. The Fe^{2+} center substitutes for In giving rise to levels about 0.65 eV below the conduction band according to Hall measurements. The resistivity of iron-doped InP ($\sim 10^{17}$ cm^{-3}) is typically 10^7–10^8 Ω cm. The energy level [8.15] for Fe^{2+} is shown in Fig. 8.5. Relatively weak absorption in the 0.27–0.59 eV spectral region is attributable to localized Fe^{2+} transitions, and gives rise to weak photoconductivity only via thermally assisted transitions. For $0.59 < h\nu < 0.7$ eV both absorption and photoconductivity are due to electronic transitions from the Fe^{2+} center to the conduction band, while for 0.7 eV $< h\nu < 1.34$ eV the transitions are due to hole excitation. The situation is then very similar to that for iron doped LiNbO$_3$, which has been

ANTIREFLECTION COATED InP:Fe ($\sim 10^{17}$ cm^{-3})

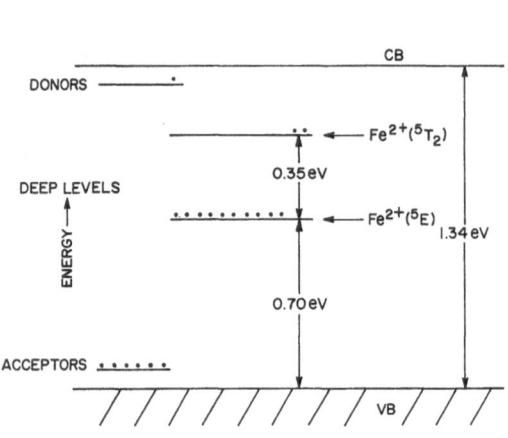

Fig. 8.5. Energy levels for divalent iron impurities in InP

Fig. 8.6. Absorption spectrum of InP : Fe ($\sim 10^{17}$ cm^{-3})

extensively studied since the discovery of the role of Fe in the photorefractive effect in LiNbO$_3$ [8.19]. At photon energies near 2.5 eV,

$$Fe^{2+} + h\nu \rightarrow Fe^{3+} + e^- \ .$$

The Fe^{3+} ions act as electron traps. The net result of excitation is a spatial redistribution of electrons amongst the Fe^{2+} and Fe^{3+} centers. The absorption spectrum of InP:Fe ($\sim 10^{17}$ cm^{-3}) is shown in Fig. 8.6. The absorption cross section remains high enough for photorefractive applications out to about 0.9 eV and the density of iron centers is satisfactory for many applications. InP:Fe permits a longer optical interaction length than GaAs:Cr for equivalent trap densities ($\gtrsim 10^{17}$ cm^{-3}) at 1.06 μm.

The understanding derived from previous studies of electrically and optically active centers in GaAs and InP allows one to design the desired density of traps as well as the desired absorption by appropriately double doping and compensating deep levels.

8.3 Transport Properties

In the later 1960s and early 1970s, the photorefractive effect was extensively investigated for optical memory applications in which the information was to be stored for long periods of time. The high resistivity of oxides made them attractive candidates for such applications. Indeed, it was the exceptionally high resistivity of LiNbO$_3$ ($\sim 10^{20}$ Ωcm) and poor photoconductivity that resulted in the discovery of the photorefractive effect in this material. With no applied field the induced index change accumulated due to the bulk photovoltaic effect [8.20] and remained over long periods of time (even more than 1 year). Since the saturation value of the electric field

$$E_{\text{sat}} = \frac{\kappa \alpha I}{\sigma} \ ,$$

where κ is the bulk photovoltaic constant and σ is the crystal conductivity, low values of σ give large electro-optic effects. During the late 1970s and 1980s, the interest in memory applications declined (mainly because of the volatility of the memory) and interest in optical processing applications, which took advantage of the volatility, increased. For such applications the memory need not be longer than milliseconds in many cases and indeed more and more attention is now being paid to higher speed applications in which much shorter relaxation times are desirable.

Most ferroelectric oxides have extremely low carrier mobilities. This is related to the strong interaction of the free carriers and the lattice via the strong optic modes of the ferroelectric, which is in turn related to the high polarizability of the oxygen ions in octahedral coordination. The semiconductors on the other hand are well known to have high free carrier mobilites. Electron mobilities fall in the range 100–10000 $\text{cm}^2/\text{V s}$, as seen in Table 8.1, which compares with values around 1 $\text{cm}^2/\text{V s}$ for even the higher mobility oxides. The higher mobility of semiconductors translates into:

1) high speed charge separation.
2) more efficient photoconductivity, and
3) rapid dielectric relaxation times in the presence and absence of light.

The first of these items depends on mobility alone. Suppose a hologram is recorded with very short optical pulses. At the instant of excitation, $t = 0$, the excited electrons are located at the same positions as the donor impurity and no space-charge field is developed. Only an index change due to absorption saturation is observed. At times $t > 0$ the electrons diffuse or drift away from the donors. As the carriers move a space-charge field, and hence an index grating, is continuously generated. If trapping sites are available the carriers are eventually trapped, giving rise to metastable space-charge fields which persist for long times determined by the lifetime of trapped carriers. If no traps are available the space-charge field decays as the free carriers rapidly return (driven by the local field) to the donor sites thereby screening the field. The optimum recording efficiency is achieved if carriers drift or diffuse a distance $\sim \Lambda/2\pi$ from the donor, where Λ is the spatial period of the hologram. The time for the carriers to travel a distance $\Lambda/2\pi \sim 0.17$ μm in a field of say 1000 V/cm with a mobility $\mu \sim 1000$ $\text{cm}^2/\text{V s}$ is ~ 17 ps! Thus holographic recording can be accomplished on a picosecond time scale [8.21, 22]. While holograms have been recorded with single picosecond optical pulses [8.23, 24] no one to date has actually measured the evolution time of the hologram to determine the ultimate speed. This should afford a direct measure of the free carrier drift velocity. For ultra-high speed processing it would be desirable not to trap carriers. In this case electron-hole excitation across the band gap of undoped semiconductors is probably the best approach. Because of the different electron and hole drift or diffusion velocities, transient gratings will be generated only for the time it takes electrons and holes to screen the local fields. Recent measurements [8.24] using a mode-locked YAG:Nd laser indicated that in undoped GaAs the response time is shorter than about 40 ps.

The second two factors b) and c) listed above depend not just on the mobility but on the mobility-carrier lifetime product $\mu\tau$. For high speed pulsed operation, a short value of τ is desirable to avoid persistent photoconductivity long after a short optical pulse has been absorbed. The values of can be greatly affected by the incorporation of dopants, particularly deep levels in semiconductors. For example, the addition of 10^{17} cm^{-3} iron impurities in InP reduces the carrier lifetime from ~ 20 ns to < 100 ps due to trapping [8.25]. The mobility is also reduced but to a much lesser extent. The addition of ionized impurities to these materials results in increased free carrier scattering and the mobility can be reduced in a controlled manner if desired. For recording information there is no advantage in having a $\mu\tau$ product so great that carriers travel distances greater than the spatial period $\Lambda/2\pi$. At low spatial frequencies, it is usually desirable to increase the material sensitivity by application of an external field E_{ext}. The maximum space-charge field which can be developed by the light is just $E_{max} = E_{ext}$ when the free carriers screen the external field. Thus the material must be designed such that for the required value of E_{max} we have

$$\mu\tau E_{max} \sim \frac{\Lambda}{2\pi} \ .$$

For InP:Fe (10^7 cm^{-3}) and for GaAs:Cr (10^{16} cm^{-3}) we find experimentally that $\mu\tau \sim 10^{-7}$ cm^2/V. Thus for an applied electric field of 2000 V/cm the carrier drift length is ~ 2 µm. This is then a suitable length of optimized recording of holograms with $\Lambda \sim 12$ µm.

For diffusion transport without an applied electric field, the maximum space-charge field for sufficiently large Λ or sufficiently high trap density is [8.26] $E_{max} = 2\pi k_B T/e\Lambda$, which is not affected by the transport properties. However, the diffusion length of carriers is $L_D = (k_B T\mu\tau/e)^{1/2}$ which affects the recording *sensitivity* (index change per absorbed photon density). For $\mu\tau \sim 10^{-7}$ we find $L_D = 0.5$ µm.

It is interesting to note that the $\mu\tau$ product of BSO is close to those of InP:Fe and GaAs:Cr. In BSO, however, μ is some three orders of magnitude smaller than that of the semiconductors while τ is some three orders of magnitude greater.

Details of the effect of diffusive and conductive transport in semiconductors on energy transfer, gain, spatial frequency response and so on are the same as those for ferroelectrics discussed in other chapters. At the present time, bulk photovoltaic effects [8.20] do not appear to play an important role in semiconductors (even though symmetry does permit them to exist). This is because the maximum carrier transport distance associated with bulk photovoltaic transport is only the mean carrier scattering length and this is always much less than the drift or diffusion length in semiconductors. The effect of high mobility on the material response time and relaxation time for holographic recording is discussed in the following section.

8.4 Material Response Times

The rate at which a photorefractive index change builds up following illumination, and the rate at which it relaxes after illumination is turned off, to a large extent determines the usefulness of a material for practical applications in optical data processing. While the ultimate speed at which information can be recorded, as discussed above, is determined by the electron and hole drift velocity, this is only true if the input energy is delivered in a time short compared with the drift times. For cw illumination, this is not the case. Photocarriers are excited and displaced to trapping sites at a rate determined by the incoming photon flux, and the space-charge field increases until a steady-state value is reached when the relaxation rate of the field due to the crystal conductivity is equal to the generation rate of the field due to carrier displacement. It is shown elsewhere in this volume that for grating periods Λ large compared with the Debye screening length that this relaxation time is just [8.27–29]

$$\tau_R = \tau_d [1 + F_D(\Lambda) + F_E(\Lambda)E^2] \, , \tag{1}$$

where τ_d is the dielectric relaxation time $\tau_d = \varepsilon\varepsilon_0/(\sigma_0 + \sigma_p)$ where σ_0 is the dark conductivity and σ_p is the photoconductivity: $\sigma_p = \alpha I e \mu \tau / h\nu$ and τ is the carrier trapping time. In this equation

$$F_D(\Lambda) = 4\pi^2 \frac{\mu r k_B T}{e\Lambda^2} \quad \text{and} \quad F_E(\Lambda) = 4\pi^2 \frac{e(\mu\tau)^2}{e\Lambda^2 + 4\pi^2 \mu\tau k_B T} \, .$$

Thus for large values of grating period Λ and small values of applied electric field E, the material response time is just τ_d, the dielectric relaxation time. For $\sigma_p \gg \sigma_0$, which is the case in semi-insulators such as GaAs even for relatively low levels of illumination, the above equations yield, for small E and large Λ

$$\tau_R = \frac{\varepsilon\varepsilon_0 h\nu}{\alpha I e \mu\tau} \quad \text{so that}$$

$$\tau_R \propto (\mu\tau)^{-1} \quad \text{and} \quad \tau_R \propto (\alpha I)^{-1} \, .$$

This indicates that for high speed applications, large values of $\mu\tau$ (with the qualifications discussed in the previous section) and high intensities I are desirable. Increasing the incident intensity during writing increases the rate at which the space-charge field develops but decreases the relaxation time in proportion. The net result is that the maximum attainable space charge field and diffraction efficiency remains constant as the intensity is increased. When the incident light is turned off, the grating relaxes at a value of τ_d determined by the dark conductivity. In semiconductors this can be controlled over a wide range by controlling the doping. In GaAs and InP this can be adjusted from the millisecond time scale to the picosecond time scale.

For small values of Λ or large applied fields, the correction factors $F_D(\Lambda)$ and $F_E(\Lambda)$ appearing in (1) become dominant. For InP:Fe $(10^{17}\,\text{cm}^{-3})$ this occurs for $\Lambda < 2\,\mu\text{m}$ and $E > 300\,\text{V/cm}$. Even for reasonable applied fields $E \sim 3\,\text{kV/cm}$ the effect is to increase the response time some 100-fold! In these limits, the material response time becomes independent of $\mu\tau$ and strongly dependent on Λ and E. Indeed, no material parameters appear in the material response at all! In many cases it is desirable to apply electric fields to obtain increased sensitivity, but it appears from these considerations that under these circumstances it is not possible to achieve fast response times even with high mobility materials unless high intensity illumination is used. After the illumination is removed, the dark relaxation time is similarly affected by the correction factor. However, semiconductors are still faster than oxides.

In pulsed operation, high intensities can be used efficiently and this offers certain advantages for optical processing operations. The incident optical pulses deliver sufficient energy in short bursts to record an index grating. Following the pulse, the hologram relaxes with the *dark* relaxation time of the material. Thus, for most purposes it is desirable to have the dark relaxation time short compared with the pulse repetition rate but long compared with the optical pulse length. As previously stated, this can be accomplished in semiconductors over a wide range by appropriate doping. Alternatively, because of the high density of free carriers created by a high intensity pulse, the gratings can be erased on a short time scale with a delayed optical pulse.

For high sensitivity recording, an electric field pulse can be applied to the crystal for the duration of the writing pulses and removed for rapid erasure. These operations are demonstrated schematically in Fig. 8.7. The writing pulses

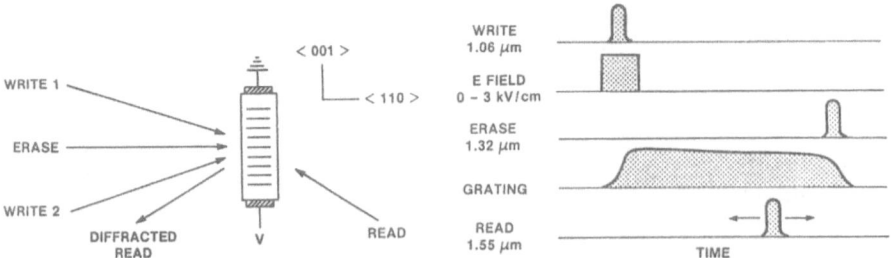

Fig. 8.7. Geometrical arrangement and timing sequences of the optical recording, reading and erasing pulses and the applied electric field in transient photorefractive recording experiments

at $t=0$ record a hologram only if both pulses arrive simultaneously. The electric field acts as an enable pulse since efficient gratings are recorded only when the electric field is present (at least for larger Λ). The delayed optical pulses can be used for reading or erasing at specified intervals, depending on their wavelength and intensity.

8.5 Experimental Work

A generic experimental arrangement for photorefractive measurements on GaAs, InP and CdTe is shown in Fig. 8.8. Light from a 1.06 µm laser is split into two beams and directed toward the semiconductor crystal (InP:Fe in Fig. 8.8) so that they intersect at an angle 2θ at the crystal. A diffraction grating of period $\Lambda = \lambda/2\sin\theta$ is recorded with this configuration. For degenerate four-wave mixing, one beam is reflected backwards along their paths and diffracted by the grating. In Fig. 8.8, the second beam is also shown reflected backwards for alignment purposes. This beam is blocked during an experiment. The diffraction efficiency in either beam can be measured with an InGaAs detector by placing a beam splitter in the path of the reflected wave. To minimize noise due to scattered light, a modulator is placed behind the crystal (with respect to the detector) in the path of the reflected light and the signal from the detector is measured with a lock-in amplifier.

For nondegenerate four-wave mixing, the diffraction efficiency is measured with a 1.32 µm laser beam incident at the Bragg angle by $\sin\theta' = \lambda/2\Lambda$ in a similar manner. In this case, the 1.06 µm laser can be modulated and there is little scattered background light at 1.32 µm.

It is essential that the semiconductor crystal be completely and uniformly illuminated by the writing beams at 1.06 µm. Any dark regions within the crystal near the electrodes act as a high series impedance, which decreases the applied electric field within the illuminated region of the crystal. This is best accomplished by having the electroded surfaces of the crystal parallel to one (the most

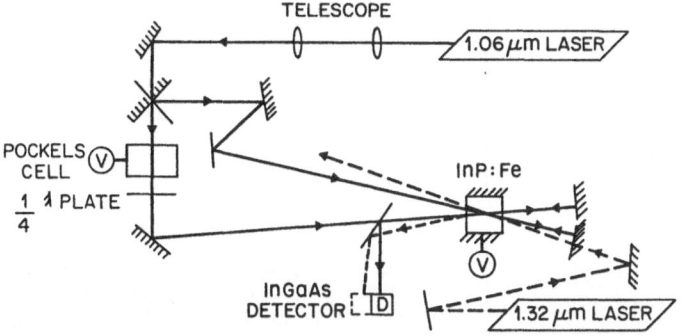

Fig. 8.8. Versatile experimental setup for photorefractive measurements

intense) of the writing beams. Because of the high refractive indices of the semiconductors, it is desirable to antireflection coat the surfaces, for instance with Al_2O_3 films, to minimize additional reflections and associated gratings within the crystal.

The crystal is electroded along ⟨100⟩ faces in the plane parallel to the planes of the grating in the crystal as shown in Fig. 8.8. The electrodes are connected to an external circuit containing a high bias voltage supply ($V = 0$ to 4 kV) and a series ammeter to monitor the dark current and photocurrent in the crystal. In some instances a series resistance R_L was placed in the external current to provide passive feedback to the bias current: The voltage across the crystal (of resistance R_c) is then $V_b = VR_c/(R_L + R_c)$. When the light intensity incident on the crystal increases, R_c decreases, and hence V_b decreases. Thus the series resistance behaves as a limiter determining the maximum diffraction efficiency.

For pulsed experiments several options are possible. Either one of the beams can be pulsed on and off with the Pockel's cell arrangement in Fig. 8.8 or the bias field can be pulsed on and off electronically. A third option is to Q-switch the 1.06 and 1.32 μm lasers in such a way that the trigger for the 1.32 μm reading beam can be delayed with respect to the 1.06 μm laser beam in a variable manner. The time of the electric field pulse and the laser pulses can then be adjusted at will.

8.5.1 Continuous-Wave Four-Wave Mixing Experiments

The diffraction efficiency η of a chromium doped GaAs crystal is shown in Fig. 8.9 as a function of applied electric field. The crystal was doped with approximately 10^{16} Cr ions per cubic centimeter and had an absorption coefficient of approximately 5 cm^{-1} at 1.06 μm.

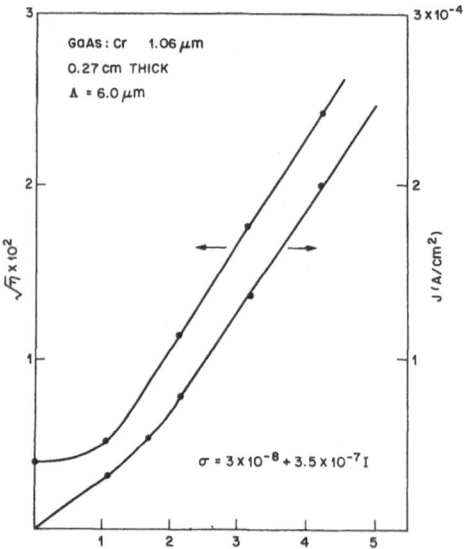

Fig. 8.9. Square root of the diffraction efficiency $\eta^{1/2}$ in GaAs : Cr (10^{16} cm^{-3}) as a function of applied electric field E. The current–field characteristic is shown for comparison

The optical path length in the crystal was 0.27 cm. The diffraction efficiency in Fig. 8.9 was measured with the degenerate four-wave mixing configuration and is defined for practical purposes as the ratio of diffracted light to undiffracted light transmitted through the crystal. The angle 2θ between the incident writing beams was 10°, corresponding to a grating period of about 6 μm. With zero applied field, the diffraction efficiency due to diffusive transport was only 1.6×10^{-5} with an incident intensity of $10\,mW/cm^2$. As the applied field was increased, the diffraction efficiency increased to values of about 0.06% for $E = 4.2\,kV/cm$. The behavior of $\eta^{1/2}$ as a function of field is linear when transport due to drift in the applied field exceeds diffusion transport. This is, of course, expected since for small η, $\eta \propto (\Delta n)^2 \alpha E^2$. The current–voltage characteristic measured at the same time as the optical behavior is linear over the same region, as shown in Fig. 8.9, but below about 1.5 kV/cm there is a departure from linearity, presumably due to non-ohmic behavior of the contacts with high resistivity GaAs, or other field homogeneities. The measured conductivity of this sample is

$$\sigma\,[\Omega\,cm]^{-1} = 3 \times 10^{-8} + 3.5 \times 10^{-7}\,I\,[W/cm^2]$$

from which the $\mu\tau$ product is calculated to be $0.9 \times 10^{-7}\,cm^2/V$. This value is some 500 times smaller than that of undoped GaAs where the mobility of electrons $\mu \sim 8500\,cm^3/Vs$ and the lifetime is ~ 20 ns. The drift length $(\mu\tau E)$ of the chromium doped crystal at field of 4 kV/cm is thus 3.6 μm, which is close to optimum for the grating period of 6 μm recorded. Further increased in applied field will not result in correspondingly increased diffraction efficiency. The reduction of carrier lifetime by the deep levels is clearly necessary for operation of the crystals with an external field. On the other hand, the diffusion length $(k_B T\mu\tau/e)^{1/2}$ is only about 0.46 μm in the doped crystal.

The diffraction of these chromium-doped GaAs crystals is low primarily because of the small crystal length used, which is dictated by the high optical absorption. A fivefold reduction of the absorption would result in increased diffraction efficiency to levels of the order of percent in crystals of centimeter dimensions.

Indium phosphide doped with $\sim 10^{17}\,cm^{-3}$ iron impurities has an appropriately small absorption coefficient of $1\,cm^{-1}$. The square root of the diffraction efficiency is shown as a function of applied field in Fig. 8.10 for an optical path length of 0.82 cm in the crystal. At 4.2 kV/cm the diffraction efficiency is approximately 1.2% as anticipated. With greater applied fields, a diffraction efficiency of over 5% has been observed, but the crystals become warm because of Joule heating by the photocurrent, and the diffraction becomes unstable. The variation of $\eta^{1/2}$ with applied field is seen to be nonlinear below about 2 kV/cm for reasons which are not understood.

Recording with 1.32 μm light in the same sample exhibited no such nonlinearity as shown in Fig. 8.11. Both the photocurrent and $\eta^{1/2}$ are linear in applied field. However, at this wavelength the absorption coefficient of the crystal was considerably smaller resulting in greatly reduced diffraction efficiency.

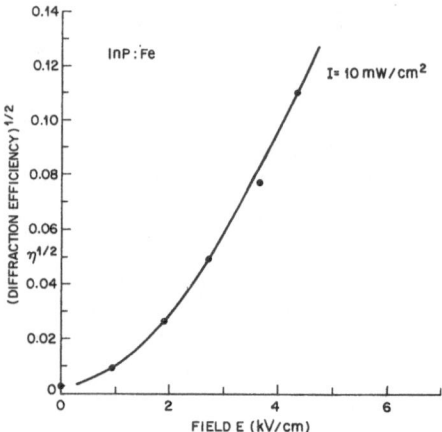

Fig. 8.10. Square root of the diffraction efficiency $\eta^{1/2}$ as a function of applied electric field E in InP:Fe ($\sim 10^{17}\,\mathrm{cm}^{-3}$) using 1.06 µm recording light

The effect of nonuniform illumination of an InP:Fe crystal is shown in Figs. 8.12 and 8.13. With no external load resistor R_L, the photocurrent in the crystal tends to saturate as the incident optical power is increased. At low intensities, the crystal resistance is high and the influence of a series resistance on nonuniform illumination (i.e., a dark region near the electrodes) is small. The effect is most pronounced at high intensities. By inserting a load resistor in the external circuit, this effect can be exaggerated as shown in Fig. 8.12.

The influence of this effect on four-wave mixing is shown in Fig. 8.13. The diffraction efficiency is seen to reach a maximum value at relatively low intensities 10–20 mW/cm².

Some examples of the transient response of an InP:Fe sample to a pulsed rectangular electric field are shown in Fig. 8.14. The diffracted beam was monitored with a boxcar integrator. The crystal illumination was constant throughout the electric field pulse. The rise time of the diffraction efficiency in this figure is twice the response time of the refractive index change τ_R for the particular values of intensity, grating period and applied electric field. When the field is turned off the decay of the diffraction has twice the value of the index relaxation time with no applied field for the same values of intensity and grating period. The slight decay observed after the initial rise in Figs. 8.14a, b is attributed to surface effects discussed earlier, which decrease the applied field in the bulk. It is immediately evident from the figure that an applied field greatly increases the response time of the material in accordance with theoretical expectations (1) within experimental error. Also in agreement with theory is the observation that higher intensities result in linearly decreased response times. The relaxation of the grating in the absence of the electric field pulses is in all cases faster than the rise time in the presence of the field, and is determined by the incident intensity and spatial frequency of the grating in accordance with (1). Those data emphasize the point made earlier that with cw illumination the increased sensitivity achieved by applying an electric field is accompanied by a dramatic slowing down of the material response.

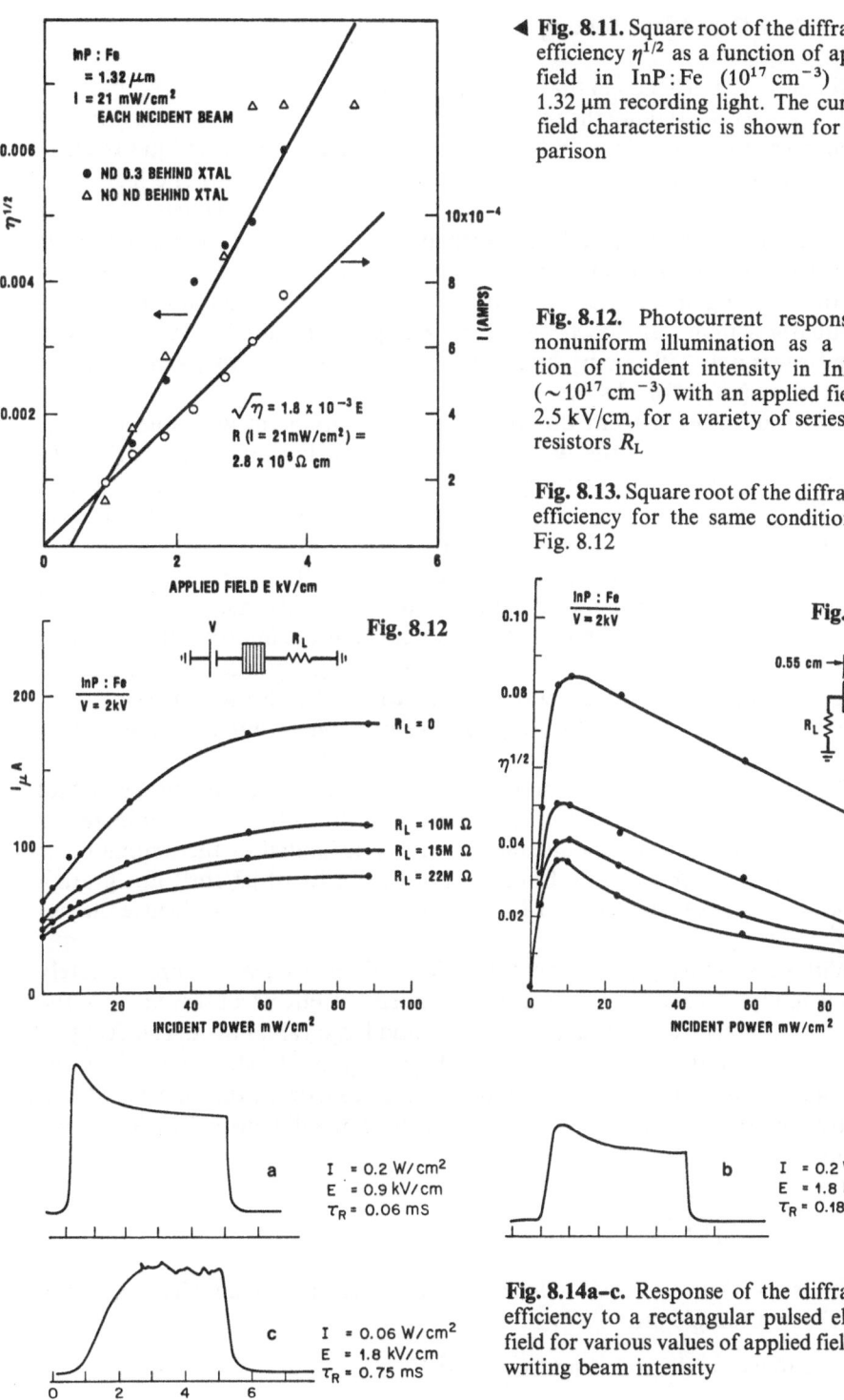

◀ **Fig. 8.11.** Square root of the diffraction efficiency $\eta^{1/2}$ as a function of applied field in InP:Fe $(10^{17}\,\mathrm{cm^{-3}})$ using 1.32 µm recording light. The current–field characteristic is shown for comparison

Fig. 8.12. Photocurrent response to nonuniform illumination as a function of incident intensity in InP:Fe $(\sim 10^{17}\,\mathrm{cm^{-3}})$ with an applied field of 2.5 kV/cm, for a variety of series load resistors R_L

Fig. 8.13. Square root of the diffraction efficiency for the same conditions as Fig. 8.12

Fig. 8.14a–c. Response of the diffraction efficiency to a rectangular pulsed electric field for various values of applied field and writing beam intensity

8.5.2 Pulsed Four-Wave Mixing

It was concluded in Sect. 8.4 that efficient high speed formation of photorefractive gratings would require the use of short optical pulses. Experiments using Q-switched writing pulses from YAG:Nd lasers at 1.06 µm and 1.55 µm reading pulses from an InGaAsP diode laser have been reported in InP:Fe [8.7]. The pulse repetition rate was 1 kHz, which permitted the photorefractive gratings in the crystal to relax completely between pulses. The writing pulse length at 1.06 µm was of approximately 300 ns duration, which is considerably longer than the trapping time of carriers in the crystal. Thus these experiments were in the "quasi-cw limit" discussed by *Valley* [8.30]. The maximum writing energy per pulse was only 280 µJ/cm^2, and intensities less than 1 W/cm^2, well below the intensities required for absorption saturation.

The reading beam from the diode laser was 0.1 µs in duration and was delayed with respect to the writing beams. The delay time was scanned to monitor the photorefractive response of the crystal. The InP:Fe crystal was insensitive to low intensity 1.55 µm radiation used for reading so the reading light did not destroy the recorded information.

To erase the photorefractive grating, a second YAG:Nd laser operating at 1.32 µm was used with a pulse duration of 1 µs. This laser could also be independently delayed with respect to the writing pulse. Thus the pulse sequence is that shown in Fig. 8.7. An electric field pulse of 3 µs duration was applied to the crystal to coincide with the writing pulse to enhance the writing sensitivity. No electric field was applied during erasure in order to minimize the grating relaxation time.

Figure 8.15a shows an example of the data taken with an InP:Fe crystal having a resistivity of 10^8 Ωcm. The intensity of the diffracted beam increases abruptly during the writing pulses with a rise time limited by the writing pulse duration. Neither the writing beam energy (down to 10 µJ/cm^2) nor grating period Λ, nor applied field (up to 3.2 kV/cm) were observed to influence the rise time.

Without an erasing pulse, the diffraction efficiency decayed approximately exponentially, except for some grating saturation effects observed near the beginning for large writing pulse intensities and large values of Λ. The decay of the diffraction efficiency is $\tau_R/2$ for small grating amplitudes where τ_R is the grating decay time discussed in Sect. 8.4. The grating decay time measured as a function of grating period is shown in Fig. 8.16. The solid line in Fig. 8.16 is the relation

$$\tau_R = \tau_d \left(1 + \frac{a}{\Lambda^2} \right)$$

derived from (1) for zero field. The values of τ_d and $\mu\tau$ calculated from data for the 10^8 Ωcm sample are

$$\tau_d = 290 \ \mu s \quad \text{and} \quad \mu\tau = 1.6 \times 10^{-7} \ cm^2/V \ .$$

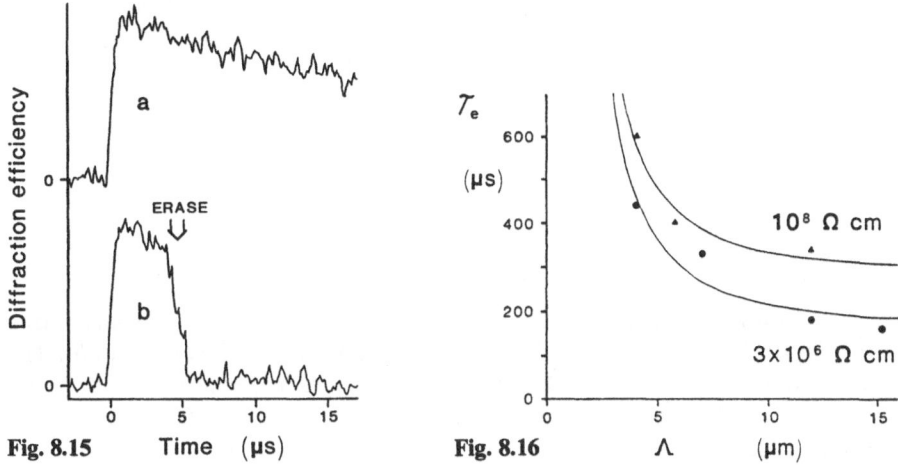

Fig. 8.15   Time   (μs)   Fig. 8.16   Λ   (μm)

Fig. 8.15. Diffraction efficiency as a function of time following photorefractive recording in InP:Fe with short (300 ns) optical pulses at 1.06 μm. (**a**) No erase pulse. (**b**) Erase pulse approximately 4 μs after the recording pulse

Fig. 8.16. Relaxation time of the photorefractive grating as a function of spatial period Λ of the grating

Optical erasure is shown in Fig. 8.15b. The grating is erased within the 1 μs duration of the erasure pulse. However, since the absorption coefficient of the InP:Fe crystal is considerably smaller at 1.32 μm than at 1.06 μm the energy required for complete erasure was greater than that for recording. By varying the delay of the erasure beam, the duration of the grating observed in Fig. 8.15b could be continuously varied over the dark relaxation time. Diffraction efficiencies (neglecting absorption) as large as 3% were achieved with a total writing energy of only 85 μJ over a 0.5 cm diameter beam, a pulsed field 3.2 kV/cm and $\Lambda = 4$ μm. The maximum efficiency which could be achieved if the space-charge field were equal to (i.e., completely screened) the external bias field is 4% in the crystal. These data are in agreement with theoretical expectations and indicate that high speed photorefractive response can be achieved using short pulses without loss in sensitivity. Since the trapping time of carriers in InP:Fe can be shorter than 10 ps, it can be anticipated that high efficiency recording is feasible with picosecond laser pulses.

8.5.3 Beam Coupling

If the phase of the index grating in photorefractive materials is displaced with respect to the incident optical interference pattern then beam coupling gives rise to energy transfer from one beam to the other. The origin of this effect is discussed in detail elsewhere in this volume. The first report of beam coupling in

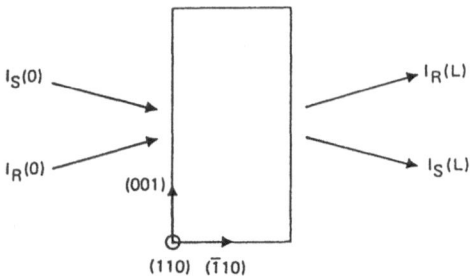

Fig. 8.17. Geometrical arrangement for studying beam coupling

semiconductors was performed in undoped GaAs which was semi-insulating ($\sigma = 1.5 \times 10^{-8}\ \Omega^{-1}\ \text{cm}^{-1}$) by virtue of uncompensated EL2 centers [8.5]. The experimental setup was a simple two-beam mixing experiment using a 1.06 µm laser, as shown schematically in Fig. 8.17. The two incident beams $I_R(0)$ and $I_S(0)$ are each diffracted by the photorefractive grating, giving rise to beam coupling, but, because of asymmetry of the electro-optic behavior of the crystal, one beam is amplified and the other attenuated. The extrinsic absorption coefficient of the crystal at 1.06 µm was approximately 1 cm^{-1}, corresponding to an occupied EL2 center concentration of about 10^{16} cm^{-3}. From temperature-dependent conductivity studies the density of ionized centers was $(1.5-2) \times 10^{15}$ cm^{-3}. The crystal dimensions were 6 mm $\langle 001 \rangle$, 5 mm $\langle 011 \rangle$ and 4 mm $\langle 110 \rangle$. The signal beam $I_S(0)$ had an intensity of 10 mW, at least an order of magnitude smaller than the reference beam $I_R(0)$. The crystal geometry (i.e., the sense of the $\langle 001 \rangle$ axis) was adjusted to provide gain of the weak signal beam.

The transmitted signal beam intensity $I_S(L)$ is then given within this small gain limit (negligible pump depletion) by

$$I_S(L) = I_S(0)e^{(\Gamma - \alpha)L}\ ,$$

where Γ is the gain coefficient, α the absorption (loss) coefficient and L the crystal length. When the reference beam is turned off

$$I_S(L) = I_S(0)e^{-\alpha L}\ .$$

Thus Γ is readily measured from the ratio. A plot of Γ as a function of spatial frequency Λ of the grating is shown in Fig. 8.18. The curve shows a maximum value for $\Lambda \sim 0.75$ µm when the grating period is equal to the Debye screening length $L_T = (\pi \varepsilon \varepsilon_0 k_B T / e^2 N_{EL2})^{1/2}$, where N_{EL2} is the density of ionized EL2 centers. This yields a value of $N_{EL2} \sim 1.3 \times 10^{15}$ cm^{-3} in agreement with conductivity data.

In these experiments no electric field was applied and carrier transport was solely by diffusion, with the index grating shifted $\pi/2$ from the incident intensity variation. For this situation

$$\Gamma = \frac{2\pi n^3 r_{41} E_{sc}}{\lambda \cos \theta}\ .$$

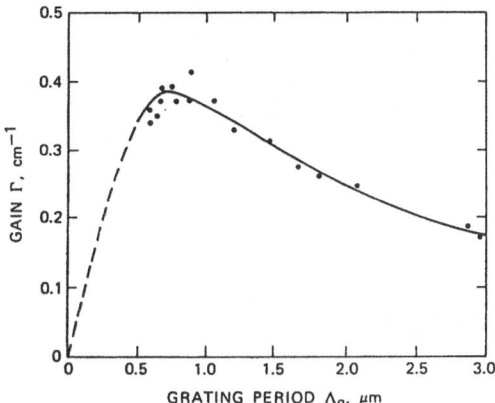

Fig. 8.18. Gain measured by two-wave mixing, as a function of grating period Λ [8.5]

For the maximum gain in Fig. 8.18, this maximum space charge field $E_{\text{sc}} = 3500$ V/cm. The magnitude of the gain observed in Fig. 8.18 is close to that observed with BSO in similar experiments with no applied electric field and with 0.5145 μm radiation. In BSO, however, by using applied electric fields and moving grating techniques, much higher values of gain have been achieved, which to a large degree compensate for the small value of the electro-optic coefficient. Similar techniques have been applied to InP : Fe and gains as high as 13 dB cm^{-1} have been observed [8.35] at 1.06 μm.

The transient response of the GaAs crystal above, was measured by modulating the reference beam with an electro-optic switch. The response time of the gain (the time for the signal beam to reach a new steady-state value after the reference beam is turned on) was 20 μs for $\Lambda = 1$ μm and a reference beam intensity of 4 W/cm^2. This value is long compared with the dielectric relaxation time during illumination because of the dominance of the factor $F_{\text{D}}(\Lambda)$ in (1) for this small value of Λ.

The gain measurements described above give a simple alternative to four-wave mixing techniques for measuring photorefractive parameters of the semiconductors. Extremely small values of gain can be easily measured by modulating the reference beam and measuring changes in the signal with phase sensitive detection.

There is, however, an important piece of information which can be obtained from gain measurements which is not achievable by four-wave mixing studies alone. This is the sign of the photocarriers involved in the photorefractive process. It was mentioned earlier that beam coupling amplifies one beam and attenuates the other. In ferroelectric crystals having a unique sense of the polar axis, the sign of the polarization and the sign of the electro-optic coefficient determines which beam is amplified. The zinc blende semiconductors have no unique polar properties. The sign of the electro-optic coefficient with respect to the crystal geometry used in the beam coupling experiment must be separately determined. The sign of the photocarriers can then be directly determined from

the beam coupling. Because of the problem associated with bipolar photocon-
duction with deep levels in semiconductors, beam coupling affords an unambiguous determination of the majority photocarrier. These experiments have been
demonstrated in InP:Fe [8.6] and GaAs:EL2 [8.5]. In both cases the photocarriers were determined to be electrons at 1.06 µm for the crystals examined,
but, as discussed earlier, this result depends on the degree of compensation of
deep levels in specific crystals.

8.5.4 Four-Wave Mixing Using Injection Lasers

Since photorefractive recording requires energies of only ~1 pJ/µm^2, injection
lasers can be used for the recording, reading and erasing of information. At
present, lasers are commercially available in the 0.85 µm spectral region
(GaAlAs technology) and the 1.3–1.55 µm spectral region (InGaAsP technology) with cw powers of a few milliwatts or higher. In the research laboratory,
laser arrays have achieved power levels of ~1 W. It has been shown that BaTiO$_3$
has sufficient sensitivity at 0.85 µm (albeit significantly lower than at shorter
wavelengths) to permit four-wave mixing with GaAs lasers [8.31]. GaAs:Cr and
InP:Fe are well matched to the longer wavelength spectral region (1.3–1.5 µm)
of current importance for optical communications. Using the compact experimental arrangement shown in Fig. 8.19, four-wave mixing in GaAs:Cr was
demonstrated with an InGaAsP laser emitting at 1.3 µm with 5 mW per facet
[8.8]. By using the laser output from opposite facets, coherent object and
reference beams could be derived from one source without the use of a beam
splitter. The laser was pulsed with a 250 µs pulse width and the incident energy
density on the crystal was 80 µJ/cm^2 [0.8 pJ/µm^2] over a 4 mm area of crystal per
pulse. The GaAs crystal was doped with about 10^{16} cm^{-3} chromium ions,
resulting in a crystal resistivity of about 10^8 Ωcm, and an absorption coefficient α
of about 4 cm^{-1}. For degenerate four-wave mixing, the optimum crystal
thickness d is given by $\alpha d = 1/2$.

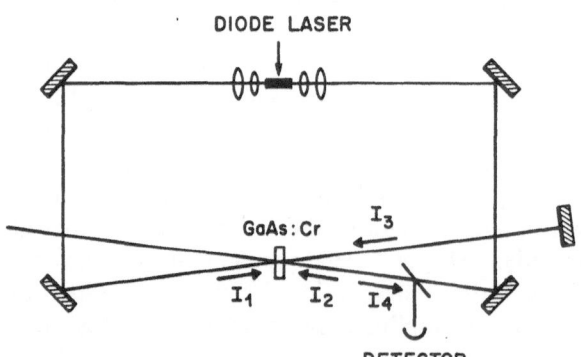

Fig. 8.19. Compact experimental arrangement for four-wave
photorefractive mixing in GaAs
using a 1.3 µm InPGaAs semiconductor laser

Using counterpropagating beams I_1 and I_2 gives the smallest possible grating period $\Lambda = 0.19$ μm in the sample. This value of Λ is 1.5 times smaller than the diffusion length of carriers and smaller than the Debye screening length and hence is not optimum for this crystal. In this configuration, the amplitude of the sinusoidal optical interference pattern is constant throughout the crystal thickness, even though the d.c. component of optical intensity varies because of optical absorption.

The maximum diffusion field which can be generated in this experimental arrangement is $E_d = 2\pi k_B T/e\Lambda = 8.6$ kV/cm, which in turn would give a diffraction efficiency of $\eta = \tanh^2(\pi\Delta nd/\lambda) = 0.4\%$. However, to achieve this value requires the displacement and trapping of 4×10^{16} cm^{-3} carriers, which exceeds both the density of Cr ions in the crystal and the absorbed photon flux in the optical pulse. The experimentally observed diffraction efficiency was $\eta \sim 0.01\%$ due to these limiations. A 20-fold improvement in η should be possible by increasing the writing energy density (i.e., reducing the beam cross section) and optimizing the density of chromium ion traps. The experiment demonstrates the feasibility of optical processing with very small energies per pulse using diode lasers. Indeed, by taking advantage of moving grating techniques, which have been shown to give enhanced optical gain in bismuth silicon oxide, it should be possible to design an image amplifier using injection lasers in this important spectral region.

8.5.5 Nonlinear Susceptibility

The photorefractive effect is only one of the nonlinear mechanisms currently under study for optical information processing. Other nonlinear mechanisms in semiconductors include index changes due to exciton saturation and absorption changes due to electro-absorption [8.32] in optically generated space-charge fields, which is conceptually similar to the photorefractive effect. These can be used to process information on a subnanosecond time scale. It may, therefore, be useful to express the photorefractive index change as a nonlinear susceptibility n_2 for comparison with other nonlinear optical materials.

since $\quad \Delta n = \dfrac{1}{2} n_i^3 r_{ij} E_j \quad$ and $\quad E_j = \dfrac{Ne\Lambda}{4\pi\varepsilon\varepsilon_0}$

then $\quad \Delta n = \left(\dfrac{n_i^3 r_{ij}\alpha\tau e\Lambda}{8\pi\varepsilon\varepsilon_0 h\nu}\right) I$.

The term in parenthesis is the photorefractive nonlinear index n_2. For 1 μm radiation, this expression gives

$n_2 \sim 250 \; \alpha\Lambda\tau$.

Thus, for the GaAs:Cr and InP:Fe used in the studies reported above with $\Lambda \sim 1\ \mu m$,

for   GaAs:Cr $n_2 \sim 2.5 \times 10^{-5}\ cm^2/W$,

for   InP:Fe $n_2 \sim 2.5 \times 10^{-6}\ cm^2/W$.

These are large nonlinearities which compare with n_2 for exciton saturation of $n_2 \sim 2 \times 10^{-5}\ cm^2/W$ for bulk GaAs at 300 K and $n_2 \sim 4 \times 10^{-3}\ cm^2/W$ for quantum well structures [8.33] based on GaAs. The nonlinear index is not, however, a useful parameter to compare materials for optical information processing. Indeed, the nonlinear index can be arbitrarily increased by increasing the dielectric relaxation time or by increasing the optical absorption! For practical purposes, the useful parameter is the *energy* required to perform an elementary processing function (i.e., switch) provided that the material responds on the required time scale. For most optimized photorefractive materials studied to date (materials in which photocarriers drift a distance $\Lambda/2\pi$), this energy is about 1 pJ/ μm^2. That is, only about one picojoule is required for a diffraction-limited optical switching event. This energy compares well with the best of other nonlinear mechanisms and materials.

8.5.6 Future Directions

The relative invariance of the ratio r_{ij}/ε_j is valid only for materials in the spectral region where the host crystal is transparent and the free carrier excitation involves extrinsic centers. In the spectral region near the band edge of a semiconductor, a resonant enhancement of the electro-optic coefficient occurs without the concomitant increase of the dielectric constant. The variation of the quadratic electro-optic effect near the band edges of InP and GaAs [8.34] are shown in Fig. 8.20. The use of the quadratic electro-optic coefficient implies a doubling of the spatial frequency of the refractive index grating relative to the intensity grating. This mechanism offers a relatively simple way in which to achieve increased photorefractive sensitivity.

As with all resonant nonlinearities, a penalty that must be paid is the increased optical absorption of the crystal, which results in a decreased optical interaction length. In addition, increased heating of the crystal by incident optical beams becomes a factor, particularly at higher intensities and short pulses and, because of the proximity of the band edge, the optical parameters are temperature sensitive. To date, there have been no experiments performed which make use of resonant enhancement of the photorefractive effect. Clearly there is a need for some study of the compromises involved with such an approach. For direct imaging applications using longitudinal electric field configurations, the decreased crystal length would translate into increase spatial resolution. To make use of resonant enhancement, only the reading beam need be near resonance. The space-charge fields can be generated with longer wavelength radiation if appropriate. Of course, resonant excitations from the valence to conduction bands of the host crystal can be the source of photocarriers as, has

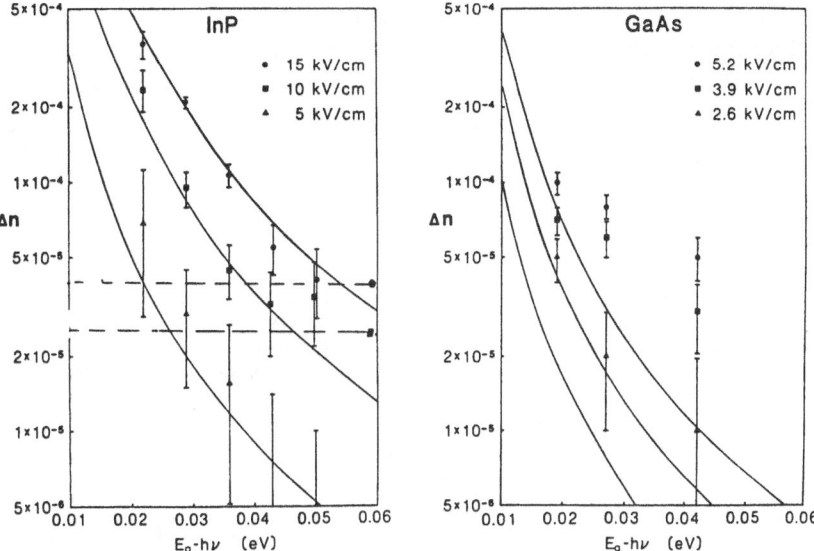

Fig. 8.20. Resonant enhancement of the electro-optic effect in GaAs and InP close to the fundamental absorption edge [8.32]

been demonstrated in KTN and $LiNbO_3$ using two-photon absorption [8.23]. To achieve ultra-high speed applications, carrier trapping is not required. The different free-carrier mobilities of electrons and holes lead to transient space-charge fields due to bipolar diffusion. Such an effect has recently been demonstrated experimentally. Extrinsic defects (or cooling) may be necessary to reduce the dark free carrier density and the resulting rapid screening of the field. Certainly for long pulse and cw studies, extrinsic defects and carrier trapping are essential.

In addition to these approaches to achieving enhanced speed and sensitivity, a search for new electro-optic semiconductors with higher electro-optic coefficient may still lead to improved photorefractive performance. Little or nothing is known about the electro-optic behavior of most semiconductors – even well-established electronic materials. Some preliminary experiments have been performed [8.7] with CdTe:In, which, according to Table 8.1, should exhibit superior photorefractive behavior. However, the centers contributing to the observation in CdTe were not identified and certainly not optimized.

8.6 Conclusion

The full potential of electro-optic semiconductors for photorefractive materials has clearly not been realized. The few experiments already performed merely represent a beginning. The combination of high sensitivity, high speed, and the ability to tailor materials to specific requirements using procedures already established for these important electronic materials make the future of this class of materials very promising for optical data processing applications. The potential for image amplification and processing in the infrared spectrum (even perhaps to the important 10.6 μm region) will surely stimulate experimental activity in this field. The approach used with bismuth silicon oxide to achieve high gain by means of moving gratings is clearly applicable to the semiconductors because of their similar electro-optic coefficients.

References

8.1   A.M. Glass, D. von der Linde: Ferroelectrics **10**, 163 (1976)
8.2   J.P. Huignard, F. Micheron: Appl. Phys. Lett. **29**, 591 (1976)
8.3   A. Ashkin, B. Tell, J.M. Dziedzic: IEEE J. QE-3, 400 (1967)
8.4   A.M. Glass, A.M. Johnson, D.H. Olson, W. Simpson, A.A. Ballmann: Appl. Phys. Lett. **44**, 948 (1984)
8.5   M.B. Klein: Opt. Lett. **9**, 350 (1984)
8.6   A.M. Glass, M.B. Klein, G.C. Valley: Electron. Lett. **21**, 220 (1985)
8.7   J. Strait, A.M. Glass: J. Opt. Soc. Am. B**3**, 342 (1986)
8.8   J. Strait, A.M. Glass: Appl. Opt. **25**, 338 (1986)
8.9   I.P. Kaminow, E.H. Turner: In *Hanbook of Lasers*, by R.J. Pressley (CRC Press, Boca Raton, Fla. 1971) p. 447
8.10 S.M. Sze: *Physics of Semiconductor Devices* (Wiley, New York 1981)
8.11 The data for InP was compiled by A.A. Ballman (unpublished)
8.12 G.M. Martin, A. Mitonneau, A. Mircea: Electron. Lett. **13**, 191 (1977)
8.13 R. Orlowski, E. Krätzig: Solid State Commun. **27**, 1351 (1978)
8.14 J. Feinberg, D. Heiman, A.R. Tanguay, R.W. Hellwarth: J. Appl. Phys. **51**, 1297 (1980)
8.15 D.C. Look: Phys. Rev. B**20**, 4160 (1979)
8.16 D.C. Look, J.S. Blakemore (eds.): *Semi-insulating III-V Materials*, Section IV (Shiva Publishing, Nantwich, UK 1984)
8.17 S. Makram-Ebid, P. Langlade, G.M. Martin: In [8.16], p. 184
8.18 S.G. Johnson, W.B. Leigh, C.K. Chan, J.S. Blakemore: In [8.16], p. 273
8.19 G.E. Peterson, A.M. Glass, T.J. Negran: Appl. Phys. Lett. **19**, 130 (1971)
8.20 A.M. Glass, D. von der Linde, T.J. Negran: Appl. Phys. Lett **25**, 233 (1974)
8.21 A.M. Glass: Science **226**, 657 (1984)
8.22 J. Strait, A.M. Glass: J. Opt. Soc. Am. B**3**, 342 (1986)
8.23 D. von der Linde, A.M. Glass, K.F. Rogers: Appl. Phys. Lett. **26**, 22 (1975); ibid. **25**, 155 (1974)
8.24 G.C. Valley, A.L. Smirl, M.B. Klein, K. Bohnert, T.F. Boggess: Opt. Lett. **11**, 647 (1986)
8.25 R.B. Hammond, N.G. Paulter, R.S. Wagner, T.E. Springer: Appl. Phys. Lett. **44**, 620 (1984)
8.26 J.J. Amodei: Appl. Phys. Lett. **18**, 22 (1971)
8.27 N.V. Kukhtarev: Sov. Tech. Phys. Lett. **2**, 438 (1976)
8.28 N.K. Kukhtarev, V.B. Markov, S.G. Odulov, M.S. Soskin, V.L. Vinetskii: Ferroelectrics **22**, 949 (1979)
8.29 G.C. Valley, M.B. Klein: Opt. Eng. **22**, 704 (1983)
8.30 G.C. Valley: IEEE J. QE-**19**, 1637 (1983)
8.31 M. Cronin-Golomb, K.Y. Lau, A. Yariv: Appl. Phys. Lett. **47**, 567 (1985)
8.32 D.A.B. Miller, D.S. Chemla, T.C. Damen, A.C. Gossard, W. Wiesman, T.H. Wood, C.A. Burrus: Appl. Phys. Lett. **45**, 13 (1984)
8.33 D.A.B. Miller, D.S. Chemla, D.J. Eilenberger, P.W. Smith, A.C. Gossard, W. Wiegman: Appl. Phys. Lett. **42**, 925 (1983)
8.34 T.E. VanEck, L.M. Walpita, W.S.C. Chang, H.H. Wieder: In Conference on Lasers and Electro-Optics May 21–24, 1985: Digest of Technical Papers (Optical Society of America, Washington DC 1985) p. 242
8.35 R.B. Bylsma, A.M. Glass, D.H. Olson: Electron. Lett. to be published

9. Nonstationary Holographic Recording for Efficient Amplification and Phase Conjugation

Sergei I. Stepanov and Mikhail P. Petrov

With 16 Figures

Conventional techniques of holographic recording in photorefractive crystals (see, e.g., [9.1–3]) involve the use of stationary external conditions, such as a fixed interference pattern, application of a steady external electric field, etc. Recent experiments using nonstationary conditions of recording (namely, a moving [9.4–6] or oscillating [9.6] interference pattern, as well as externally applied alternating field [9.7, 8]) in cubic crystals of the $Bi_{12}SiO_{20}$ (BSO) type have succeeded, however, in producing a much greater efficiency of holographic recording. The high diffraction efficiency and the shifted nature of the phase holograms make the nonstationary mechanisms especially attractive for use in holographic two- and four-wave mixing geometries. For instance, application of an alternating field across a cubic $Bi_{12}TiO_{20}$ (BTO) crystal in an efficient four-wave mixing scheme with positive feedback [9.9, 10] yielded a phase conjugate reflectivity of $R \gtrsim 30$ [9.7, 11], which is 3–4 orders of magnitude greater than that typically observed in cubic crystals [9.12, 13] and nearly equal to the highest reflectivity obtained in unique samples of the ferroelectric $BaTiO_3$ [9.14, 15].

Efficient nonstationary holographic recording relies on the long drift lengths of photoelectrons in an external electric field E, or, to be more precise, on fulfilment of the basic condition

$$Kr_E \gtrsim 1 , \tag{9.1}$$

where $r_E = \mu\tau E$ (μ is mobility and τ the lifetime of a photoelectron) and $K = 2\pi/\Lambda$ is the spatial frequency of the recorded hologram (Λ is the fringe spacing). This also allows the nonstationary recording mechanisms under consideration to be called the long drift length (LDL) mechanisms.

In particular, the hologram buildup on recording with a moving interference pattern [9.4–6] is due to the drift of holograms in a photorefractive crystal in a steady external electric field [9.5, 6]. This implies that the hologram (recorded earlier in a photorefractive crystal) drifts as a whole in the direction of the external field upon uniform illumination. It should be particularly emphasized that the drift of the hologram in this case is in no way determined by any effects of self-diffraction [9.4, 16] and is observed during incoherent illumination of the crystal as well.

In order to explain the origin of this phenomenon, let us assume that a simple hologram with a given sinusoidal electric field distribution $E_{sc}(x)$ is uniformly illuminated by an external incoherent light source. If there is no external field

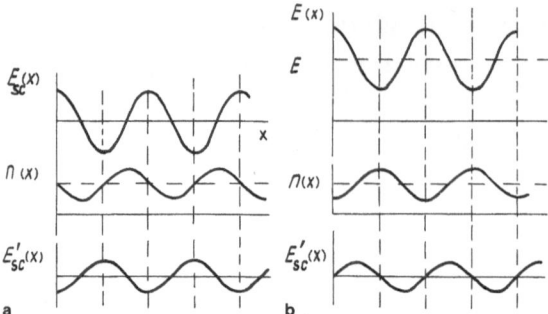

Fig. 9.1a, b. Spatial redistribution of uniformly photogenerated electrons induced by the hologram electric field $E_{sc}(x)$. (a) With no external field; (b) in an external steady electric field E, the long drift length approximation ($Kr_E \gg 1$)

($E=0$), the uniformly excited photoelectrons are concentrated on the left slopes of the distribution $E_{sc}(x)$, thereby giving rise to a new (secondary) hologram with the field distribution $E'_{sc}(x)$ that is opposite in phase to $E_{sc}(x)$ (Fig. 9.1a). The amplitude of the initial hologram therefore decreases, and this is how the usual incoherent optical erasure proceeds.

In the short drift length approximation ($Kr_E \ll 1$), application of a steady external field does not appreciably change the photoelectron grouping and produces no qualitative change in the hologram erasure process. In the other extreme case ($Kr_E \gg 1$), however, the photoelectrons manage to pass through several fringe spacings Λ during their mean lifetime τ. They will no longer be grouped in the initial hologram field $E_{sc}(x)$ as efficiently as before and their peak concentration will occur at the minima of the total field $E(x) = E + E_{sc}(x)$, where the average velocity of the photoelectrons is lowest. So, the secondary field $E'_{sc}(x)$ turns out to be shifted by $1/4$ of the spatial period with respect to the initial distribution $E_{sc}(x)$ (Fig. 9.1b). This leads to a shift of the hologram along the external field E (or in the opposite direction to E in the case of hole photoconductivity) rather than to relaxation of the initial hologram, that is, the hologram drifts.

Thus, the simplified analysis given above shows that the use of an ordinary stationary interference pattern to record a hologram appears to be inappropriate if long drift lengths of photoelectrons are involved. The stationary hologram amplitude will apparently reach its maximum when the condition of phase synchronism is satisfied, i.e., when the recording pattern moves synchronously with the hologram being recorded at a resonance velocity v_0.

There is also an alternative method of synchronization, namely, to "stop" the hologram. This can be accomplished by recording in an external alternating field with the time period $\tau_\sim$ much smaller than the characteristic time (τ_{sc}) of hologram formation. In this case, the hologram, while moving in opposite directions during two successive half periods of the field oscillation, remains immobile on average and is thereby synchronized with the fixed recording interference pattern.

The last-mentioned nonstationary recording mechanism can also be considered to be a further development of diffusion recording. In fact, there is a

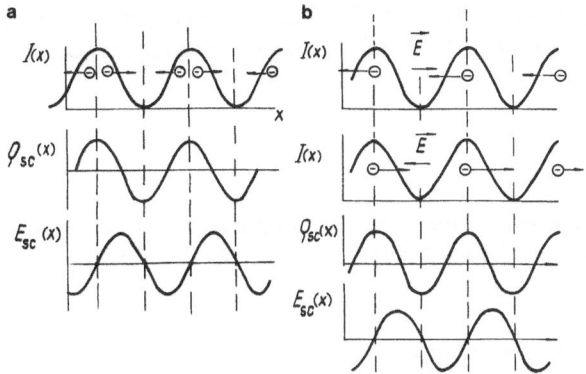

Fig. 9.2. Formation of a shifted hologram via (**a**) diffusion mechanism (without an external field), and (**b**) nonstationary mechanism (in an external alternating electric field). Also shown: light intensity distribution $I(x)$ in the interference pattern, preferred directions of electron motion (for nonstationary recording for two successive half-periods of an alternating field), the resulting space charge density $\varrho_{sc}(x)$, and shifted electric field distributions in the holograms $E_{sc}(x)$

symmetric blurring of the photoinduced electrons around the interference pattern maxima (Fig. 9.2) resulting in a shifted phase hologram [9.1–3] in both cases. However, in nonstationary recording, the nonuniformly generated electrons are redistributed by their drift in an external electric field; the process is much more efficient than thermal diffusion.

9.1 Theoretical Analysis of Nonstationary Recording Mechanisms

9.1.1 Basic Equation for a Hologram Complex Amplitude

In the following analysis we shall use complex amplitudes for all space-varying periodic values (see, e.g., [9.3, 17, 18]). In particular, the sinusoidal distribution of light intensity in the recorded interference pattern is

$$I(x) = I_0 [1 + |m| \cos(Kx + \Psi_0)]$$

$$= I_0 + \frac{I_0}{2} (m e^{iKx} + m^* e^{-iKx}) , \tag{9.2}$$

where $I_0 m = I_0 |m| \exp(i\Psi_0)$ is the complex amplitude of the spatially periodic part of the interference pattern. The density of the photoelectrons in the conduction band is

$$n(x) = n_0 + \frac{n_0}{2} (a e^{iKx} + a^* e^{-iKx}) ,$$

where n_0 is the average steady state density of photoelectrons under uniform illumination with light of intensity I_0. The density of positively charged acceptor centers is

$$N_A(x) = N_A + \frac{N_A}{2}\left(A\,e^{iKx} + A^*e^{-iKx}\right) ,$$

where N_A is the average acceptor density, considered here to be well below the average donor concentration N_D [9.2, 3, 19]. The electric field distribution throughout the sample is

$$E(x) = E + E_{sc}(x) = E + \tfrac{1}{2}\left(E_{sc}\,e^{iKx} + E_{sc}^*e^{-iKx}\right) ,$$

where E is the external electric field applied to the sample. We neglect higher spatial harmonics in the distributions $n(x)$, $N_A(x)$, and $E(x)$, which approximation is known to be valid only for low initial contrasts of the recorded pattern ($|m| \ll 1$) and when the modulation depths of the parameters considered are also small [9.3, 17, 20].

The basic equations employed in the theoretical analysis of the holographic recording processes, namely, the balance equation for mobile electrons in the conduction band, the balance equation for the charge at deep trapping levels, the law of charge conservation, and Poisson's equation, reduce in this approximation to linear relations between the complex amplitudes of relevant parameters. Specifically, Poissons's equation assumes the form

$$iK\varepsilon\varepsilon_0 E_{sc} = eAN_A . \tag{9.3}$$

Here we ignore the contribution of mobile electrons to the amplitude of the hologram electric field, which can be done when the lifetime of a photoelectron τ is much shorter than the characteristic time of hologram recording τ_{sc}.

The balance equation for concentration of acceptor centers has the form

$$\frac{\partial A}{\partial t}\,N_A = mg_0 - (a+A)n_0\tau^{-1} . \tag{9.4}$$

Here the first term on the right-hand side corresponds to spatially nonuniform generation of photoelectrons in the interference pattern ($g_0 = n_0\tau^{-1}$ is the average photoelectron generation rate resulting from the average light intensity I_0), and the second term corresponds to their nonuniform trapping by acceptor centers.

In the approximation $\tau \ll \tau_{sc}$ assumed here, the photoelectron distribution can be treated as quasi-stationary (for the light and field distributions and trap population at any given moment) and is described by

$$mg_0 - (a+A)n_0\tau^{-1} = -iK\mu n_0(aE + E_{sc}) + K^2 Dan_0 . \tag{9.5}$$

Here the left-hand side of the expression describes the electron exchange between the conduction band and trapping levels [see also (9.4)], and the right-hand side is the redistribution throughout the conduction band due to drift in the electric field $E(x)$ and diffusion (D is the diffusion coefficient).

A self-consistent solution of (9.3–5) yields the following final equation for the hologram complex amplitude [9.3]:

$$\frac{\partial E_{sc}}{\partial t} = -\frac{m(E + iE_D) + E_{sc}(1 + K^2 l_s^2 - iK l_E)}{\tau_M (1 + K^2 r_D^2 - iK r_E)}, \tag{9.6}$$

where $E_D = K(D/\mu) = K(k_B T/e)$ is the diffusion field, $r_D = \sqrt{D\tau}$ is the photoelectron diffusion length, $l_s = \sqrt{\varepsilon \varepsilon_0 k_B T/e^2 N_A}$ is the Debye screening length, $l_E = \varepsilon \varepsilon_0 E/e N_A$ is the length of electron tightening by the electric field E, and $\tau_M = \varepsilon \varepsilon_0/e \mu n_0$ is the Maxwell relaxation time.

Note that (9.6) immediately predicts practically all major features of holographic recording in photorefractive crystals. In particular, for the steady-state regime of recording (i.e., at $dE_{sc}/dt = 0$), the hologram amplitude turns out to be equal to a well-known value [9.2, 3, 18]

$$E_{sc} = -\frac{m(E + iE_D)}{1 + K^2 l_s^2 - iK l_E}. \tag{9.7}$$

Equation (9.7) also predicts saturation of the acceptor traps (i.e., violation of the quasi-neutrality condition) and a stationary hologram amplitude in the case when $K l_E \gtrsim 1$ or $K l_s \gtrsim 1$ for the usual drift or diffusion holographic recording.

On the other hand, by taking $m = 0$ in (9.6) we obtain an equation describing incoherent optical erasure

$$\frac{\partial E_{sc}}{\partial t} = -\frac{E_{sc}(1 + K^2 l_s^2 - iK l_E)}{\tau_M (1 + K^2 r_D^2 - iK r_E)} \tag{9.8}$$

that predicts the exponential decay of the hologram

$$\frac{E_{sc}(t)}{E_{sc}(t=0)} = \exp\left[-\frac{t(1 + K^2 l_s^2 - iK l_E)}{\tau_M (1 + K^2 r_D^2 - iK r_E)}\right]$$

$$= \exp\left\{-\frac{t[(1 + K^2 r_D^2)(1 + K^2 l_s^2) + K^2 r_E l_E]}{\tau_M[(1 + K^2 r_D^2)^2 + K^2 r_E^2]}\right\}$$

$$\times \exp\left\{-\frac{it[(1 + K^2 l_s^2) K r_E - (1 + K^2 r_D^2) K l_E]}{\tau_M[(1 + K^2 r_D^2)^2 + K^2 r_E^2]}\right\} \tag{9.9}$$

with a decay time [9.2, 3, 19]

$$\tau_{sc} = \frac{\tau_M[(1 + K^2 r_D^2)^2 + K^2 r_E^2]}{(1 + K^2 r_D^2)(1 + K^2 l_s^2) + K^2 r_E l_E}. \tag{9.10}$$

The presence of an additional exponential multiplier with a purely imaginary argument in (9.9) points to a uniform drift of the hologram being erased along the external field E at a characteristic velocity

$$v_0 = \frac{r_E(1 + K^2 l_s^2) - l_E(1 + K^2 r_D^2)}{\tau_M[(1 + K^2 r_D^2)^2 + K^2 r_E^2]} \ . \tag{9.11}$$

As can be seen from the derivation, the basic equation (9.6) is nothing but simply a balance equation for the electric charge at deep trapping levels obtained under the assumption of low contrast ($|m| \ll 1$) and a quasi-stationary distribution of photoelectrons ($\tau \ll \tau_{sc}$). Therefore (9.6) is valid not only for holographic recording of a fixed interference pattern in a steady external field, but also for the description of hologram recording under nonstationary conditions with the characteristic time of variation much longer than τ. Thus we can use it to analyze further the recording of moving interference patterns, as well as holographic recording in time-varying electric fields.

9.1.2 Recording a Moving Interference Pattern in a Steady Field

The most convenient way of using (9.6) to analyze holographic recording of a moving interference pattern

$$I(x, t) = I_0 \{1 + |m| \cos[K(x - vt)]\} \tag{9.12}$$

is to employ a coordinate system that moves synchronously with the interference pattern at a velocity v [9.6, 21]. In this coordinate system, the amplitude of the steady-state hologram that drifts synchronously with the pattern being recorded turns out to be

$$E_{sc} = \frac{-m(E + iE_D)}{(1 + K^2 l_s^2 - iKl_E) - iKv\tau_M(1 + K^2 r_D^2 - iKr_E)} \ . \tag{9.13}$$

It attains its maximum absolute value

$$E_{sc} = -\frac{m(E + iE_D)(1 + K^2 r_D^2 + iKr_E)}{(1 + K^2 r_D^2)(1 + K^2 l_s^2) + K^2 r_E l_E} \tag{9.14}$$

at a resonance velocity v_0 (9.11).

In normal conditions, i.e., $Kl_s \ll 1$ (no saturation of traps during recording in the absence of field E), for an electric field amplitude E such that $Kr_E \gtrsim (1 + K^2 r_D^2)$, the steady-state grating transforms to a shifted one (purely imaginary E_{sc}):

$$E_{sc} \approx -imE \frac{Kr_E}{1 + K^2 r_D^2} \ ,$$

$$v_0 \approx (K^2 \tau_M r_E)^{-1} \ , \qquad \text{at} \quad K^2 r_E l_E \gtrsim (1 + K^2 r_D^2) \tag{9.15}$$

and

$$E_{sc} \approx -im E_q \, ,$$

at $K^2 r_E l_E \gtrsim (1 + K^2 r_D^2)$.   (9.16)

$$v_0 \approx \frac{r_E - l_E (1 + K^2 r_D^2)}{\tau_M K^2 r_E^2} \, ,$$

The latter means that the regime of trap saturation is reached, since $E_q = e N_A / K \varepsilon \varepsilon_0$ is equal to the maximum possible hologram amplitude for given values of N_A and K.

Analogous drifting holograms with stationary amplitudes comparable to (9.15, 16) can be recorded by a periodically oscillating (in amplitude or phase) interference pattern. In particular, in the case of a more efficient phase modulation of the interference pattern

$$I(x, t) = I_0 \{ 1 + |m| \cos [Kx + \Delta \cos(\Omega t)] \} \qquad (9.17)$$

there is a whole set of drifting space-time components with corresponding resonance velocities $v_n = n\Omega/K$ and complex amplitudes $I_0 m i^n J_n(\Delta)$ (where n is any integer and J_n is the nth order Bessel function). Thus, if any one of the resonance conditions ($v_0 = v_n$) is satisfied, the corresponding drifting hologram is also efficiently recorded.

It is possible that the well-known instability of traditional holographic recording of a fixed interference pattern in BSO-type crystals in a steady external electric field results from beating of such a resonantly amplified drifting component and the main fixed part of the recorded steady-state hologram. The necessary phase modulation of the interference pattern (9.17) arises in this case from occasional mechanical vibrations and air currents in the holographic setup.

9.1.3 Recording a Stationary Interference Pattern in an Alternating Electric Field

For the case of interest, when the external field varies with period $\tau_\sim : \tau \ll \tau_\sim \ll \tau_{sc}$, the general expression for the time-average amplitude of the hologram electric field is obtained by averaging the left- and right-hand sides of (9.6) over a period $\tau_\sim$. For a square-wave field (Fig. 9.3a), this leads to [9.7, 8]

$$\frac{\partial E_{sc}}{\partial t} = - \frac{im [E_D (1 + K^2 r_D^2) + E K r_E] + E_{sc} [(1 + K^2 r_D^2)(1 + K^2 l_s^2) + K^2 r_E l_E]}{\tau_M [(1 + K^2 r_D^2)^2 + K^2 r_E^2]} \, ,$$

(9.18)

which yields the following expression for the steady-state ($dE_{sc}/dt = 0$) amplitude of a purely shifted hologram:

$$E_{sc} = -im \left(\frac{E_D}{1 + K^2 l_s^2} \right) \frac{1 + K r_E E / [E_D (1 + K^2 r_D^2)]}{1 + K^2 r_E l_E / [(1 + K^2 r_D^2)(1 + K^2 l_s^2)]} \, . \qquad (9.19)$$

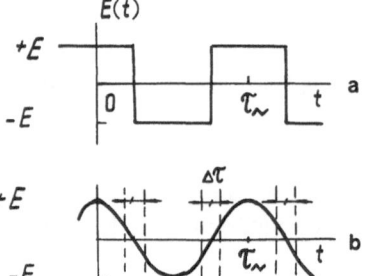

Fig. 9.3a, b. Typical alternating electric fields used in nonstationary hologram recording experiments: (a) square wave; (b) sinusoidal. For (b) the time periods $\Delta\tau$ when the recorded hologram is effectively erased are shown

Usually $Kl_s \ll 1$, and then for $E=0$

$$E_{sc} = -imE_D \ . \tag{9.20}$$

At larger amplitudes of the external field E, such that $Kr_E \gtrsim (E_D/E)(1+K^2 r_D^2)$, (9.19) gives values similar to (9.15, 16):

$$E_{sc} \approx -imE \frac{Kr_E}{1+K^2 r_D^2} \quad \text{at} \quad K^2 r_E l_E \gtrsim 1 + K^2 r_D^2 \ , \tag{9.21}$$

and

$$E_{sc} \approx -imE_q \qquad \text{at} \quad K^2 r_E l_E \gtrsim 1 + K^2 r_D^2 \ . \tag{9.22}$$

Finally, we give the theoretical expression for E_{sc} for holographic recording in a sinusoidal alternating field (Fig. 9.3b) obtained from (9.6) using a similar averaging procedure:

$$E_{sc} = -im \frac{E_D}{(1+K^2 l_s^2)}$$
$$\times \frac{1 + E[\sqrt{(1+K^2 r_D^2)^2 + K^2 r_E^2} - (1+K^2 r_E^2)]/(Kr_E E_D)}{1 + l_E[\sqrt{(1+K^2 r_D^2)^2 + K^2 r_E^2} - (1+K^2 r_E^2)]/[r_E(1+K^2 l_s^2)]} \ . \tag{9.23}$$

For the long drift length approximation ($Kr_E \gtrsim 1 + K^2 r_D^2$), this leads to

$$E_{sc} = \begin{cases} -imE & \text{at} \quad E \gtrsim E_q \ , \\ -imE_q & \text{at} \quad E \gtrsim E_q \ . \end{cases} \tag{9.24}$$

So, for a not too large electric field amplitude, when $K^2 r_E l_E \gtrsim 1 + K^2 r_D^2$, the steady-state hologram recorded in a square-wave alternating field (9.21) turns out to be much more efficient than that recorded in a sinusoidal alternating field of the same amplitude E. This is attributable to comparatively long time intervals $\Delta\tau$ when the main condition (9.1) is not satisfied for a sinusoidal field (Fig. 9.3b).

Because of a quadratic dependence on E, the characteristic time τ_{sc} (9.10) during $\Delta\tau$ turns out to be much smaller than its maximum value, and this is mainly responsible for the dramatic decrease in the hologram steady-state amplitude.

9.1.4 Discussion

The expressions (9.15, 16, 21, 22) derived above indicate that by using non-stationary recording methods it is possible to get a shifted hologram with the amplitude exceeding not only that of the diffusion, but also that of a normal drift hologram (i.e., in a steady field E).

The following problem is of great practical significance: What spatial frequency K is optimal for the nonstationary recording method at given values of N_A and E (the latter is typically ≈ 10–15 kVcm^{-1} and is limited by surface breakdowns)? From (9.19) we find

$$K_{opt} \approx r_D^{-1} \frac{k_B T}{eE} \left[\left(\frac{k_B T}{eE} \right)^2 + l_s^2 \right]^{-1/2} , \tag{9.25}$$

which yields the following maximum hologram amplitude:

$$|E_{sc}|^{max} = \frac{1}{2} mE K_{opt} r_E = \frac{1}{2} mE r_D \left[\left(\frac{k_B T}{eE} \right)^2 + l_s^2 \right]^{-1/2} . \tag{9.26}$$

The product of the hologram amplitude and spatial frequency band ($\Delta K \sim K_{opt}$)

$$\Delta K |E_{sc}|^{max} m^{-1} = \frac{1}{2} \left(\frac{k_B T}{e} \right) \left[\left(\frac{k_B T}{eE} \right)^2 + l_s^2 \right]^{-1} \tag{9.27}$$

is independent of the photoelectron parameters (μ, D, τ) and is affected by N_A and E alone.

One more general remark about the nonstationary holographic recording methods that we have discussed should be made. Both of them (recording a drifting pattern, as well as applying an alternating field) enhance not the sensitivity (determined by the initial stage of recording) but the stationary hologram amplitude (with an initial low contrast $|m| \ll 1$). Actually this means an increase in the gain factor of the photorefractive crystal [9.2, 3, 16]

$$\Gamma = -\mathrm{i} 2\pi E_{sc} m^{-1} \frac{n^3 r}{\lambda} , \tag{9.28}$$

where λ is the light wavelength, n is the average refractive index, and r is the linear electro-optic coefficient of the crystal. In practice, utilization of non-stationary methods of holographic recording in cubic photorefractive crystals (BSO, BGO, BTO) rapidly brings about saturation (9.16, 22). However, in this case Γ turns out to be proportional to the r/ε ratio (approximately the same for all presently known photorefractive crystals [9.22]) and hence it depends practically on concentration and type of trapping centers only.

9.2 Light Diffraction and Degenerate Four-Wave Mixing in Cubic Photorefractive Crystals

9.2.1 Introduction to Four-Wave Mixing in Photorefractive Crystals

Phase conjugation in photorefractive crystals is known to occur by degenerate four-wave mixing (Fig. 9.4a) (see [9.12, 14, 23, 24] and also the following volume in this series). Since the cubic crystals involved exhibit typically a fairly low efficiency of reflection hologram recording [9.25], the four-wave mixing process can be regarded as the result of two simultaneous oppositely directed processes of two-wave mixing (Fig. 9.4b, c), which are united by one common volume transmission hologram. In the approximation of constant-amplitude reference beam R_1, for the steady-state hologram, the complex amplitude of the object wave S_1 in a two-wave mixing geometry (Fig. 9.4b) changes throughout the sample (see, e.g., [9.2, 16, 18, 26, 27]) as

$$\frac{\partial S_1}{\partial z} = \frac{\Gamma}{2} S_1 , \qquad (9.29)$$

where Γ is the gain factor of the photorefractive crystal (9.28). Neglecting the optical absorption of the sample and Fresnel reflections from its surfaces, we arrive at the transmittivity

$$T = \left| \frac{S_1(d)}{S_0} \right|^2 = \exp(\mathrm{Re}\{\Gamma d\}) , \qquad (9.30)$$

where S_0 is the amplitude of the signal wave at the front and $S_1(d)$ at the back surface of the sample. Thus, in the case of formation of shifted gratings [a diffusion or nonstationary recording mechanism when Γ is real (9.28)], the object beam experiences gain ($\Gamma > 0$) or loss ($\Gamma < 0$) depending on the phase mismatch between the hologram being recorded and the interference pattern. It should be remembered that, for the shifted phase holograms, it is the light wave in whose propagation direction the refractive index distribution $n(x)$ maxima

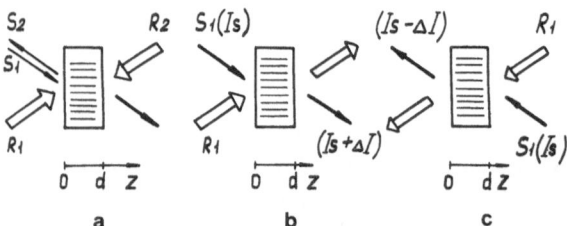

Fig. 9.4. Holographic schemes for four-wave (**a**) and two-wave (**b, c**) mixing via a transmission dynamic hologram. For the two-wave mixing geometry, the change of the energy transfer direction resulting from reversing the light beams' directions is shown

are shifted relative to the distribution $I(x)$ that always experiences gain. This implies that the effect automatically changes its direction (gain or loss in the signal wave S_1) when the direction of both waves S_1 and R_1 (Fig. 9.4c) is reversed.

For discussion of the four-wave mixing geometry, (9.29) given above should be supplemented with a similar expression describing the diffraction of the pump wave R_2 into a conjugate signal wave S_2, as well as their contributions to the hologram recording process. The set of these equations has the simplest form if the plane pump waves are strictly collinear and their intensities are equal ($|R_1| = |R_2|$):

$$\frac{\partial S_1}{\partial z} = \frac{\Gamma}{4} (S_1 + S_2^*) \ , \qquad \frac{\partial S_2^*}{\partial z} = -\frac{\Gamma'}{4} (S_1 + S_2^*) \ . \tag{9.31}$$

The minus sign in the second equation represents the opposite direction of propagation of R_2 and S_2 with respect to R_1 and S_1. The coefficients Γ, Γ' are of opposite signs in the usual situation [9.28], because S_1 and S_2 are counterpropagating (Fig. 9.4a) and one of them experiences gain (S_1 at $\Gamma > 0$, S_2 at $\Gamma < 0$) and the other loss.

Solution of the set of equations (9.31) for $\Gamma' = -\Gamma$ and typical boundary conditions $S_1(0) = S_0$ and $S_2(d) = 0$ yields the following expressions for transmittivity and reflectivity [9.28]:

$$T^- = \left|\frac{S_1(d)}{S_0}\right|^2 = \left|\frac{2\exp(\Gamma d/2)}{1 + \exp(\Gamma d/2)}\right|^2 \ ,$$

$$R^- = \left|\frac{S_2(0)}{S_0}\right|^2 = \left|\frac{1 - \exp(\Gamma d/2)}{1 + \exp(\Gamma d/2)}\right|^2 \ . \tag{9.32}$$

For the case of real $\Gamma > 0$, the reflectivity therefore appears always to be smaller than unity and the transmittivity is less than the analogous value (9.30) in the two-wave mixing geometry.

The latter is not surprising, since introduction of a counterpropagating pump beam R_2 results in additional feedback in the crystal volume. By this we mean an additional hologram recording mechanism near the input face of the crystal due to interference of R_2 and S_2. In this case, however, we can speak of negative feedback, because the hologram that is recorded by S_2 and R_2 is opposite in phase to that recorded by S_1 and R_1 (coherent erasure of the hologram [9.29] by R_2 and S_2).

The effect of the negative feedback in a four-wave mixing geometry in photorefractive crystals with shifted holograms can be offset by appreciably decreasing the intensity of the counterpropagating pump beam R_2 [9.28, 30]. However, photorefractive crystals offer the possibility of a radical solution of this problem by employing positive feedback, which considerably increases T and R. The latter is based on some features of volume phase gratings in cubic crystals to be briefly discussed in the following section.

9.2.2 Light Diffraction from Volume Phase Holograms in Cubic Photorefractive Crystals

Major specific features of light diffraction from volume phase holograms in photorefractive crystals arise from the anisotropic nature of the recorded phase hologram with a tensor amplitude $\hat{\varepsilon}_{sc}$ [9.3, 31]

$$(\varepsilon_{sc})_{nm} = - \sum_{i,k,l=1}^{3} \varepsilon_{ni} r_{ikl} (E_{sc})_l \varepsilon_{km} \ . \tag{9.33}$$

Analysis of diffraction of an arbitrarily polarized light beam in such crystals should be carried out by independent consideration of two orthogonally polarized eigenmodes of diffraction. Because of the tensor nature of $\hat{\varepsilon}_{sc}$, each eigenmode is characterized by its own eigenstate of linear polarization and diffraction efficiency. For not too large Bragg angles ($\theta \ll 1$) and a symmetric holographic experimental setup, the polarization eigenvectors in cubic crystals of the BSO type coincide with the principal axes of the intersection of the index ellipsoid $\hat{\varepsilon} + \hat{\varepsilon}_{sc}$ with the front face of the crystal [9.3, 31].

In particular, for a typical holographic orientation of a cubic crystal [9.32] ([110] is perpendicular to the sample surface), the eigenstates of diffraction are E- and H-polarized at $K \parallel [001]$ (Fig. 9.5a) [9.3, 33]. In this case the diffraction efficiency of the grating for the first eigenmode of diffraction is fairly high (η_E), whereas it goes to zero for the second eigenmode ($\eta_H = 0$). The latter apparently indicates that the H-polarized light waves are not modulated by the hologram electric field $E_{sc} \parallel K \parallel [001]$.

For the orthogonal crystal orientation, when $K \parallel [1\bar{1}0]$, the eigenstates of diffraction turn out to be polarized at $\pm 45°$ to the incidence plane [9.3, 33]. All other conditions being equal, the diffraction efficiencies of the grating for the above two eigenmodes of diffraction are equal to the efficiency of the grating in the previous orientation when read-out is accomplished with E-polarized light ($\eta_{+45°} = \eta_{-45°} = \eta_E$). In this situation, however, the signs of the refractive index changes for light beams polarized at $+45°$ and $-45°$ become opposite

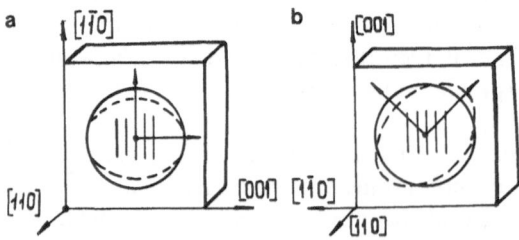

Fig. 9.5. Typical orientations of the cubic BSO-type crystals ((110) cut) in the holographic setup. The solid circles and dashed ellipses show the initial index ellipsoid and that distorted by the hologram electric field ($E_{sc}(x) \parallel K$), respectively. Polarization of the eigenmodes of diffraction is shown by arrows

(Fig. 9.5b) and there is a 180° phase shift between the $\pm 45°$ polarized waves diffracted from the same hologram. This feature, namely, the contrast inversion of the same phase grating with the change of the light wave linear polarization (from $+45°$ to $-45°$) will be shown in the following section to be extremely useful for increasing the efficiency of four-wave mixing.

Another important factor that leads to serious complications of the diffraction phenomena in crystals of the silenite family arises from their natural optical activity (for BSO $\varrho \approx 42$ degrees/mm at $\lambda = 0.5$ μm [9.34]). This effect can apparently be ignored only in crystals of small thickness d, where the angle of rotation of the polarization plane of the transmitted light wave does not exceed 90°. We should merely assume that the real light beam polarization is at the center of the sample, not at its front surface (they differ by the angle $\varrho d/2$).

For thicker crystals ($\varrho d \gtrsim 90°$), the diffraction phenomena in cubic crystals are even more complicated. In particular, the Bragg maximum splits into several differently polarized peaks [9.35], which results in a considerable reduction of the gain factor Γ [9.36].

9.2.3 Four-Wave Mixing with Positive Feedback via Shifted Gratings

As follows from the preceding section, cubic photorefractive crystals offer a means of investigating an unusual four-wave mixing geometry with opposite signs of the hologram for the light beams passing through the sample in opposite directions. To accomplish this, in the orientation presented in Fig. 9.5b the polarization directions of the counterpropagating light beams must be ortho-gonal and oriented at $\pm 45°$ to the incidence plane. As a consequence, the direction of self-diffraction of the reference beam R_2 from the recorded hologram will be opposite to that in a conventional four-wave mixing scheme. In this case we should take $\Gamma' = \Gamma$ in the set of equations (9.31), which yields the following transmittivity and reflectivity for the usual boundary conditions [9.9, 10]:

$$T^+ = \left| \frac{4}{4 - \Gamma d} \right|^2 ,$$

$$R^+ = \left| \frac{\Gamma d}{4 - \Gamma d} \right|^2 . \tag{9.34}$$

These parameters peak at $\Gamma > 0$, that is, when the two object waves S_1 and S_2 experience gain due to self-diffraction of R_1 and R_2 from the recorded hologram. At $\Gamma d \to 4$, the coefficients (9.34) go to infinity (Fig. 9.6), and this corresponds to the regime of self-oscillation or generation in the crystal. Note that recently [9.37] equations analogous to (9.34) were also obtained for four-wave mixing via shifted phase gratings recorded using the circular photogalvanic effect.

It is worth noting that the expressions for T^+ and R$^+$ given above yield finite values above the generation threshold, i.e., at $\Gamma d > 4$, as well. A thorough

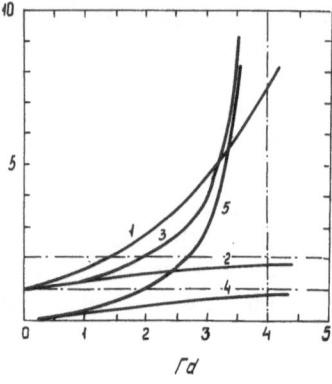

◀ **Fig. 9.6.** Transmittivities and reflectivities for two- and four-wave mixing processes via a shifted phase transmission hologram (theory, undepleted pumps approximation). Two-wave mixing: (*1*) *T*, and four-wave mixing: (*2*) T^-, (*3*) T^+, (*4*) R^-, (*5*) R^+

Fig. 9.7. (a) Reflectivity R^+ for four-wave mixing with positive feedback as a function of the pump beam intensity ratio (r, r^{-1}). [Theory, undepleted pumps approximation, $\Gamma d = 2\,(1); 3\,(2); 3.5\,(3).$] **(b)** Theoretical dependence of $R^+(\theta)$ half-width on the product Γd

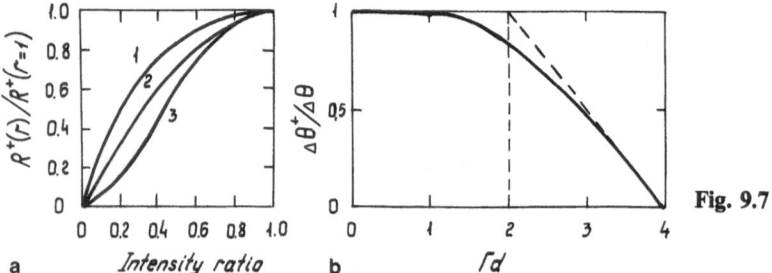

a   *Intensity ratio*   b   **Fig. 9.7**

analysis involving consideration of the extended set of equations (including an additional expression describing the temporal evolution of the recorded hologram) shows that the stationary solutions above the generation threshold are unstable.

We also studied the below-threshold stationary states so as to find the optimum ratio between the intensities of pump beams, similar to that described in [9.28] for a conventional four-wave mixing geometry. The maximum reflectivity was found to occur in this case for equal powers of these beams. For an arbitrary ratio ($I_{R_2}/I_{R_1} = r$) the transmittivity and reflectivity are given by

$$R^+ = \frac{\sinh^2\left(\dfrac{1-r}{1+r}\dfrac{\Gamma d}{4}\right)}{\sinh^2\left(\dfrac{1-r}{1+r}\dfrac{\Gamma d}{4} + \dfrac{\ln r}{2}\right)},$$

$$T^+ = \frac{\sinh^2\left(\dfrac{\ln r}{2}\right)\exp\left(\dfrac{1-r}{1+r}\dfrac{\Gamma d}{2}\right)}{\sinh^2\left(\dfrac{1-r}{1+r}\dfrac{\Gamma d}{4} + \dfrac{\ln r}{2}\right)}.$$

(9.35)

See Fig. 9.7a.

A more detailed analysis [9.10] (Fig. 9.7b) also gives a fairly simple relation between the maximum possible magnitude of the reflectivity R^+ and the allowable deviation of the backward pump beam R_2 from collinearity with the direction of R_1. Indeed, for the most interesting interval $2 < \Gamma d < 4$, the following simple expression turns out to be valid for the case in question:

$$\sqrt{R^+} \left(\frac{\Delta\theta^+}{\Delta\theta} \right) \approx 2 , \qquad\qquad (9.36)$$

where $\Delta\theta^+$ is the allowable deviation of the backward pump beam direction from collinearity and $\Delta\theta$ is the half-width of the maximum of Bragg diffraction from the volume hologram which is uniform throughout the sample thickness d and has the same spatial frequency K. The latter imposes rather severe restrictions on both the quality and accuracy of alignment of plane pump waves, on the phase homogeneity and on the quality of the polished crystal faces.

9.2.4 Discussion

The expressions derived above for the transmittivity and reflectivity hold also for an unshifted grating (Γ is imaginary) and a grating with an arbitrary shift (Γ is complex). However, the maximum reflectivity R^+ at a given absolute value of the product (Γd) is achieved for a purely shifted grating.

It follows from Sect. 9.1, 3 that rather efficient shifted holograms can be obtained in cubic photorefractive crystals via nonstationary methods of hologram recording. Fairly high gain factors for shifted holograms can also be achieved in other photorefractive crystals and, in particular, in $BaTiO_3$ [9.14, 15] by the diffusion recording mechanism. In these birefringent (noncubic) crystals, it is more appropriate to employ a four-wave mixing geometry without feedback, which occupies an intermediate position between the two limiting cases considered above. For this purpose, we have to eliminate one of the self-diffraction processes suppressing the hologram, for instance, to use either the polarization of the waves R_1 and S_1, for which the effect of self-diffraction is negligible, or the polarization of R_2, for which diffraction accompanied by rotation of the plane of polarization is observed [9.31]. For the former case one should take $\Gamma = 0$ in (9.33), which gives, for standard boundary conditions,

$$T^0 = 1 ,$$
$$R^0 = |e^{\Gamma d/4} - 1|^2 . \qquad\qquad (9.37)$$

A similar R^0 is obtained in the second case as well. So, elimination of one of two oppositely directed processes (in the case of negative feedback) of self-diffraction yields here reflectivities exceeding unity even for equal intensities of pump beams.

In conclusion, it should be emphasized that the geometries discussed in this section differ markedly from a conventional four-wave mixing scheme in that the original S_1 and conjugate S_2 waves are orthogonally polarized. In a number of situations (for example, in holographic real-time vibrometry [9.38] and lensless imaging schemes [9.39]) this feature can be used to reduce light losses. For this purpose, the traditional beam splitter in the signal shoulder of the interferometer is replaced by a birefringent Glan prism.

9.3 Nonstationary Holographic Recording Mechanisms in Cubic Photorefractive Crystals

9.3.1 Recording Moving Interference Patterns in $Bi_{12}SiO_{20}$ (BSO) Crystals

Experimental techniques of recording moving interference patterns in photorefractive crystals were first reported in [9.4, 40], where two-wave mixing in BSO and BGO crystals was studied. In these experiments, an appreciable steady-state amplification of a weak signal beam for a certain resonance velocity of the interference pattern along the steady external electric field E was observed. This effect was attributed to a shifted component of the phase hologram, resulting from its mismatch with the recorded pattern, similar to that observed earlier in dynamic holographic recording experiments using nonlinear liquids [9.16].

Another experimental investigation of the nonstationary holographic recording in BSO via a moving interference pattern was described in [9.5, 6]. Special steps were taken in order to eliminate a possible intensity transfer between the writing beams [9.4, 40] and to study the effect of the resonance amplification of the hologram by itself (for a given $|m|$). Samples of small thickness ($d \approx 1$ mm) and equal initial intensities of the recording beams were used, and the crystallographic orientation of the crystal and the writing beams' polarization were chosen to ensure diffraction with rotation of the plane of polarization. The necessary low contrast of the grating was provided by the external uniform illumination of the sample with incoherent green light.

The experimental curve for the steady-state diffraction efficiency of the grating ($\eta \sim |E_{sc}|^2$) as a function of the interference pattern velocity v (Fig. 9.8) displays a pronounced resonance behavior, with a peak at $v = v_0 \neq 0$, thus verifying the basic conclusion of the above analysis. Direct observation of the holographic grating in a polarization microscope showed that, in the absence of the recording interference pattern, the direction of the drift of the holographic grating coincides with the external field E direction. This confirms the electronic origin of photoconductivity in BSO in the blue-green spectral region [9.41]. According to the theoretical expression (9.11), v_0 turned out also to be proportional to the average intensity $I_0 (v_0 \sim \tau_M^{-1} \sim I_0)$ and decreased with increasing external field $E [v_0 \sim E^{-1}$ at $Kr_E \gtrsim (1 + K^2 r_D^2)]$ (Fig. 9.9a). The drift of a

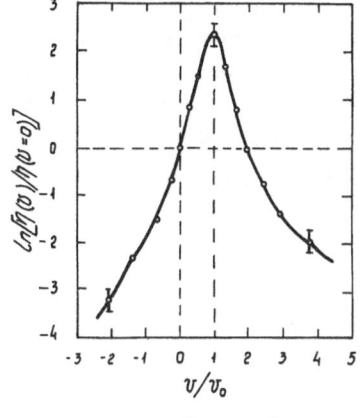

◀ **Fig. 9.8.** Steady-state hologram diffraction efficiency η as a function of the interference pattern drift velocity v [9.5, 6]. (Experiment, BSO, $\lambda = 0.51$ μm, $E \approx 9$ kV cm^{-1}, $\Lambda \approx 2.5$ μm, resonance velocity $v_0 \approx 1$ μm s^{-1})

Fig. 9.9. (a) Resonance velocity of the hologram drift v_0 as a function of steady electric field amplitude E (1) and average light intensity I_0 (2) [9.5, 6]. **(b)** Hologram optical erase time τ_{sc} as a function of steady external electric field E [9.5, 6]. (Experiment, BSO, $\lambda = 0.51$ μm, $\Lambda = 2.5$ μm)

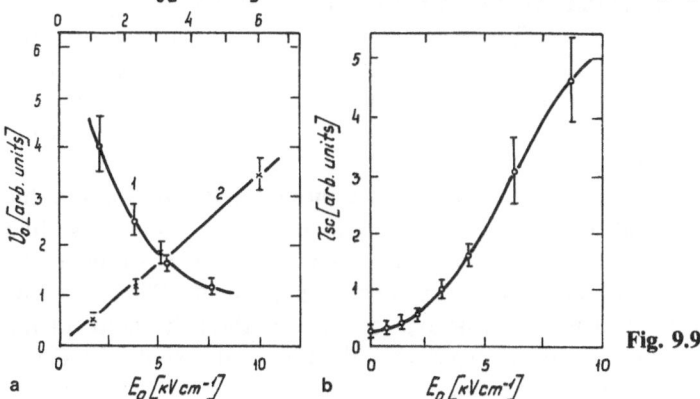

Fig. 9.9

grating erased in BSO under an external steady field E was also confirmed in [9.42] using a standard interferometric technique.

Vital parameters of the crystal studied (with respect to the nonstationary recording mechanism) r_D and l_s can be obtained from the experimental plot of the hologram erase time τ_{sc} against E (Fig. 9.9b). Fitting of the theoretical expression (9.10) to the experimental curve yields $r_D \approx 0.27$ μm and $l_s \approx 0.024$ μm ($N_A \approx 10^{17}$ cm^{-3}). A similar dependence of v_0 and τ_{sc} on E in BSO was also reported in a recent paper [9.43].

In accordance with the theoretical expression (9.26) the maximum increase of the hologram diffraction efficiency (κ) on recording a resonantly drifting pattern is

$$\kappa = \frac{\eta(v_0)}{\eta(v=0)} \approx \frac{1}{4}\, r_D^2 l_s^{-2} \approx 30 \ . \tag{9.38}$$

Taking into account an inhomogeneous broadening of the resonance curve $\eta(V)$ (Fig. 9.8) induced by variations in the average sample illumination, a spatially nonuniform external field distribution E, and photoconductive properties of the crystal itself, the discrepancy between the obtained estimate for κ and the experimental value ($\kappa = 10$, Fig. 9.8) appears to be quite natural. Indeed, an additional analysis reveals that in order to achieve $\kappa \approx 30$ experimentally, the spatial homogeneity of I_0 along the x axis should be kept accurate to $\gtrsim 1/2\sqrt{\kappa} \approx 0.1$. The active spatial frequency band of the signal beam will also in this case be no more than $1/2\sqrt{\kappa} \approx 0.1$ of the carrier spatial frequency of the hologram.

9.3.2 Nonstationary Holographic Recording in an Alternating Field in $Bi_{12}TiO_{20}$ (BTO) Crystals

In the experiments [7–9, 11], 4-mm thick BTO samples grown at the Laboratory of Quantum Radioelectronics of the A.F. Ioffe Physical Technical Institute, USSR, were used (Table 9.1). The geometry similar to that utilized in BSO experiments presented above (Fig. 9.5b) was employed with HeNe laser radiation ($\lambda = 0.63$ μm). In order to provide the optimum conditions for two-wave mixing, the light beams were polarized at $\pm 45°$ to the incidence plane in the middle of the crystal thickness.

In contrast to nonstationary recording via a moving pattern, holographic recording in an external alternating electric field results in a purely shifted hologram (9.19, 23). Thus, the two-wave mixing geometry was found to be suitable for studying the efficiency of this recording mechanism (Fig. 9.4b). The basic parameter to be measured in this experiment was the gain factor of the crystal Γ (9.28) evaluated from the experimental value of transmittivity (for a weak signal beam) according to (9.30).

A typical experimental $\Gamma(E)$ curve obtained for BTO in such an experiment is shown in Fig. 9.10a. Along with this, the first experiments also revealed the

Table 9.1. Data of crystals used in the experiments[a]

	BSO ($Bi_{12}SiO_{20}$)	BTO ($Bi_{12}TiO_{20}$)
Point group	23	23
Density [g cm^{-3}]	9.2	9.1
Lattice parameter [Å]	10.103 ± 0.0005	10.176 ± 0.0005
Melting point [°C]	895	875
Optical absorption [cm^{-1}]	0.1–0.3	0.4–0.6
Refractive index	2.54	2.58
Half-wave voltage $\dfrac{1}{2}\left(\dfrac{n^3 r_{41}}{\lambda}\right)^{-1}$ [kV]	3.9–4.4	3.2–3.3
Natural optical activity [deg. mm^{-1}]	22.4	6.3
Static dielectric constant	56	47

[a] The optical data were measured at $\lambda = 633$ nm in the crystals grown at the laboratory for Quantum Radioelectronics of the A. F. Ioffe Physical Technical Institute

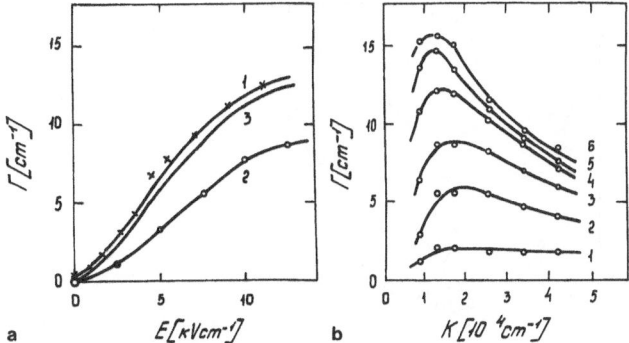

Fig. 9.10. (a) Gain factor Γ vs alternating electric field amplitude obtained for a two-wave mixing scheme with a probe signal beam (*1*), and from optically induced noise measurements (*2, 3*). [Curve 2 is calculated from (9.30) and curve 3 from (9.39)]. (b) Gain factor Γ vs spatial frequency K for an alternating electric field amplitude $E = 2.5$ kV cm^{-1} (*1*); 5 kV cm^{-1} (*2*); 7.5 kV cm^{-1}(*3*); 10 kV cm^{-1} (*4*); 12.5 kV cm^{-1}(*5*); 15 kV cm^{-1} (*6*)[9.8]. (BTO, $\lambda = 0.63$ μm, $\Lambda^{-1} = 300$ mm^{-1}, a square-wave alternating electric field, $\tau_\sim \approx 25$ ms)

presence of a strong photoinduced noise (see, e.g., [9.3, 44]), thus suggesting a simple way to obtain the dependence $\Gamma(E, K)$ by comparing the optical noise level with and without an alternating external field. However, the estimates for Γ obtained in a standard two-wave mixing experiment with a probe signal beam and those defined from the photoinduced noise calculated from (9.30) showed a drastic discrepancy (Fig. 9.10a). This is attributable to the volume nature of the distribution of the scattering centers responsible for the optical noise in the crystal. In this model, the output noise signal intensity grows as

$$I_N = I_N^0 \, \frac{e^{\Gamma d} - 1}{\Gamma d} \,, \tag{9.39}$$

where I_N^0 is the output intensity of the noise signal with no alternating applied field E.

The set of curves $\Gamma(K)$ plotted in accordance with (9.39) from the experimental curves for the photoinduced noise buildup in an external square-wave alternating field is presented in Fig. 9.10b. The curves are qualitatively consistent with the theoretical expressions (9.12–22), i.e., $\Gamma \sim K$, if the condition of quasi-neutrality is fulfilled, and $\Gamma \sim K^{-1}$ when it is violated. The data obtained give the following estimates: $r_D = 0.25$ μm and $l_s \approx 0.08$ μm ($N_A \approx 10^{16}$ cm^{-3} for $\varepsilon \approx 47$ [9.45]).

The theoretical analysis given in Sect. 9.1.3 also predicts appreciably different efficiencies for recording in square-wave and sinusoidal alternating fields (9.19, 23). In order to verify this conclusion, we carried out two-wave mixing experiments with a probe signal beam using BTO for the above two types of alternating field. The experimental $\Gamma(E)$ curves calculated from (9.30) are depicted in Fig. 9.11a. As the theoretical analysis predicts, the gain factor Γ for

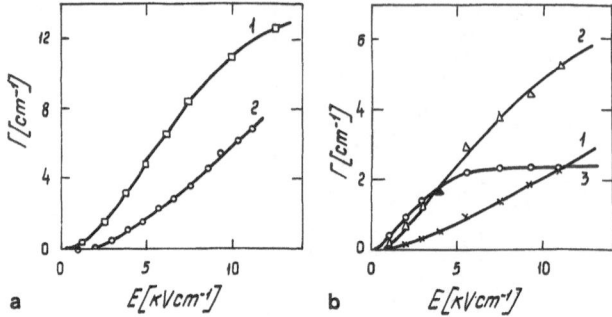

Fig. 9.11. (a) Gain factor Γ vs alternating electric field amplitude for a square-wave (1) and sinusoidal (2) alternating voltage. (BTO, $\lambda = 0.63$ μm, $\tau_\sim \approx 20$ ms, $\Lambda^{-1} \approx 200$ mm^{-1}, initial signal/pump intensity ratio about 10^{-4}). (b) Gain factor Γ vs square-wave alternating electric field amplitude E. [BSO, $\lambda = 0.63$ μm, $\tau_\sim \approx 25$ ms, $\Lambda^{-1} = 10$ mm^{-1}(1), 40 mm^{-1}(2), 70 mm^{-1} (3)]

the sinusoidal field turns out to be much lower and displays a longer flat portion of nearly linear growth.

In conclusion, we shall give the experimental curves for $\Gamma(E)$ obtained for the BSO crystal in a square-wave field at $\lambda = 0.63$ μm (Fig. 9.11b). The two-wave mixing experiments using BSO were performed under conditions similar to those for BTO crystals, with the same orientations and typical sample sizes. The maximum gain factor Γ observed in BSO on nonstationary holographic recording in an alternating field proves to be well below that for BTO and is achieved at appreciably lower spatial frequencies, which is consistent with the results reported in [9.46] for two-wave mixing BSO experiments using running interference patterns at $\lambda = 0.568$ μm.

The curves shown in Fig. 9.11b are in reasonable agreement with (9.19) and give the following estimates for the main characteristic lengths in these crystals: $r_D \approx 0.6$ μm and $l_s \approx 0.4$ μm ($N_A \approx 3 \times 10^{14}$ cm^{-3}). It should be noted that the estimate obtained for N_A is well below the concentration of acceptor centers that contribute to the hologram buildup in BSO at $\lambda = 0.5$ μm as measured in a holographic experiment under other conditions (see [9.21, 25, 36] and also Sect. 9.3.1). More-detailed studies of specific features of nonstationary holographic recording in BSO in an alternating field are required.

9.3.3 Discussion

As the experiments described above and the results reported in [9.4, 40, 46] have shown, nonstationary holographic recording in an externally applied alternating field and recording of moving interference patterns in a steady external field can succeed in forming highly efficient shifted gratings. However, it should be pointed out that in practice the use of an alternating field is often favored because of both easier technical implementation and the nonresonant character of the recording mechanism. Indeed, the time period $\tau_\sim$ (if condition $\tau_{sc} \gg \tau_\sim \gg \tau$ is met) turns out to be independent of light intensity, hologram spatial frequency, magnitude of the field, or local variations of the crystal parameters. Moreover, high-quality holographic recording appears to be possible even under nonuniform illumination of the sample, since at $\tau_\sim \ll \tau_M$ no marked redistribution of the average electric field during a half-period of its variation $\tau_\sim/2$ occurs.

Of great interest also is the possibility of using the alternating field for holographic recording in other photorefractive crystals where condition (9.1) can be fulfilled, namely, in $KNbO_3$ [9.47], InP, and GaAs [9.48a]. First experiments on two-wave mixing in GaAs ($\lambda = 1.06 \, \mu m$) using sinusoidal alternating electric fields were reported in [9.48b].

9.4 Image Amplification and Phase Conjugation in BTO via Nonstationary Recording in an Alternating Electric Field

9.4.1 Image Amplification in BTO Crystals

Efficient recording of shifted phase holograms in an alternating electric field allows application of BTO crystals for the stationary amplification of complex wave fronts via two-wave mixing (Fig. 9.4b). Experiments using (110)-cut samples of a 4-mm thickness were performed at the wavelength of a HeNe laser ($\lambda = 0.63 \, \mu m$) [9.11]. Typical photographs of amplified images are presented in Figs. 9.12a, b.

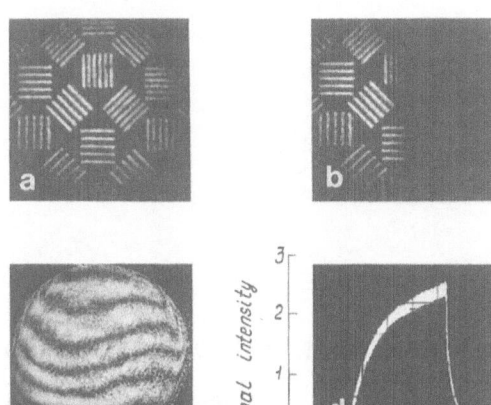

Fig. 9.12a–d. Two-wave mixing amplification of signals in BTO via shifted holograms recorded in a square-wave electric field. ($\lambda = 0.63 \, \mu m$, $d = 4 \, mm$, $E \approx 12 \, kV \, cm^{-1}$, $\tau_\sim \approx 25 \, ms$, $I_0 \approx 1 \, mW \, mm^{-2}$). (a) $20^\times$ amplified image. (b) The same image with the right part of the pump beam blocked. (c) Double-exposure interferogram of a thin BSO plate. (d) Oscilloscope traces of the signal intensity response to pulse switching on the pump beam

The same holographic geometry can be used to record real-time holographic interferograms (e.g., [9.49]). An example of a double-exposure interferogram of phase inhomogeneities in a thin BSO plate [9.50] is given in Fig. 9.12c. Dynamic characteristics of the holographic geometry under discussion are illustrated by the experimental plot of the amplified beam intensity as a function of time for typical recording conditions (Fig. 9.12d).

The experiments considered above were carried out for a typical orientation of the cubic photorefractive crystal (Fig. 9.5b) with the light beams' polarization oriented at $\pm 45°$ in the middle of the crystal thickness. However, such an orientation of the (110)-cut sample did not prove to be optimum for two-wave mixing experiments and this can be easily seen from the angular distribution of the photoinduced noise behind the BTO sample (Figs. 9.13a–c). Indeed, though the ac field orientation along the x axis provides the most efficient formation of noise holograms with wave vectors lying in the incidence plane, the noise lobe deviates from it by 20–30°.

Detailed consideration reveals that the most efficient phase grating (for a given amplitude E_{sc}) in a cubic photorefractive crystal is in fact achieved for $K \parallel [111]$, i.e., when [111] (or any other crystallographic axis of threefold symmetry) lies in the plane of incidence [9.51]. The optimum light polarization in this case appears to be of the H type. However, the gain is not very high; the

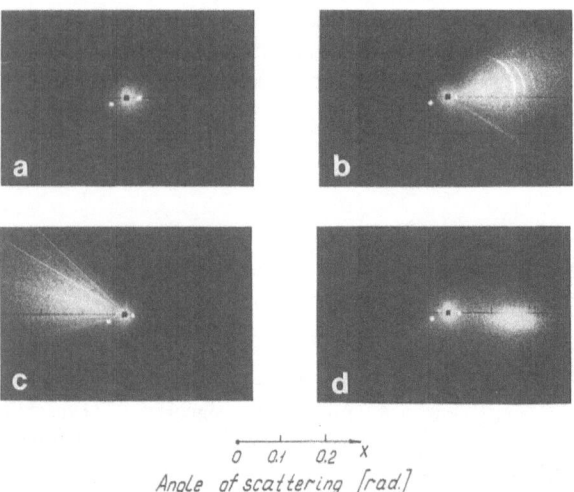

0   0.1   0.2   x
Angle of scattering [rad.]

Fig. 9.13a–d. Angular distribution of light noise behind the (110)-cut BTO sample. (**a**) With no external field. (**b–d**) In a square-wave alternating electric field ($\tau_\sim = 20$ ms, $E \approx 15$ kV cm^{-1}). (**b, c**) Only one pump beam with $+45°$ and $-45°$ orientation of the light polarization in the middle of the sample thickness. [The ring structure in (**b**) is a result of efficient noncollinear four-wave mixing (with positive feedback) of the pump beam with its Fresnel reflection from the back surface of the crystal repeated due to the wedge of the sample]. (**d**) Two counterpropagating plane pump waves. ($\lambda = 0.63$ μm, $d \approx 4$ mm, the black square at the center of each picture is a hole in the screen through which the pump beams pass)

grating amplitude is only 15% larger than that for typical orientations (Fig. 9.5). An adequate theoretical analyis of the phenomenon observed is bound to include some other effects and, in particular, piezoelectric and eleasto-optic properties of the crystal [9.52].

9.4.2 Phase Conjugation in BTO Crystals

The experiments employed a typical four-wave mixing geometry used for photorefractive crystals (see, e. g., [9.23–25]) with two counterpropagating plane pump beams (Fig. 9.4a). However, a square-wave alternating electric field was applied to the BTO crystal and the polarizations of the light waves crossing the sample in opposite directions were orthogonal and oriented at $\pm 45°$ relative to the plane of incidence. A low-power HeNe laser was used as a source of coherent radiation. A major portion of the laser light power ($P \gtrsim 5$ mW) was divided between two pump beams of nearly equal intensity, which were about 1 mm in diameter on the sample surface. The proper light beam polarization was set by quartz plates. Total losses (caused by optical absorption and Fresnel reflections from the sample faces) in the intensities of the light beams as they propagated through the 4-mm thick sample amounted to about 50%.

Simultaneously with the conjugate wave front intensity ($I_{S_2}^+$), we measured the intensity of the transmitted signal wave ($I_{S_1}^+$) and the intensity of the transmitted signal wave (I_{S_1}) that was amplified by two-wave mixing with a reference wave R_1 (the counterpropagating wave R_2 being blocked at this moment). The experimental curves for the above-mentioned light beam intensities normalized to the intensity of the original signal wave at the front face of the sample I_{S_0} (i.e., the experimental transmittivity and reflectivity) are presented in Fig. 9.14. A typical amplified image produced by the conjugate wave front and also the original image (slide) are shown in Fig. 9.15 [9.7, 11].

Thus the experimentally obtained value of reflectivity ($R_{exp} = I_{S_2}^+/I_{S_0} \gtrsim 30$) is comparable to the highest values of R achieved in BaTiO$_3$ [9.14, 15] and is 3–4

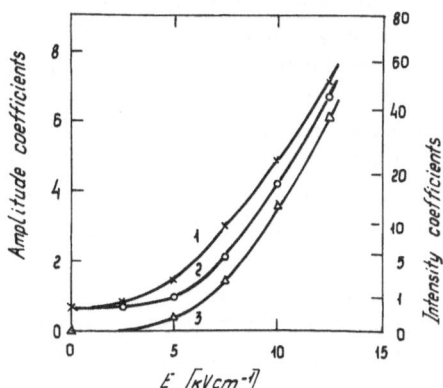

Fig. 9.14. Two-wave transmittivity (1), positive feedback four-wave mixing transmittivity (2), and reflectivity (3) vs alternating electric field amplitude [9.7]. (Experiment, BTO, $\lambda = 0.63$ μm, $d \approx 4$ mm, $\Lambda^{-1} \approx 200$ mm^{-1}, square-wave alternating electric field, $\tau_\sim \approx 25$ ms, initial signal/pump intensity ratio $\approx 2 \times 10^{-3}$)

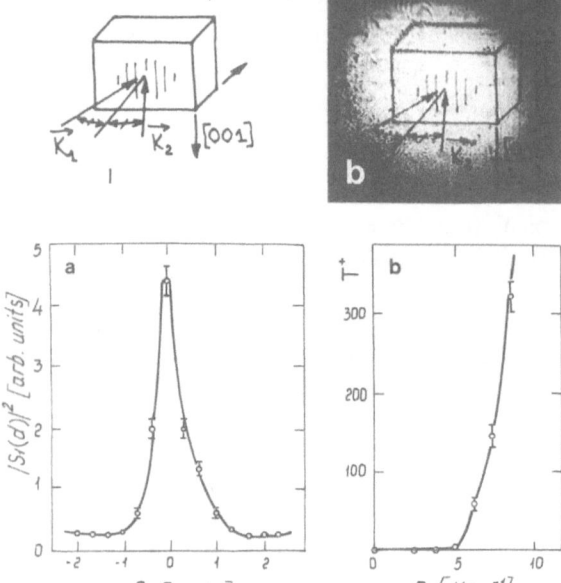

Fig. 9.15. Part of the initial object (**a**) and ×20 amplified image reconstructed from a phase conjugate wave (**b**). (BTO, $\lambda = 0.63 \,\mu\text{m}$, $d \approx 4\,\text{mm}$, $\Lambda^{-1} \approx 200\,\text{mm}^{-1}$, square-wave alternating electric field, $\tau_{\sim} \approx 25\,\text{ms}$, $E \approx 12.5\,\text{kV cm}^{-1}$)

Fig. 9.16. (**a**) Experimental plot of the transmitted signal beam intensity against misalignment ($\Delta\theta$) of pump plane waves in a four-wave mixing experiment with positive feedback [9.9]. (**b**) Experimental dependence of the optically induced noise gain at the center of the generation spot on the external alternating electric field amplitude [9.9]. (BTO, $\lambda = 0.63 \,\mu\text{m}$, $d = 4\,\text{mm}$, $\Lambda \approx 6\,\mu\text{m}$, square-wave alternating electric field, $\tau_{\sim} \approx 25\,\text{ms}$, $E \approx 6\,\text{kV cm}^{-1}$)

orders of magnitude greater than that measured experimentally in a conventional four-wave mixing geometry using BSO [9.12, 13]. It is worth mentioning that the same recording conditions in BTO in a four-wave mixing geometry with negative feedback (the same polarizations of all the beams involved) yielded a conjugate wave intensity of about two orders of magnitude lower.

According to the theoretical expression (9.34), the generation regime must be reached in the geometry under consideration at $\Gamma d \to 4$, which requires application of an external alternating field with an amplitude of $E \approx 8\,\text{kV cm}^{-1}$ to the 4-mm thick BTO crystal (Fig. 9.10b). And, in fact, beginning from $E \approx 6\,\text{kV cm}^{-1}$ we observed experimentally formation of a generation noise spot (Fig. 9.13d). In spite of the relatively wide spectral range of the optically induced noise in a two-wave mixing geometry (Figs. 9.13b,c), the spot appears to be rather narrow and lies almost in the incidence plane, i.e., at the $[1\bar{1}0]$ axis along which the external field is directed. It should be noted that once the spot is formed, the photoinduced noise outside its area goes down.

The need for precise adjustment of the setup to achieve the generation regime, which follows from the theoretical analysis (9.36), is supported by the experimental dependence of the light intensity at the center of the noise spot on the pump beams' misalignment (Fig. 9.16a) [9.9]. If the pump beams are carefully aligned, the growth of the photoinduced noise intensity at the center of the generation spot exhibits a threshold behavior (Fig. 9.16b) [9.9]. On passing to the above-threshold field amplitudes E, the resonance peak (Fig. 9.16a) is appreciably broadened.

9.4.3 Discussion

We have demonstrated in Sects. 9.4.2, 1 that it is possible to get a high gain for weak signal waves and phase conjugate reflectivity in experiments using non-stationary recording in BTO. But, in practice, the pump-to-signal conversion efficiency and the optical noise level are also of great significance.

For the BTO samples studied, the intensity of the initial (unamplified) noise was proportional to K^{-2} in the spatial frequency range 50–500 lines/mm, and the light scattering coefficient per unit solid angle at a spatial frequency of 200 lines/mm was about $\alpha \approx 0.1$. To achieve high fidelity of amplified and phase-conjugate signal waves (Figs. 9.12, 15) with this noise level, the holographic setups included conventional schemes for optical filtering of the output image. They consisted of a pinhole placed behind the sample in the object beam focusing plane. Its diameter was matched with the spatial frequency spectrum of the image being processed. Such a pinhole gave a considerable increase of the signal-to-noise ratio in the amplified image plane due to reduction in the photoinduced noise level and removal of amplified multiple signal waves arising from Fresnel reflections inside the sample.

It should also be pointed out that the angular density of noise ($\sim \alpha$) sets a minimum illumination level for diffuse scattering objects needed to achieve a signal-to-noise ratio of unity in the reconstructed image. For the BTO crystals involved, it may be only a factor of $\alpha^{-1} \approx 10$ lower than the intensity of the reference beam within the hologram volume.

Optical noise also limits the maximum gain for weak signal beams in a similar manner. For instance, for the BTO crystals studied, when an alternating field with an amplitude of $15\,\mathrm{kV\,cm^{-1}}$ is applied to the sample (Fig. 9.10b), as much as 25% of the reference beam energy is transferred to the photoinduced noise lobe. The two-wave gain $T \approx 200$ obtained in this case is likely to be close to the maximum gain for the BTO crystals investigated.

9.5 Conclusion

In this chapter we theoretically analyzed the nonstationary mechanisms of holographic recording in photorefractive crystals, such as recording a moving (oscillating) interference pattern in a steady external field and recording in an alternating field. The efficiency of the new methods was demonstrated using the cubic photorefractive crystals $Bi_{12}SiO_{20}$ (BSO) and $Bi_{12}TiO_{20}$ (BTO) as examples. The major aspects of the theory of light diffraction from anisotropic phase holograms in cubic photorefractive crystals were briefly discussed. We also analyzed the operation of an efficient four-wave mixing geometry that differs from a conventional scheme in that the light waves propagating through the sample in opposite directions are orthogonally polarized. It is shown that experiments employing the new four-wave mixing geometry and nonstationary holographic recording in BTO in an external alternating electric field succeed in achieving extremely high reflectivities of conjugate wave fronts.

Acknowledgements. The authors wish to express their thanks to Dr. A. A. Petrov and Dr. M. V. Krasin'kova for growing BTO and BSO crystals, and to Mrs. N. N. Nazina for her help in translating and typing the manuscript.

References

9.1   D.L. Staebler: "Ferroelectric Crystals", in *Holographic Recording Materials*, Topics Appl. Phys., Vol. 20 (Springer, Berlin, Heidelberg 1977) pp. 101–132
9.2   P. Günter: Phys. Rep. **93**, 201–299 (1982)
9.3   M.P. Petrov, S.I. Stepanov, A.V. Khomenko: *Photosensitive Electro-optic Media in Holography and Optical Information Processing* (Nauka, Leningrad 1983)
9.4   J.P. Huignard, A. Marrakchi: Opt. Commun. **38**, 249–254 (1981)
9.5   S.I. Stepanov, V.V. Kulikov, M.P. Petrov: Sov. Tech. Phys. Lett. **8**, 229–230 (1982)
9.6   S.I. Stepanov, V.V. Kulikov, M.P. Petrov: Opt. Commun. **44**, 19–23 (1982)
9.7   S.I. Stepanov, M.P. Petrov: Sov. Tech. Phys. Lett. **10**, 572–573 (1984)
9.8   S.I. Stepanov, M.P. Petrov: Opt. Commun. **53**, 292–295 (1985)
9.9a   S.I. Stepanov, M.P. Petrov, M.V. Krasin'kova: Sov. Phys. – Tech. Phys. **29**, 703–705 (1984)
9.9b   S.I. Stepanov, M.P. Petrov: Opt. Acta **31**, 1335–1343 (1984)
9.10   S.I. Stepanov, M.P. Petrov: Opt. Commun. **53**, 64–68 (1985)
9.11   M.P. Petrov, S.I. Stepanov: Proc. ICO-13 Conf., Sapporo, Japan, August 20–24, 1984, p. 430–431
9.12   J.P. Huignard, J.P. Herriau, P. Aubourg, E. Spitz: Opt. Lett. **4**, 21–23 (1979)
9.13   O. Ikeda, T. Suzuki, T. Sato: Appl. Opt. **22**, 2192–2195 (1983)
9.14   J. Feinberg, R.W. Hellwarth: Opt. Lett. **5**, 519–521 (1980)
9.15   F. Laery, T. Tschudi, J. Alberts: Opt. Commun. **47**, 387–390 (1983)
9.16   V.L. Vinetskii, N.V. Kukhtarev, S.G. Odulov, M.S. Soskin: Sov. Phys. – Usp. **22**, 742 (1979)
9.17   M.G. Moharam, T.K. Gaylord, R. Magnusson, L. Young: J. Appl. Phys. **50**, 5642–5651 (1979)
9.18   N.V. Kukhtarev, V.B. Markov, S.G. Odulov, M.S. Soskin, V.L. Vinetskii: Ferroelectrics **22**, 949–960 (1979)
9.19   M. Peltier, F. Micheron: J. Appl. Phys. **48**, 3683–3690 (1977)
9.20   G.A. Alphonse, R.C. Alig, D.L. Staebler, W. Phillips: RCA Rev. **36**, 213–229 (1975)
9.21a   G.C. Valley: J. Opt. Soc. Am. B**1**, 868–873 (1984)
9.21b   P. Refregier, L. Solymar, H. Rajbenbach, J.P. Huignard: J. Appl. Phys. **58**, 45–57 (1985)
9.22   A.M. Glass: "Holographic Storage", in *Photonics*, ed. by M. Balkanski, P. Lallemand (Gauthier-Villars, Paris 1975) pp. 163–192
9.23   J. Feinberg, D. Heiman, R.W. Hellwarth: J. Opt. Soc. Am. **68**, 1367–1368 (1978)
9.24   N.V. Kukhtarev, S.G. Odulov: JETP Lett. **30**, 4–8 (1979)
9.25   J.P. Huignard, J.P. Herriau, G. Rivet, P. Günter: Opt. Lett. **5**, 102–104 (1980)
9.26   Y. Ninomiya: J. Opt. Soc. Am. **63**, 1124–1130 (1973)
9.27   D.W. Vahey: J. Appl. Phys. **46**, 3510–3515 (1975)
9.28   B. Fisher, M. Cronin-Golomb, J.O. White, A. Yariv: Opt. Lett. **6**, 519–521 (1981)
9.29   T.K. Gaylord, T.A. Rabson, F.K. Tittel, C.R. Quick: J. Appl. Phys. **44**, 896–897 (1973)
9.30   H. Rajbenbach, J.P. Huignard, P. Refregier: Opt. Lett. **9**, 558–560 (1984)
9.31   M.P. Petrov, T.G. Pencheva, S.I. Stepanov: J. Opt. **12**, 287–292 (1981)

9.32   J.P.Huignard, F.Micheron: Appl. Phys. Lett. **29**, 591–593 (1976)
9.33a M.P.Petrov, S.V.Miridonov, S.I.Stepanov, V.V.Kulikov: Opt. Commun. **31**, 301–305 (1979)
9.33b A.Marrakchi, J.P.Huignard, P.Günter: Appl. Phys. **24**, 131–138 (1981)
9.34   S.C.Abrahams, C.Svensson, A.R.Tanguay, Jr.: Solid State Commun. **30**, 293–295 (1979)
9.35   T.G.Pencheva, M.P.Petrov, S.I.Stepanov: Opt. Commun. **40**, 175–178 (1981)
9.36   P.Refregier, L.Solymar, H.Rajbenbach, J.P.Huignard: Electron. Lett. **20**, 656–657 (1984)
9.37   A.D.Novikov, V.V.Obuchovskii, S.G.Odulov, B.I.Sturman: JETP Lett. **44**, 538–542 (1986)
9.38   J.P.Huignard, J.P.Herriau, T.Valentin: Appl. Opt. **16**, 2796–2798 (1977)
9.39   M.D.Levenson, K.M.Johnson, V.C.Hanchett, K.Chiang: J. Opt. Soc. Am. **71**, 737–743 (1981)
9.40   P.Günter: Opt. Commun. **41**, 83–88 (1981)
9.41a S.L.Hou, R.B.Lauer, R.E.Aldrich: J. Appl. Phys. **44**, 2652–2658 (1973)
9.41b T.G.Pencheva, S.I.Stepanov: Sov. Phys. – Solid State **24**, 687–688 (1982)
9.42   G.Hamel de Montchenault, B.Loiseaux, J.P.Huignard: Electron. Lett. **22**, 1030–1032 (1986)
9.43   J.-M.C.Jonathan, R.W.Hellwarth, G.Roosen: IEEE J. QE-**22**, 1936–1941 (1986)
9.44a G.A.Alphonse, W.Phillips: RCA Rev. **37**, 184–205 (1976)
9.44b J.Feinberg: J. Opt. Soc. Am. **72**, 46–51 (1982)
9.45   A.J.Fox, T.M.Bruton: Appl. Phys. Lett. **27**, 360–362 (1975)
9.46   H.Rajbenbach, J.P.Huignard, B.Loiseaux: Opt. Commun. **48**, 247–252 (1983)
9.47   P.Günter: J. de Phys., Colloq. **44**, C2, 141–147 (1983)
9.48a A.M.Glass, A.M.Johnson, D.M.Olson, W.Simpson, A.A.Ballman: Appl. Phys. Lett. **44**, 948–950 (1984)
9.48b J.Kumar, G.Albanese, W.H.Steier, M.Ziari: Opt. Lett. **12**, 120–122 (1987)
9.49   J.P.Huignard, A.Marrakchi: Opt. Lett. **6**, 622–624 (1981)
9.50   G.S.Trofimov, S.I.Stepanov: Sov. Tech. Phys. Lett. **11**, 256–257 (1985)
9.51   S.I.Stepanov: Proc. ICO-14 Conf., Quebec, Canada, August 24–28, 1987
9.52   A.A.Izvanov, N.D.Hat'kov, S.M.Shandarov: Digests of V All-Union School on Optical Information Processing (Kiev 1984) p. 53–54

Additional References

Sochava, S.L., Stepanov, S.I., Petrov, M.P.: Ring oscillator using photorefractive $Bi_{12}TiO_{20}$ crystal. Pis'ma Zh. Tekh. Fiz. **13**, 660–664 (1987), to be translated into English in Sov. Tech. Phys. Lett. **13**

Stepanov, S.I., Shandarov, S.M., Hat'kov, N.D.: Photoelastic contribution to the photore-fractive effect in cubic crystals. Fiz. Tverd. Tela **29**, 3054–3058 (1987), to be translated into English in Sov. Phys. – Solid State **29**

Stepanov, S.I., Sochava, S.L.: Efficient energy exchange in two-wave mixing in $Bi_{12}TiO_{20}$. Zh. Tekh. Fiz. **57**, 1763–1766 (1987), to be translated into English in Sov. Phys. – Tech. Phys. **57**

Subject Index